# The VC-1 and H.264 Video Compression Standards for Broadband Video Services

# MULTIMEDIA SYSTEMS AND APPLICATIONS SERIES

*Consulting Editor*

**Borko Furht**
*Florida Atlantic University*

***Recently Published Titles:***

*SIGNAL PROCESSING FOR IMAGE ENHANCEMENT AND MULTIMEDIA PROCESSING* edited by E. Damiani, A. Dipanda, K. Yetongnon, L. Legrand, P. Schelkens, and R. Chbeir; ISBN: 978-0-387-72499-7

*MACHINE LEARNING FOR MULTIMEDIA CONTENT ANALYSIS* by Yihong Gong and Wei Xu; ISBN: 978-0-387-69938-7

*DISTRIBUTED MULTIMEDIA RETRIEVAL STRATEGIES FOR LARGE SCALE NETWORKED SYSTEMS* by Bharadwaj Veeravalli and Gerassimos Barlas; ISBN: 978-0-387-28873-4

*MULTIMEDIA ENCRYPTION AND WATERMARKING* by Borko Furht, Edin Muharemagic, Daniel Socek: ISBN: 0-387-24425-5

*SIGNAL PROCESSING FOR TELECOMMUNICATIONS AND MULTIMEDIA* edited by T.A Wysocki,. B. Honary, B.J. Wysocki; ISBN 0-387-22847-0

*ADVANCED WIRED AND WIRELESS NETWORKS* by T.A.Wysocki,, A. Dadej, B.J. Wysocki; ISBN 0-387-22781-4

*CONTENT-BASED VIDEO RETRIEVAL: A Database* Perspective by Milan Petkovic and Willem Jonker; ISBN: 1-4020-7617-7

*MASTERING E-BUSINESS INFRASTRUCTURE* edited by Veljko Milutinović, Frédéric Patricelli; ISBN: 1-4020-7413-1

*SHAPE ANALYSIS AND RETRIEVAL OF MULTIMEDIA OBJECTS* by Maytham H. Safar and Cyrus Shahabi; ISBN: 1-4020-7252-X

*MULTIMEDIA MINING: A Highway to Intelligent Multimedia Documents* edited by Chabane Djeraba; ISBN: 1-4020-7247-3

*CONTENT-BASED IMAGE AND VIDEO RETRIEVAL* by Oge Marques and Borko Furht; ISBN: 1-4020-7004-7

*ELECTRONIC BUSINESS AND EDUCATION: Recent Advances in Internet Infrastructures* edited by Wendy Chin, Frédéric Patricelli, Veljko Milutinović; ISBN: 0-7923-7508-4

*INFRASTRUCTURE FOR ELECTRONIC BUSINESS ON THE INTERNET* by Veljko Milutinović; ISBN: 0-7923-7384-7

*DELIVERING MPEG-4 BASED AUDIO-VISUAL SERVICES* by Hari Kalva; ISBN: 0-7923-7255-7

***Visit the series on our website: www.springer.com***

# The VC-1 and H.264 Video Compression Standards for Broadband Video Services

by

Jae-Beom Lee
*Sarnoff Corporation*
*USA*

Hari Kalva
*Florida Atlantic University*
*USA*

*Authors:*
Jae-Beom Lee
Sarnoff Corp.
Video, Communications and Networking Systems Division
201 Washington Road
Princeton, NJ 08540
jlee@sarnoff.com

Hari Kalva
Florida Atlantic University
Dept. Computer Science & Engineering
777 Glades Road, PO Box 3091
Boca Raton, FL 33431
hari@cse.fau.edu

*Series Editor:*
Borko Furht
Florida Atlantic University
Department of Computer Science & Engineering
777 Glades Road, PO Box 3091
Boca Raton, FL 33431
borko@cse.fau.edu

ISBN-13 978-1-4419-4376-7 e-ISBN-13: 978-0-387-71043-3



Printed on acid-free paper

9 8 7 6 5 4 3 2 1

springer.com

# Contents

# Preface

Probably the most interesting and influential class to the authors about video compression was EE E6830 (Digital Image Processing and Understanding) at Columbia University in 1995, offered by adjunct Professors Dr. Netravali, Dr. Haskell and Dr. Puri at AT&T. In the class, they impressed the authors with how such difficult and mysterious statements in video standards could be interpreted/ understood in plain human languages. Since then, the authors had had a dream that similar services could also be provided to interpret difficult video subjects into reasonable level of explanations in the future.

The VC-1 standard is fundamentally the same as WMV-9. WMV-x video compression technologies of Microsoft have long been the most popular over the Internet due to popularity of Microsoft Operating Systems. The technologies were published in August 2005 for the first time in a formal SMPTE document in the name of VC-1, and the official standard then was finalized in April 2006. In contrast, the MPEG committee recently standardized the MPEG AVC (H.264) video coding standard, whose first version was officially published in May 2003, and several subsequent amendments and corrigenda then followed until recently. These two are highly efficient compression standards that can make high-quality video services possible for Digital Storage Media (e.g., Blu-ray DVD or HD DVD) and/or broadband networks applications (e.g., IPTV).

In the industry, on the other hand, video compression text/ reference books have become less useful due to the advance of bitstream analyzer tools such as Interra or Vprove. The tools cross-link statements in the standards in the middle of bitstream verification. In other words, documents explaining in low level are not useful very much any more. Therefore, the focus on the text/

reference books might need to shift from definitions of bits and pieces to ideas/ philosophies about technologies/ tools. This book is designed to present the readers with reasonable understanding and reasoning about why such tools are devised in such ways – as was once done by Dr. Netravali, Dr. Haskell and Dr. Puri. Only the domain is shifted in this book from MPEG-2 to VC-1/ H.264.

The authors are grateful to Professors Anastassiou, Chang and Eleftheriadis (now with the University of Athens, Greece) in the department of Electrical Engineering at Columbia University who helped to shape our understanding about video compression more than a decade ago with the ADVENT project at Center for Telecommunications Research.

A companion website for this book is available at: www.cse.fau.edu/~hari/vc1-h264. The web site will be updatated with useful resources, software tools, and errata. The authors hope that the readers find this book enjoyable and useful.

Dr. Jae-Beom Lee
Princeton, NJ

Dr. Hari Kalva
Boca Raton, FL

April, 2008

# Acknowledgements

The authors would like to thank the Series Editor Dr. Borko Furht and Publishing Editor Susan Lagerstrom-Fife for their encouragement and support. The authors would also like to thank Dr. Bill Lin at Sarnoff Corporation and Dr. Gary Sullivan at Microsoft Corporation for their technical and general advice/ comments to improve the quality of this book.

Dr. Jae-Beom Lee also expresses gratitude to colleagues at Sarnoff Corporation for their kind interactions and discussions in deepening video compression knowledge: Arkady Kopansky, Yanjun Xu, Ric Conover, Dennis McClary, Norm Hurst, Mike Isnardi, Hans Baumgartner, Iris Caesar, Jun Hu, Azfar Inayatullah, Joe Frank, Indu Kandaswami, Sandip Parikh, Lin Her, Yumin Zhang, Mike Patti, Khaleel Udyawar, Saurabh Shandilya, Prashant Laddha, Bedarakota Madhu Sudhan, Anup Mankar, Vishvanath Deshpande, Ramanan Narayanan, Mattamari Seshagiri Srividya, Veena Parashuram, Penmestsa Raju and Iyengar Sridhar.

*To Bong-Gum Lee, Yun-Hee Jang and Da-En Lee – the three most important women in my life*

*Jae-Beom Lee*

*To my parents*

*Hari Kalva*

# 1. Multimedia Systems

## 1.1 Overview of MPEG-2 Systems

### Systems and Synchronization

The video compression system is a part of a multimedia system that provides a means of multiplexing several types of multimedia information into one single stream [haskell:MPEG2, yu:MPEG2systems, ITU:MPEG2systems]. Summarizing a global picture of such multimedia systems helps further understanding VC-1 and H.264/ AVC video compression topics. This section describes issues regarding general aspects of system timing and media synchronization. It then discusses solutions and their practical implications including sender/ receiver system timing synchronization (i.e., Transport synchronization), video/ audio synchronization (i.e., Inter-media synchronization) and encoder/ decoder resource synchronization (i.e., Resource synchronization). It further details practical examples to implement such synchronization mechanisms on MPEG Systems.

SMPTE RP227 recommends VC-1 bitstream encoding provisions that define a minimum set of rules for the carriage of a VC-1 elementary stream in an MPEG-2 Transport Stream with additional intention to provide a generic means of carrying a VC-1 video elementary stream in an MPEG-2 Program Stream as used by the DVD Forum. In addition, Amendment 3 of ITU-T Recommendation H.222.0 recommends H.264/ AVC bitstream encoding provisions that define a minimum set of rules for the carriage of a H.264/ AVC elementary stream in an MPEG-2 Transport Stream, with additional intention to provide a generic means of carrying a H.264/ AVC video elementary stream in an MPEG-2 Program Stream as used by the DVD Forum. In other words, both VC-1 and H.264/ AVC standards adopt MPEG-2 Systems as a major system encapsulation/ transport mechanism due to its popularity in the real world [SMPTE:VC1systems, ISO:MPEG2systems.amd].

MPEG-2 Systems have been used for an extraordinary number of applications that require solid transport delivery or local playback mechanisms, where strict transport / inter-media/ resource synchronizations are recommended. However, mobile applications like cellular video streaming do not require strict transport timing synchronization. Streaming video applications using the TCP-IP protocol, where no Quality of Service (QoS) is provided, do not have a fixed delay (or a constant incoming bitrate). In such a case, transport synchronization could be ignored. Local timer-based inter media synchronization might still need to be performed though.

In general, for any video compression/ transmission standard, the inter media synchronization is performed using a Presentation Time Stamp (PTS) that dictates when a particular media unit (for example, video picture or audio frame) should be played back. The nature in which PTS is decided for audio and video units varies from standard to standard. The PTS may be described at the media level (for example, within the video bitstream in H.264/ AVC), Packetized Elementary Stream (PES) level (for example, within PES Header in MPEG-2 or VC-1) or transport level (for example, MPEG-2 video on RTP packets).

The following sections describe in detail mechanisms for transport, media and resource synchronizations in any multimedia systems.

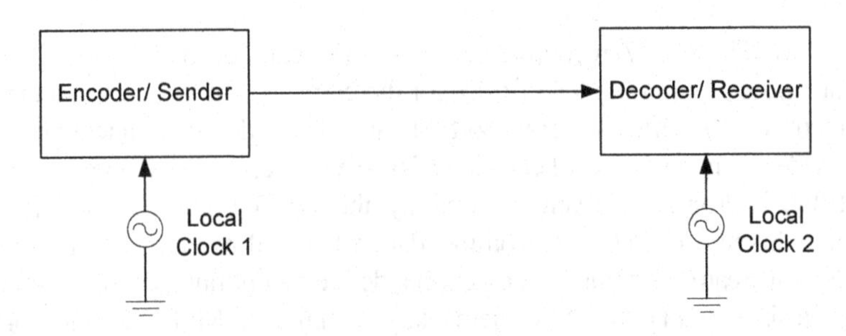


**Figure 1-1 Sender Receiver Synchronization in Communication Systems**

## Transport Synchronization

Compressed multimedia generally has of two forms in applications – networked playback or local playback. Local playback is much more relaxed in terms of timing since all data is already available in the player. On the other hand, networked playback requires more strict system timing as any loose synchronization of sender/ receiver ends up with playback jitter. Therefore, the discussion in this section mainly covers the networked scenario. Figure 1-1 depicts a general situation of a sender and a receiver, where typical multimedia communication systems assume delivery mechanisms on local clocks.

In standards like MPEG-2, the operating clock is defined as 27MHz. However, there are no perfectly matched two local clocks – one in the sender, the other in the receiver – in terms of speed in real life. Since the data transmission rate in modern communication systems is extremely high, constantly faster or slower (even with the slightest imaginable mismatch) clock at receiver side makes the receiver buffer underflow or overflow. For example, 20Mbps bandwidth communication accumulates 20 million bits in the receiver side buffer when two clocks are out of synchronization only for 1 sec.

To manage receiver side buffer for a long time (for example, two to three hours movie time), the time-average speed of two local clocks should be exactly the same. Note that slight jittering between two clocks is not a problem as long as the time-averages of two local clocks are exactly matched and a reasonable size of receiver side buffer is allowed.

To implement the same speed of two local clocks in terms of time-average, PLL and Voltage Controlled Oscillator (VCO) are used in MPEG Systems with a special type of Time Stamp called Program Clock Reference (PCR), where Time Stamp is defined to be nothing but a sampling value of a counter that runs on a local clock as shown in Figure 1-4. The MPEG-2 standard specifies PCR as being driven by a 27MHz clock.

In the delivery of MPEG multimedia information, sender and receiver synchronization is achieved with PCR in Transport Stream (TS) packets. In the TS packet header, PCR Time Stamp is written based on the local clock of the sender (i.e., system encoder).

When a receiver receives the first PCR in a TS packet, it copies PCR as an initial value into a counter increased by the local clock of the receiver. With any PCR received later, the receiver compares it with its local counter to determine whether the receiver local clock is faster or slower than that of the sender. The differential value of the received PCR and the local counter is fed back to adjust the speed of the local clock of the receiver side with VCO. For example, let's say that a received PCR is 2000, while local counter at the moment is 2001. This implies that the operating speed of the local clock at the receiver is faster than that of the sender. Such difference can be adjusted with VCO at the receiver. Once transport synchronization is achieved with Time Stamps in the transport layer, inter-media synchronization is next to be considered with Time Stamps in the media layer.

This synchronization mechanism between two close-speed running local clocks is a fundamental means in Asynchronous Transfer Mode (ATM) communications. ATM was supposed to work best for point-to-point communications and MPEG-2 TS packet was designed to be accommodated to four ATM packets with the fixed length of 188 bytes while at MPEG-2 System standardization effort. In contrast, Synchronous Transfer Mode (STM) requires nation-wide distribution of a single master clock to all nodes, thus making the time-average of the speed of the clocks the same with slight jittering. This jittering is not a big problem as aforementioned as long as buffer size is enough. Therefore, STM does not require PLL-VCO. SONET from AT&T is a famous STM backbone network that allows direct add/ drop multiplexing without intermediate demux necessity down to any layer.

MPEG-2 Systems define a provision about deviation of system clock frequencies among individual implementations as follows:

$$27000000 - 810 \leq system\_clock\_frequency \leq 27000000 + 810$$

and rate of change of *system_clock_frequency* with time $\leq 75 \times 10^{-3} Hz/s$. (1-1)

Therefore, the speeds of two local clocks should be reasonably close at the beginning.

The *system clock frequency* 27MHz is chosen for historical reasons. A common frequency from which NTSC (525/60) and PAL (625/50) line and field rates can be derived is 4.5MHz as shown in 4.5MHz/143 (a.k.a., 2fH for NTSC) and 4.5MHz/144 (a.k.a., 2fH PAL), respectively. The luma sampling rate was chosen as 13.5MHz (a.k.a., $4.5 \times 3 MHz$ ), three times of the common frequency. To apply luma sampling rate to YUV4:2:2 type format, the sampling rate is chosen as 27MHz (a.k.a., 13.5MHz + 6.75MHz + 6.75MHz).

### Inter-Media Synchronization with PTS

In general applications, more than two medias are captured/ compressed in the encoder (sender) side. Even when only two medias (video and audio) are captured simultaneously, the absolute time order cannot be maintained due to time division packet multiplexing in the delivery mechanism.

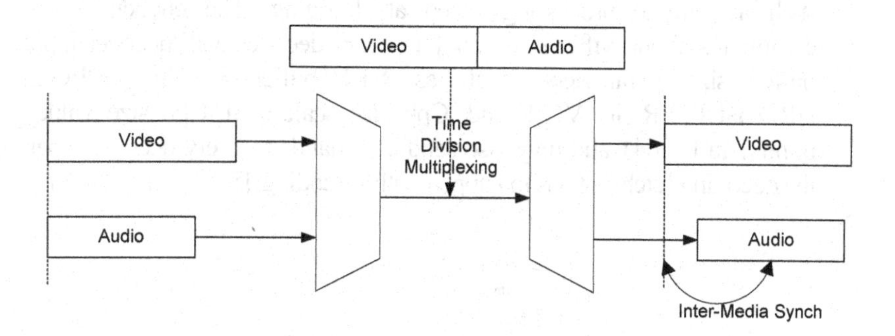


**Figure 1-2 Time Division Multiplexing**

Packet multiplexing generally causes different delay between media due to serialization as shown in Figure 1-2. On top of the aforementioned "packet multiplexer jitter," "network delivery jitter" is also added. Since concurrent media should play back simultaneously, an inter-media synchronization mechanism is to be devised.

To this end, Access Units (AU) for different media and Presentation Time Stamp (PTS) are defined. Two framed-media captured at the same time share the same PTS. And, PTS is ideally attached to each AU. If a PTS is not found in a certain AU, the value should be inferred based on the values recently received at the receiver. The encoder commands every decoder when a specific AU should be displayed with a specific PTS command. At receiver side, inter-media synchronization can be achieved with two media being buffered and played back based on PTS. Note that network delivery jitter and packet multiplexer jitter can be eliminated through buffering at the receiver side. PTS is a 90KHz-based Time Stamp locked to the 27MHz-based PCR in MPEG-2.

## Resource Synchronization with DTS

The Video Buffer Verifier (VBV), which is the Hypothetical Reference Decoder (HRD) in VC-1 or H.264, is the MPEG-2 hypothetical buffer model for a video decoder. VBV is meant to connect to the output of an encoder virtually while encoding. As bitstreams are created, the VBV fullness must be checked to ensure that it does not overflow or underflow. A dummy decoder is assumed with certain predefined behavior such as infinite processing speed at decoding. The encoder sends commands about buffer size policy to every decoder with decoder input buffer size parameters (such as VBV_buffer_size in MPEG-2, HRD_BUFFER in VC-1 and Cpb_size_scale and Cpb_size_value_minus1 in H.264), and may also send commands to every decoder when the decoding fetch-out has to happen with Decoding Time Stamp (DTS).

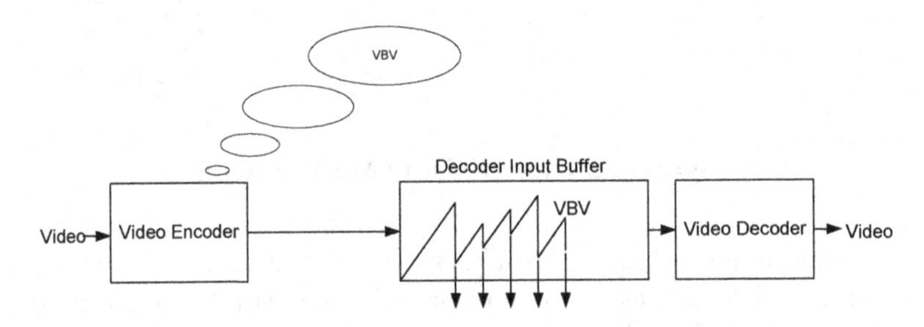


**Figure 1-3 VBV Model and Resource Synchronization**

The encoder takes full advantage of the remaining buffer resource in the VBV for bit allocation policy at GOP-level, Frame-level and MB-level.

If a decoder doesn't follow the encoder's commands regarding DTS, the remaining buffer resource assumed by the encoder can be different from that in actual decoder input buffer – this could cause an unexpected behavior resulting in overflow or underflow at the decoder input buffer. Therefore, a decoder has to do its best to conform to DTS commands of the encoder. DTS is ideally attached to each AU, as is PTS. If a DTS is not found in a certain AU, the value should be inferred based on the values recently received at the receiver. DTS is a 90KHz-based Time Stamp locked to the 27MHz-based PCR in MPEG-2.

### DTS/ PTS Locking Mechanism to PCR

A PCR is a sampling value of a counter that runs on a 27MHz-driven local clock. As such, a DTS and a PTS are sampling values of a counter that runs on a 90KHz-driven local clock. PCR is present in a 42-bit Time Stamp, while DTS/ PTS is present in a 33-bit Time Stamp in MPEG-2 as shown in Figure 1-4. For example, 0 and 2100 as shown in Figure 1-4 are two sampling values of the counter that runs at 27MHz-driven local clock – PCR Time Stamps. Also, 0 and 7 are two sampling values of the counter that runs at 90KHz-driven local clock – DTS/ PTS Time Stamps. Since a Program is defined to share a common time-base in MPEG Systems, all media streams (videos and audios) in a single program are captured and time-tagged based on a single common clock. Note that 27MHz is divisible by 90KHz with a factor of 300. In other words, a 90KHz clock can be generated from a single source of 27MHz clock and a DTS/ PTS can be directly compared/ locked to PCR by multiplying it by 300 as shown in Figure 1-4.

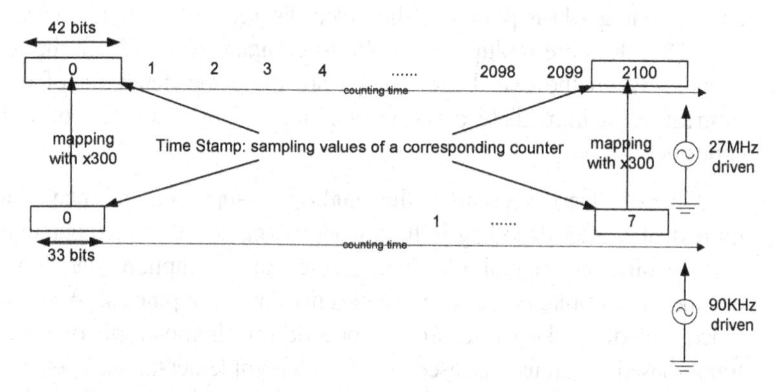


**Figure 1-4 DTS/ PTS Locking to PCR**

In summary, MPEG-2 Systems synchronization is performed in two levels – Transport level and Media level. First, Transport synchronization is carried out with a PCR Time Stamp to prevent overflow or underflow in communication buffers. Second, Inter-media synchronization and Resource synchronization are performed at Media level with PTS/ DTS locked on Transport clock already synchronized between the sender and the receiver (or potentially multiple receivers).

### General MPEG System Architecture

A Program can contain multiple media streams of video and audio. To handle multiple streams in a synchronized manner, the cores consist of three blocks – system demux, video decoders and audio decoders as shown in Figure 1-5. The system demux de-packetizes and routes video, audio and systems data to the appropriate buffers. PCR data is extracted from system data for Transport synchronization. The first PCR data is copied to a local counter of the receiver as an initial Time Stamp and the counter is increased by a PLL-VCO rectified 27MHz local clock.

The AUs of video and audio are all called "frames." Video decoders are controlled by DTS and PTS together since sometimes display order and decoding order are different. However, audio decoders are controlled only by PTS since playback order and decoding order are the same. In the video controller, DTS values are multiplied by 300 to compare with PCR value at the local counter. When the two Time Stamps are the same, fetch-out of video bitstream and immediate decoding is initiated by the video decoding controller. When the PCR hits the time of $PTS \times 300$, the corresponding AU is played in the video display. In the audio controller, the PTS values are multiplied by 300 to compare with PCR at the local counter. When the two Time Stamps are the same, fetch-out of audio bitstream and immediate decoding are initiated by the audio decoding controller.

Unlike video decoding, the audio is supposed to play back immediately after decoding in theory. However, actual implementation is slightly different in real life. The theoretical assumption that dummy decoder has infinite processing speed is not correct in practice. Audio and video decoders take some time to decode one frame-worth of data. If frame-based pipelining is used for decoder implementation, even more than one frame time delay is to be expected. In such a case, the delay is mainly dependent on how many frame-based pipelining stages are used internally in the design. Therefore, PTS values are modified to accommodate such delay. Since the video decoder requires more computation, PTS values of audio are incremented to accommodate video processing delay. In other words, audio play-out buffers are needed to hold up data until the concurrent video data is ready for inter-media synchronization.

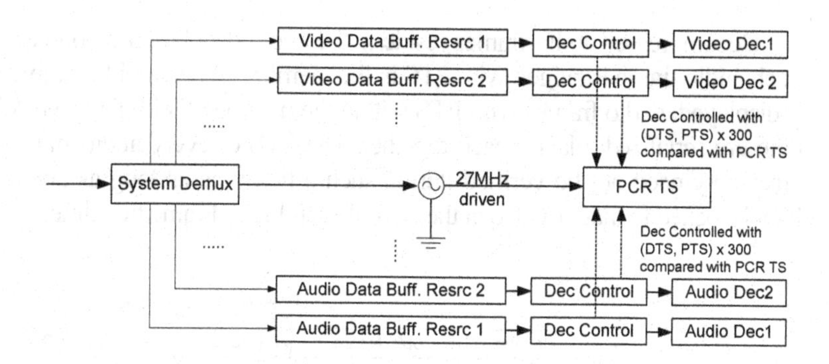


**Figure 1-5 MPEG System Architecture**

## Processor Mapping of MPEG System

Two important functional blocks of MPEG Systems are, first, System Demux and Clock Recovery and, second, Presentation Scheduler as shown in Figure 1-6.

System Demux and Clock Recovery function handles depacketization and extraction of PCR information with rectification. The Presentation Scheduler handles Inter-media synchronization and Resource synchronization based on DTS/ PTS.

One implementation of System functions can be to put the two functions on the audio processor. Audio processing is less computationally intensive, compared with video processing. This is mainly because the amount of 2-D video data handled is much more than that of 1-D audio data. Therefore, a certain amount of computational room can be available to System implementation.

One popular practical implementation of the video Presentation Scheduler function is based on Interrupt Service Routine (ISR), which is based on an interrupt signal (Field and/or Frame time) generated by the display processing device as shown in Figure 1-7. At another interrupt signal point (Direct Memory Access(DMA)-done, synchronization for decoding start), System Demux and Clock Recovery function can be performed since typically the outside Host delivers bitstream based on Field or Frame time. The key idea behind this is to guarantee secure availability of resources at the decoder – one picture decoding is performed only when one decoded picture is displayed/ released from the display buffer at PTS time.

Decompressed audio frames reside at the audio data buffer resources. Each audio frame is tagged with PTS in the memory. When a video frame is displayed, audio frames with PTS within the next one Field/ Frame tick time are serialized/ played back together. The PTS of every audio frame practically need not be verified during such a tick time as long as audio PTS is locked to that of video at the start of each Field/ Frame tick time.

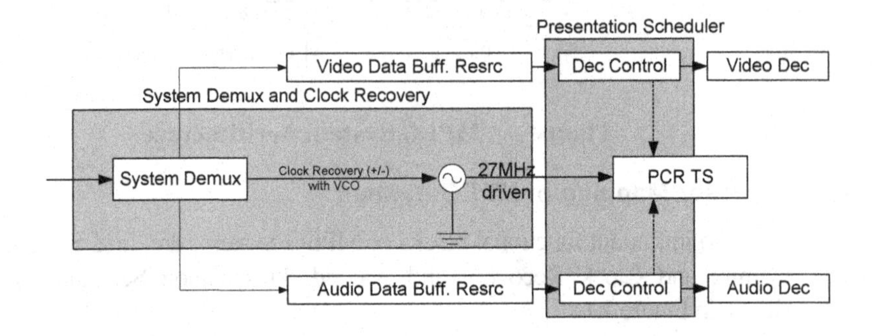


**Figure 1-6 A Typical Processor Mapping of MPEG System**

Mobile applications like cellular video streaming don't require strict timing. Still, a lot of streaming video applications depend on TCP-IP protocol where no Quality of Service is provided. So, the server and clients have a side channel to negotiate in a timely manner. In such a case, PCR Clock Recovery could be ignored and Timer-based interrupt signals (for DTS and PTS) are sufficient. The implementation mentioned above can be slightly modified to accommodate different application scenarios.

## Display and Decoder Interlocking Mechanism

Display processors can be defined in various ways. A basic function of a display processor is to perform DMA from picture memories and to convert from YUV format to certain display signals physically pre-defined. It could generate outputs for analog TVs and/or for digital displays. It could also generate interrupt signals at Field and/or Frame time. The input interface has a couple of Luma/ Chroma data pointers and Go-bit for an immediate action.

Once a display processor starts working, it cannot typically be stalled or delayed by any means (except clock rate rectification based on PLLed local clock). It is continuously displaying empty or meaningful data at the very regular rate of 30 frames/second or any other pre-defined rate. Since the display unit or processor works almost independently, other parts of the SOC need to be synchronized with the display processor operation.

Interlocking of display and decoder is the major means to secure buffer resources at the decoder. Two interrupts generated by the display processor are generally required to inter-lock between display processor and the decoder – 1) Field and/or Frame time interrupt and 2) DMA-done interrupt. The first interrupt tells about the time (Field and/or Frame) point when the display scanning line reaches those points. The second interrupt indicates the time when DMA from external memory to the display processor is done. How to use these two interrupts for interlocking is an implementation issue. However, the decoder is generally allowed to go ahead for next unit (Field or Frame) decoding at the DMA-done interrupt. Note that the decoder stalls for a while after a data unit (Field or Frame) decoding is finished.

```
While (1) { //infinite loop++
                display and decoder synch point;
                main decoding part;
                buffer reordering for display;
} //infinite loop--
```

**Figure 1-7 Display and Decoder Inter-locking**

The “buffer reordering for display” is performed at the end of the decoding stage. Generally, display order comes after a couple of frame/field ticks. The only exceptions are I and P pictures. When the next reference picture is reached, the previous reference picture can be displayed. This is not true generally for advanced video codecs such as H.264, though.

## 1.2 System Target Decoders and Encapsulations

### TS-System Target Decoder vs. PS-System Target Decoder

There are two main different applications in multimedia systems – networked application vs. local playback. Different applications require different information in the header and data encapsulation methods. In MPEG-2 Systems, "Transport Stream (TS)" and "Program Stream (PS)" are defined in this sense. TSs are defined for transmission networks that may suffer from errors and jitters, while PSs are defined for local playback environment with negligible errors. Therefore, important parts of TSs are about error protection capability over the networks and splicing of multiple programs, while an important part of PSs is to represent/ store bitstreams in Digital Storage Media (DSM) in compact and efficient manners.

Two System Target Decoders (STD) – Transport Stream STD (T-STD) and Program Stream STD (P-STD) – are defined to describe basic behaviors of systems decoders in networked or DSM applications so that system encoders consider them during construction or verification on such system streams. Actual correspondence of this part in a video decoder is HRD model in a sense that a systems encoder should operate not to violate pre-defined STD behaviors. When a TS is generated through an MPEG-2 Systems encoder for network applications, T-STD model should be carefully considered in terms of system parameters such as buffer size, buffer output rate, etc. For example, TS packet generation biased for a specific media through multiplexer would cause underflow in buffers for other media and systems. In addition, P-STD model should be carefully considered to generate PS for DVD or DSM streams.

The T-STD is a hypothetical system decoder that is synchronized exactly with the system encoder. A demux, system control, buffers and media decoders altogether define T-STD as shown in Figure 1-8. The MPEG-2 Systems standard specifies certain rules for multiplexers in the form of T-STD for compliant system bitstreams to be decodable.

The input of T-STD is bytes of multiplexed TSs. The T-STD input data is assumed to be piecewise constant rate between PCRs. There are two types of buffers defined in T-STD – Transport Buffer (TB) and Multiplexing Buffer (B). The main function of TB is to handle network delivery jitter, while that of B is to alleviate packet multiplexer jitter. In case of video data, B is broken down further to two sub-buffers –

Multiplexing Buffer (MB) and Elementary stream decoder Buffer (EB) as shown in Figure 1-8.

The purpose of "TB" is to eliminate network delivery jitter. The size of TB is proposed to be 512 bytes. The input comes at the full rate of TS, while the output leaves at a rate Rx defined based on data types.

In such a model, the full rate of TS can be determined with two consecutive PCR values and the byte count between those two PCR fields as follows:

$$transport\_rate(i) = \frac{((i'-i'') \times system\_clock\_frequency)}{PCR(i') - PCR(i'')}$$

(1-2)

where $i$ is the index of any byte in the TS for $i'' < i < i'$.

$i'$ is the index of the byte containing the last bit of the immediately following program_clock_reference_base field applicable to the program being decoded.

$i''$ is the index of the byte containing the last bit of the most recent program_clock_reference_base field applicable to the program being decoded.

$PCR(i'')$ is the time encoded in the program clock reference base and extension fields in units of the system clock.

In addition, the input arrival time, *t(i)*, of i-th byte shall be computed from a PCR value and the *transport_rate(i)* as follows:

$$t(i) = \frac{PCR(i'')}{system\_clock\_frequency} + \frac{i - i''}{transport\_rate(i)}$$

(1-3)

where $i$ is the index of any byte in the TS for $i'' < i < i'$.

$i''$ is the index of the byte containing the last bit of the most recent program_clock_reference_base field applicable to the program being decoded.

$PCR(i'')$ is the time encoded in the program clock reference base and extension fields in units of the system clock.

These two values are fundamental parameters through which the delivery speed of TS packets is controlled at a broadcast or unicast server.

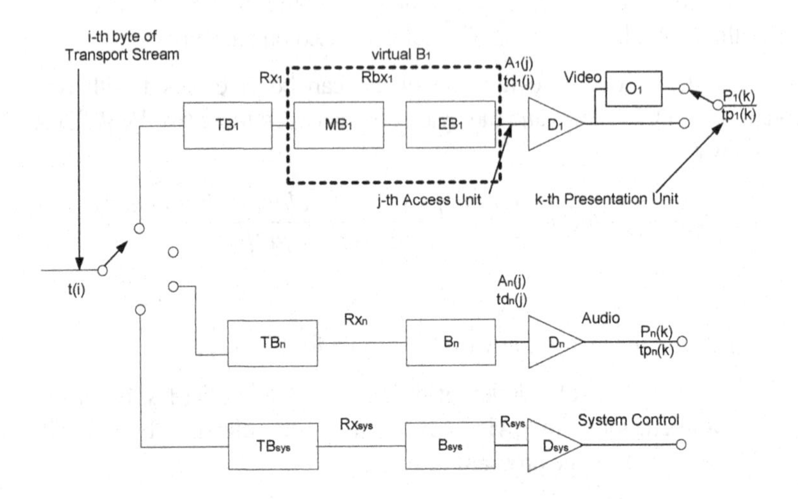


**Figure 1-8 Transport Stream System Target Decoder Model**

The output rate Rx is defined for T-STD based on data types as follows:

$Rx_n = 1.2 \times R_{\max}[profile, level]$ for video data. (1-4)

$Rx_n = 2 \times 10^6\, bits/\sec$ for audio data. (1-5)

$Rx_{sys} = 1 \times 10^6\, bits/\sec$ for systems data. (1-6)

For ISO/IEC 13818-7 ADTS audio, the number of channels and corresponding bitrates is shown in Table 1-1. Audio channels require their own decoder buffer in the elementary stream. The purpose of "B" or "virtual B" (video case) is to eliminate packet multiplexer jitter. However, virtual B is broken down into two units of buffers for video – MB and EB. The main reason for this is that video requires huge resource at the decoder due to sizable 2D image data, and Resource usage schedule should be synchronized between an encoder and a decoder in terms of VBV as was explained in previous sections. In this sense, EB is nothing but VBV buffer while MB is the prior stage to handle multiplexer jitter.

**Table 1-1 Output Rate Rxn for T-STD Model for Audio**

| Number of Channels | $Rx_n[bits/s]$ |
|---|---|
| 1~2 | 2000000 |
| 3~8 | 5529600 |
| 9~12 | 8294400 |
| 13~48 | 33177600 |

The size of MB (a.k.a., MBS) is proposed based on Levels as follows:

- Low/Main levels

$$MBS_n = BS_{mux} + BS_{oh} + VBV_{\max}[profile, level] - vbv\_buffer\_size \quad (1\text{-}7)$$

where PES packet overhead buffering

$$BS_{oh} = (1/750)\sec\times R_{\max}[profile, level], \quad (1\text{-}8)$$

and additional multiplex buffering

$$BS_{mux} = (0.004)\sec\times R_{\max}[profile, level]. \quad (1\text{-}9)$$

- High 1440/ High levels

$$MBS_n = BS_{mux} + BS_{oh} \quad (1\text{-}10)$$

- Constrained Parameters ISO/ IEC 11172-2 (MPEG-1 video) bitstreams

$$MBS_n = BS_{mux} + BS_{oh} + vbv\_\text{max} - vbv\_buffer\_size \quad (1\text{-}11)$$

where $vbv\_\text{max}$ is max vbv_buffer_size for MPEG-1 video.

The size of B (a.k.a., BS) is proposed based on media types as follows:

- Audio

$$BS_n = BS_{mux} + BS_{dec} + BS_{oh} = 3584 \text{ bytes} \quad (1\text{-}12)$$

where the size of the AU decoding buffer and the size of PES packet overhead buffer constraints $BS_{dec} + BS_{oh} \leq 2848$ bytes. (1-13)

For ISO/IEC 13818-7 ADTS audio, the number of channels and corresponding bitrates is shown in Table 1-2.

**Table 1-2 Size of buffer B for T-STD Model for Audio**

| Number of Channels | $BS_n[bytes]$ |
|---|---|
| 1~2 | 3584 |
| 3~8 | 8976 |
| 9~12 | 12804 |
| 13~48 | 51216 |

- Systems

$$BS_{sys} = 1536 \text{ bytes} \quad (1\text{-}14)$$

There are two ways to fetch the content out of the MB to the EB – Leak method and Vbv_delay method. The Leak method is basically CBR filling up the EB buffer with peak or higher-than-ES-target bit rate. If there is a PES packet payload data in the MB and the EB is not full, the PES packet payload is transferred from MB to EB at a rate equal to Rbx. If the EB is full, data is not removed from the MB. When there is no PES packet payload data present in the MB, no data is removed from the MB. The Rbx rate is proposed based on Levels as follows:

- Low/Main levels

$$Rbx_n = R_{max}[profile, level]\ . \qquad (1\text{-}15)$$

- High 1440/ High levels

$$Rbx_n = Min\{1.05 \times R_{es}, R_{max}[profile, level]\}\ . \qquad (1\text{-}16)$$

- Constrained Parameters ISO/ IEC 11172-2 (MPEG-1 video) bitstreams

$$Rbx_n = 1.2 \times R_{max} \qquad (1\text{-}17)$$

where $R_{max}$ is the maximum bit rate for a Constrained Parameters bitstream in ISO/IEC 11172-2.

Vbv_delay method is piecewise-CBR filling up of the EB buffer in vbv_delay advance. The vbv_delay method specifies precisely the time at which each byte of coded video data is transferred from MB to EB, using the vbv_delay values coded in the video elementary stream. When the vbv_delay method is used, the final byte of the video picture start code for picture j is transferred from MB to EB at the time $td_n(j) - vbv\_delay(j)$, where $td_n(j)$ is the decoding time of picture j and $vbv\_delay(j)$ is the delay time, in seconds, indicated by the vbv_delay field of picture j.

The piecewise constant rate Rbx is proposed as follows:

$$Rbx(j) = \frac{NB(j)}{vbv\_delay(j) - vbv\_delay(j+1) + td_n(j+1) - td_n(j)} \qquad (1\text{-}18)$$

where NB(j) is the number of bytes between the final bytes of the picture start codes (including the final byte of the second start code) of pictures j and j+1, excluding PES packet header bytes.

When data is removed from EB or B, the unit of the processing is typically AUs. From each elementary stream buffer EB or B, the access unit $A_n(j)$ that has been longest in the buffer is removed with stuffing bytes attached to it at $td_n(j)$. The decoding time $td_n(j)$ is specified in the DTS with corresponding AUs. In contrast, systems data are not in units of AU. Instead, data is defined to be removed from $B_{sys}$ at a rate of $R_{sys}$ whenever there is at least one byte available in the buffer $B_{sys}$.

$$R_{sys} = \max(80000 bits/s, transport\_rate(i) \times 8 bits/500) \quad (1\text{-}19)$$

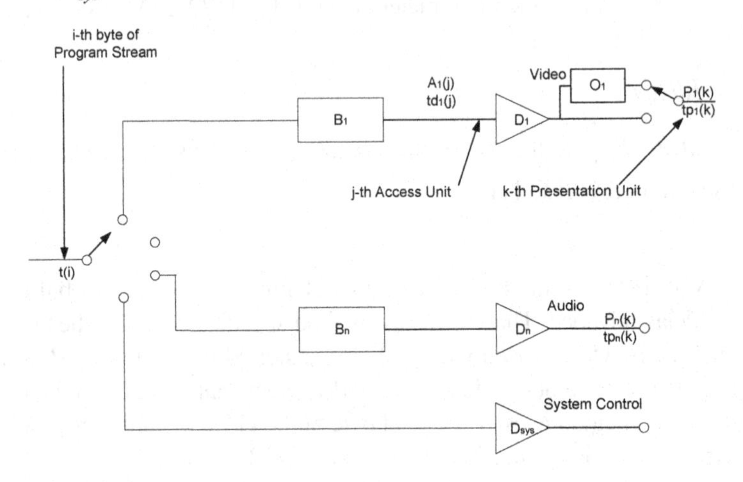


**Figure 1-9 Program Stream System Target Decoder Model**

The P-STD is a hypothetical system decoder that is synchronized exactly with the system encoder. A demux, system control, buffers and media decoders altogether define P-STD as shown in Figure 1-9. The MPEG-2 Systems standard specifies certain rules for multiplexers in the form of P-STD for compliant system bitstreams to be decodable.

The input of P-STD is bytes of multiplexed PSs at the rate specified in the Pack header. The input data of P-STD is assumed to be piecewise constant rate along with Packs. There is only one type of buffer defined in P-STD –Multiplexing Buffer (B). TB is not necessary in P-STD since network delivery jitter would not happen. However, B is still defined to reside at P-STD.

The purpose of B is not to alleviate packet multiplexer jitter any more, but to be used for local data buffer for depacketized elementary streams. The processing speed is infinite and local bandwidth is assumed to be enough in P-STD. Since a local copy of the media is already available, certain system characteristics can be somewhat ignored in practice. For example, multiplexer jitter can be ignored in P-STD. In other words, there can be practically no multiplexer jitter when a local copy of the media is always available and the processing speed is boundless. When multiplexer jitter can be ignored, B is not necessary to be broken down further into two sub-buffers – Multiplexing Buffer (MB) and Elementary stream decoder Buffer (EB) – for video streams.

In terms of system implementations, output buffer rates are not that critical since data is always available on local storage media and the bandwidth is enough internally. However, the size of buffers is more important to plan in advance for resource policy at the decoder. For example, given a DVD bitstream, the player has to decide how much memory should be allocated in advance for the systems decoder and decoder bitstream buffers.

In such a model, the theoretical input arrival time, *t(i)*, of i-th byte, through pumping from local storage media shall be constructed from a SCR value and the rate at which data arrives (program_mux_rate in Pack header) is as follows:

$$t(i) = \frac{SCR(i')}{system\_clock\_frequency} + \frac{i - i'}{program\_mux\_rate \times 50} \quad (1\text{-}20)$$

where $i$ is the index of any byte in the Pack, including the Pack header

$i'$ is the index of the byte containing the last bit of the *system_clock_reference_base* field in the Pack header.

$SCR(i')$ is the time encoded in the *system clock reference base* and extension fields in units of the system clock.

The PES packet data from elementary stream n is passed to the input buffer for stream n, $B_n$. Transfer of byte i from the system target decoder input to $B_n$ is instantaneous, so that byte i enters the buffer for stream n, of size $BS_n$, at time t(i). Bytes present in the Pack headers, Program Stream Maps, Program Stream Directories, or PES packet headers of the Program Stream such as SCR, DTS, PTS and packet_length fields, are not delivered to any of the buffers, but may be used to control the system. However, it is possible to construct a stream of data as a contiguous stream of PES packets with the same elementary stream and the same stream_id. Such a stream is called a PES stream. The PES-STD model for a PES stream is identical to P-STD. The only difference is that Elementary Stream Clock Reference (ESCR) is used in the place of the SCR and ES_rate is used in the place of program_mux_rate.

The size of B (a.k.a., BS) for PES-STD is proposed based on media types as follows:

- ITU-T Rec. H.262 | ISO/IEC 13818-2 video

$$BS_n = VBV_{\max}[profile, level] + BS_{oh} \quad (1\text{-}21)$$

where PES packet overhead buffering

$$BS_{oh} = (1/750)\sec\times R_{\max}[profile, level]. \quad (1\text{-}22)$$

- ISO/IEC 11172-2 video

$$BS_n = VBV_{\max} + BS_{oh}. \quad (1\text{-}23)$$

- ISO/IEC 11172-3 or ISO/IEC 13818-3 audio

$$BS_n = 2848 \text{ bytes}. \quad (1\text{-}24)$$

### Elementary Streams and Packetized Elementary Streams

Elementary streams (ES) consist of compressed data from a single source such as audio, video, data, ancillary data, etc. For video or audio, ES is nothing but syntax compliant compressed streams defined in the standards. The ancillary data are mainly used for synchronization, identification, and characterization of the source information. However, ES itself is a little bit risky to use since it does not have any DTS/ PTS information. Without DTS/ PTS, the decoder might crash with the ES bitstream in certain systems when a random access occurs. Therefore, it is generally the responsibility of content providers to make a complete PS or TS with all necessary information. The ESs themselves are first packetized into either constant-length or variable length packets to form Packetized Elementary Streams (PES) as shown in Figure 1-10. Figure 1-11 depicts detail syntax diagram for PES packets. One of the optional fields contains PTS/ DTS, ESCR and ES_rate, which are very important in system timing as shown in Figure 1-11.

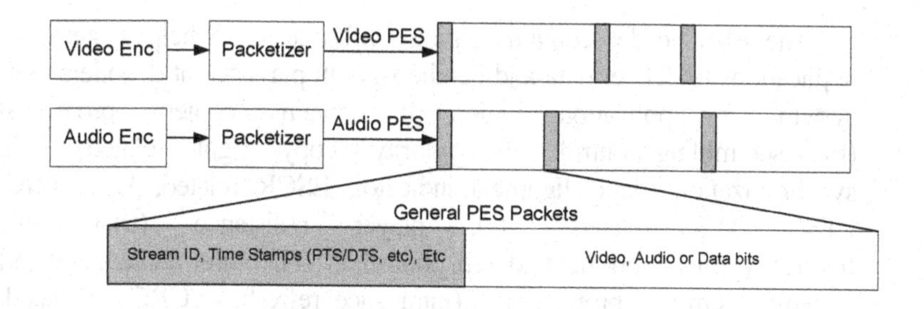


**Figure 1-10 General PES**

The stream_id is a basic means to look ahead about what the content actually is in the PES payload. There are mainly three types of payload contents –media (audio or video), system data (program_stream_map, ECM, EMM, program_stream_directory, DSMCC_stream, ITU-T RE. H.221.1 type E-stream, etc.) and padding_stream. This can be told easily from higher layer just by peeking into the stream_id data as shown in Table 1-3.

PES_packet_length data means the number of bytes in the PES packet following the last byte of the field. A value "0" indicates that the PES packet length is neither specified nor bounded and is allowed only in PES packet whose payload consists of bytes from a video elementary stream contained in TS packets. This PES_packet_length syntax element is mainly used when PES encapsulation is done with VC-1 video streams. Note that this field is used to indicate VC-1 ES as discussed in Chapter 2.

**Table 1-3 Stream_id Assignment Examples (MPEG IS: Table 2-18)**

| Stream_id | Stream coding |
|---|---|
| 1011 1100 | Program_stream_map |
| 110x xxxx | ISO/IEC 13818-3 or ISO/IEC 11172-3 audio stream number x xxxx |
| 1110 xxxx | ISO/IEC ITU-T Rec. H.262\| ISO/IEC 13818-2 or ISO/IEC 11172-2 video stream number xxxx |
| 1111 1111 | Program_stream_directory |
| ……… | ……… |

The PES header contains various information to help systems or applications to deliver data and handle smooth playback at decoders over system or application issues. It contains content protection (PES_scrambling_control, PES_priority, copy right related, etc), synchronization (data_alignment_indicator, ESCR related, ES related, PTS/ DTS related, etc), player assistance (trick_mode, frequency_truncation, field_id, rep_control, P-STD buffer related, etc) and system/ error protection (intra_slice_refresh, CRC related, program_packet_sequence_counter, etc). Note that it is not mandatory for decoders/ players to make use of all of these fields. Some might want to use specific subset of them.

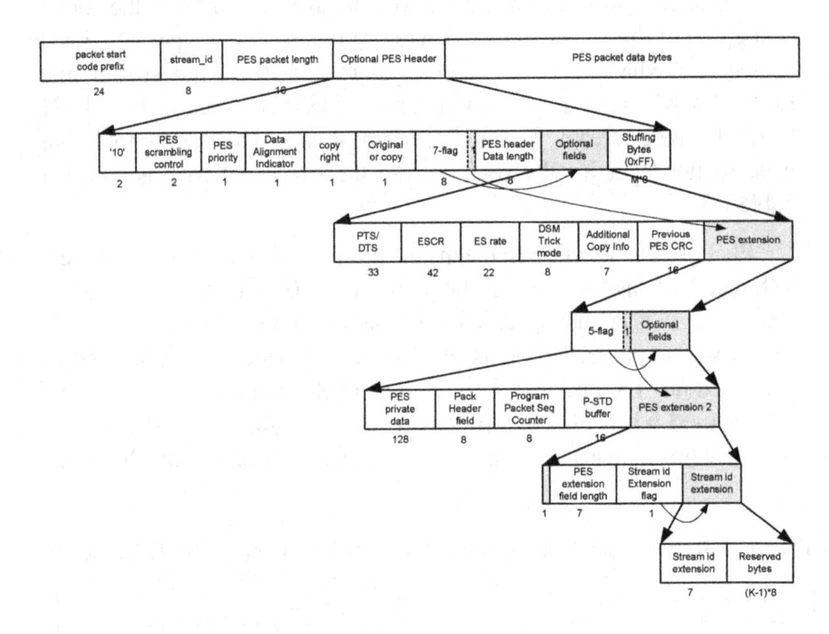


**Figure 1-11 PES Packet Syntax Diagram**

## Program Stream Map PES

Program Stream Map PES (PSM-PES) shows the relationship between a PS and its own ESs. Since this packet is a special case of PES packet, packet_start code prefix is the same as that of PES. Figure 1-12 depicts the syntax of PSM-PES. The stream_id position of the PES contains map_stream_id (0xbc) in Table 1-3. PES packet length is renamed as "program_stream_map_length." The first N-loop descriptors are about PS information, and are called Program Descriptors (PD). The second N-loop descriptors are about ES information, and are called Program Element Descriptors (PED). PEDs in PESs are nothing but additional information that an encoder wants to pass to decoders in the media layer, while PEs in PSs (i.e., PSM) or TSs (i.e., PMT) are additional information in the program or transport layer. Figure 1-12 shows also about what would be the list of ESs and additional information for each ES with elementary_stream_id and stream_type fields as shown in Table 1-3 and Table 1-4, respectively.

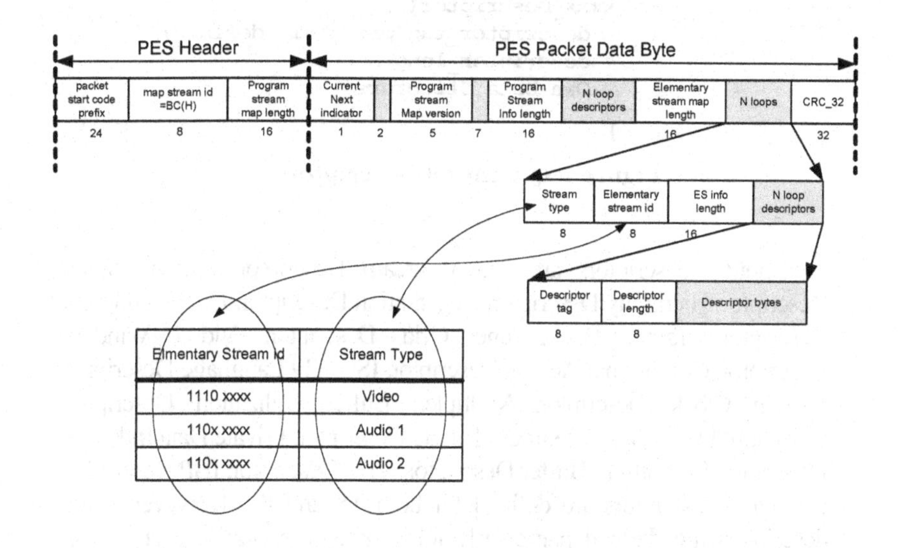


**Figure 1-12 PSM-PES Packet Syntax Diagram**

PD and PED are structures that may be used to extend the definitions of Programs and Program Elements. All descriptors have a format that begins with an 8-bit tag value that is followed by an 8-bit descriptor length and data fields as shown in Figure 1-13.

**Table 1-4 Stream_type Assignment (MPEG IS: Table 2-29)**

| Value | Description |
|---|---|
| 0x00 | ITU-T ISO/IEC Reserved |
| 0x01 | ISO/IEC 11172-2 Video |
| 0x02 | ITU-T Rec. H.262\| ISO/IEC 13818-2 or ISO/IEC 11172-2 video stream |
| 0x03 | ISO/IEC 11172-3 Audio |
| ………. | ………. |

```
xxxx_Descriptor( ){
  descriptor_tag(privately defined)
  descriptor_length
  Own Data Structure

}
```

**Figure 1-13 General Descriptors**

Popular descriptors are Video Stream Descriptor, Audio Stream Descriptor, Hierarchy Descriptor, Registration Descriptor, Data Alignment Descriptor, Target Background Grid Descriptor, Video Window Descriptor, Conditional Access Descriptor, ISO 639 Language Descriptor, System Clock Descriptor, Multiplex Buffer Utilization Descriptor, Copyright Descriptor, Maximum Bitrate Descriptor, Private Data Indicator Descriptor, Smoothing Buffer Descriptor, STD Descriptor, IBP Descriptor, etc. Most descriptors are defined for both TS and PS. However, some descriptors are defined particularly for a specific scenario as shown in Table 1-5. Descriptors reside at PMT in TS packets, while they are put in PSM-PES for PS packets.

**Table 1-5 PD and PED (MPEG IS: Table 2-39)**

| Descriptor_tag | TS | PS | Identification |
|---|---|---|---|
| .......... | | | .......... |
| 2 | x | x | Video_stream_descriptor |
| 3 | x | x | Audio_stream_descriptor |
| 9 | x | x | CA_descriptor |
| 10 | x | x | ISO_639_language_descriptor |
| 35 | x | | MultiplexeBuffer_descriptor |
| .......... | | | .......... |

### Program Stream Directory PES

The directory for an entire stream is made up of all directory data carried by Program Stream Directory PES (PSD-PES) packets with the directory_stream_id. Since this packet is a special case of PES packet, packet_start code prefix is the same as that of PES, thus making it a PSD-PES. Basically, PSD-PES provides directory services about where specific AUs are in the PES bitstreams as shown in Figure 1-14. Figure 1-15 depicts the syntax of PSD-PES. The stream_id position of the PES contains directory_stream_id (0xee) in Table 1-3. PES packet length is interpreted as in regular PES packets to indicate the total number of bytes in the program_stream_directory immediately following this field. The number_of_access_units value is the number of AUs that are referred to in this PSD-PES packet. The prev_directory_offset value tells the byte address offset of the packet start code of the previous PSD-PES packet, while the next_directory_offset value indicates for the next PSD-PES packet. The PES_header_position_offset value gives byte offset address of the first byte of the PES packet containing the AU referenced. The reference_offset value indicates the position of the first byte of the referenced AU, measured in bytes relative to the first byte of the PES packet containing the first byte of the referenced AU. The PTS value is the PTS of the AU referenced. The bytes_to_read value is the number of bytes in the PS after the byte indicated by reference_offset that are needed to decode the AU completely.

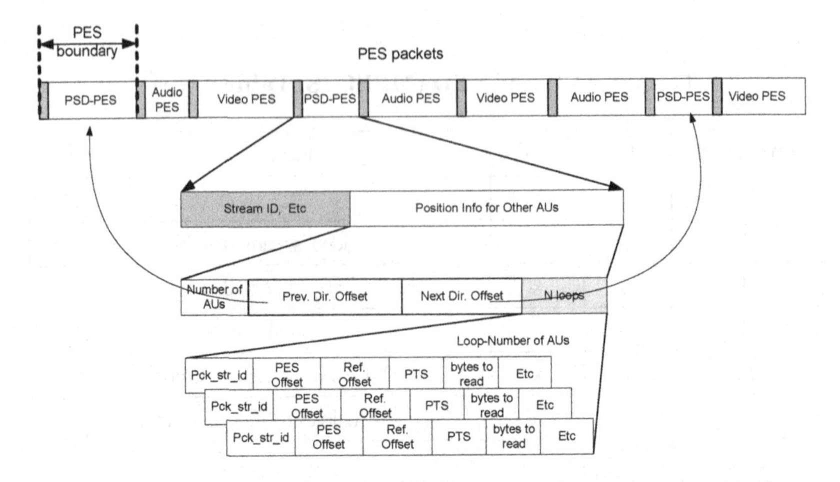


**Figure 1-14 PSD-PES Directory Services**

The intra_coded_indicator specifies whether the referenced AU is not predictively coded. The coding_parameters_indicator indicates the location of coding parameters that are needed to decode the referenced AU. For example, "00" means that all coding parameters are set to their default values, while "01" implies that all coding parameters are specified in this referenced AU. PSD-PES is particularly designed for handling PSs, while the same functionality is included in Program Specific Information (PSI) in TSs.

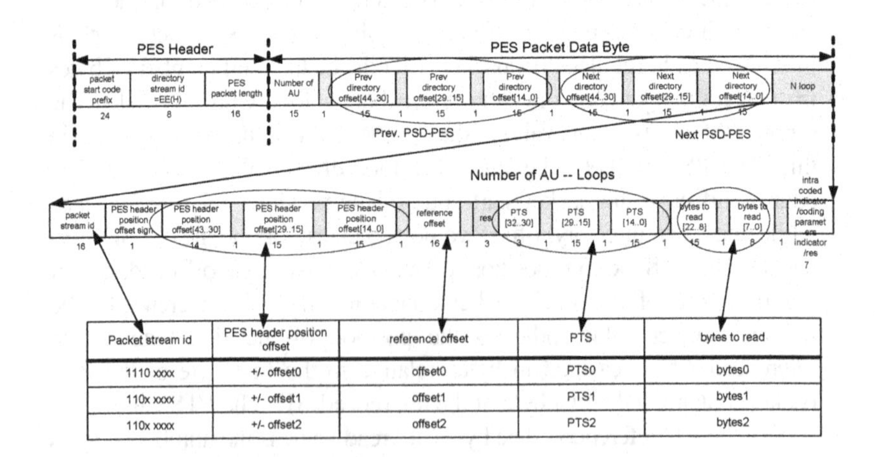


**Figure 1-15 PSD-PES Packet Syntax Diagram**

### Transport Stream

A Program Stream carries on Program and is composed of Packs. A Pack consists of a pack header followed by a variable number of multiplexed PES packets as shown in Table 1-4. Note that both TS and PS packets are based on PES packets as shown in Figure 1-16. However, PS needs one more layer of encapsulation called "Pack." TSs are defined for transmission networks that may suffer from errors and jitters. Therefore, an important part of TS is error protection capability over the networks and splicing of multiple programs.

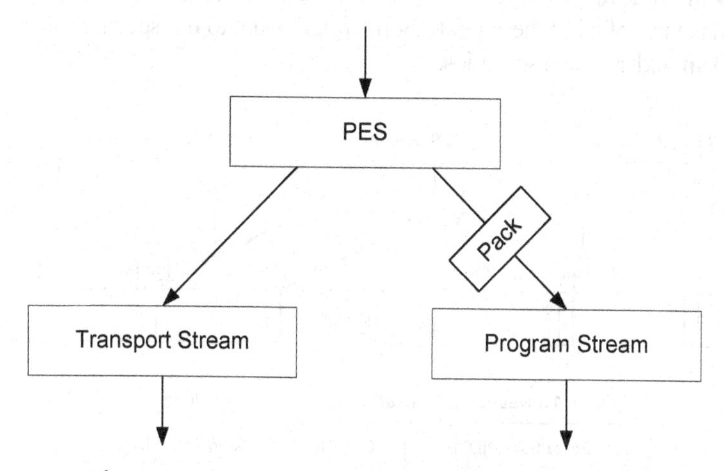


**Figure 1-16 PES, a common base for TS and PS**

The size of a TS packet is fixed at 188 bytes as shown in Figure 1-17 and Figure 1-18. The first data in the TS packet header is the sync_byte (0x47) that is byte aligned as shown in Figure 1-18. This implies that sync_byte should be examined to find the start point of a TS packet only at byte-aligned boundaries and sync_byte emulation should be avoided in any part of TS packets. One more note is that the left-over TS part is filled with stuffing data when the PES packet is finished as shown in Figure 1-17. The TS packet header contains various information to help networks and systems to deliver data smoothly to decoders. It contains transport assistance (transport_priority, PID, transport_scrambling_control, continuity_counter, discontinuity_indicator, ltw_offset related, etc), synchronization (PCR/OPCR related, etc), player assistance (random_access_indicator, splicing_point related, transport_private_data related, splice_countdown, seamless_splice/splice_type related, DTS_next_AU, etc) and system/ error protection (transport_error_indicator, elementary_stream_priority, etc). The two most important aspects of all to support are error handling and splicing between multi-programs. For example, certain important TS packets such as those of I-reference frame content can improve delivery quality with use of transport_priority, elementary_stream_priority, random_access_indicator, etc. When program switch is expected, splice downcounter and continuity_counter can be used to let decoders know an accurate switch point in advance. Note that it is not mandatory for networks/ systems to make use of all of these fields. Some might want to use specific subsets of them under certain scenarios.

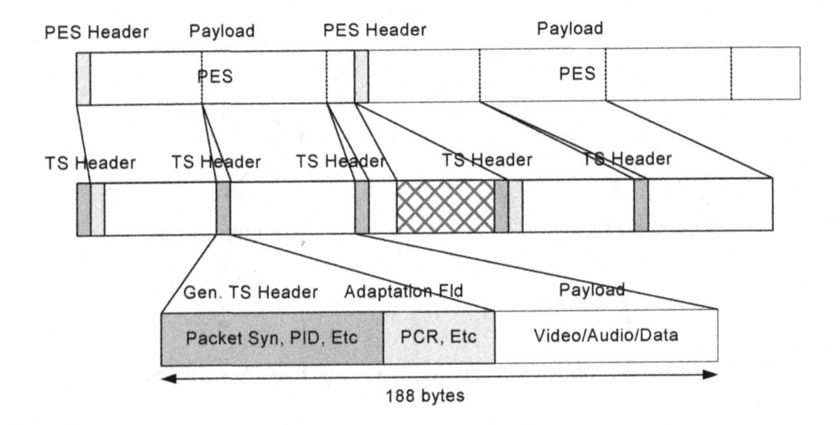


**Figure 1-17 Encapsulation with TS on PES**

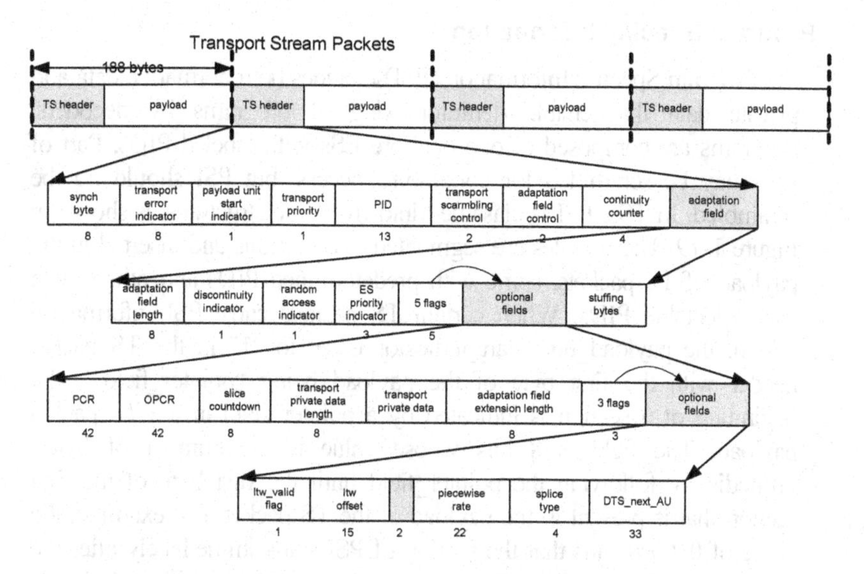


**Figure 1-18 TS Packet Syntax Diagram**

The PID is a basic means to look ahead about what the content actually is in the TS payload. There are mainly three types of payload contents –media (audio or video), system data (Program Association Table, Conditional Access Table, etc.) and null data. This can be determined easily from a higher layer just by peeping the PID as shown in Table 1-6.

**Table 1-6 PID Table (MPEG IS: Table 2-3)**

| Value | Description |
|---|---|
| 0x0000 | Program Association Table |
| 0x0001 | Conditional Access Table |
| 0x0002 | Transport Stream Description Table |
| 0x0003-0x000f | reserved |
| 0x00010..0x1ffe | May be assigned as network_PID, Program_map_PID, elementary_PID, or for other purposes |
| 0x1fff | Null packet |

## Program Specific Information

Program Specific Information (PSI) includes both normative data and private data that enable demultiplexing of programs by decoders. Programs are composed of one or more ESs with labeled PIDs. Part of ESs may be scrambled for conditional access, but PSI should not be scrambled. In TSs, PSI is classified into five table structures as shown in Figure 1-19. These tables are segmented into sections and inserted in the payload of TS packets, some with predetermined PIDs and others with user selectable PIDs. When certain TS packets carry PSI information within, the payload_unit_start_indicator is set to "1" in the TS packet header with the first byte of the payload being "pointer_field." The beginning of a section is indicated by a pointer_field in the TS packet payload. The field is 8 bits whose value is the number of bytes, immediately following the pointer_field until the first byte of the first section that is present in the payload of the TS packet. For example, the value of 0x00 means that the section of PSI starts immediately after the pointer_field. Generally, PSI can start at any byte-position with the pointer_field in the payload. The use of payload_unit_start_indicator is depicted in Figure 1-20 for PSI section insertion in TS payload as explained earlier.

| Table Name | PID # | Main Description |
|---|---|---|
| Program Association Table | 0 | Associates Program # with PMT |
| Program Map Table | Assigned in PAT | Associates PID with Program |
| Network Information Table | Assigned in PAT | Contains Physical Network Params |
| Conditional Access | 1 | Associates PID's with private streams |
| Transport Stream Description Table | 2 | Associates one or more descriptors from pre-defined descriptor IS Table 2-39 to an entire Transport Stream |

**Figure 1-19 PSI in TS Payload**

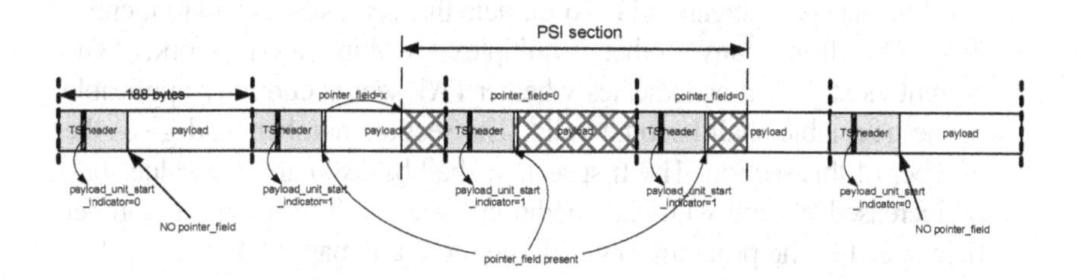


**Figure 1-20 Section in TS Payload for PSI**

The payload formats of Program Association Table (PAT), Program Map Table (PMT) and Conditional Access Table (CAT) are called Program Association section, Program Map section and Conditional Access section as shown in Figure 1-21, Figure 1-22 and Figure 1-23, respectively.

PSI sections contain various information to help systems tune decoders for specific programs. PSI is nothing but table entry information required to tune a decoder up to a specific media channel like CNBC. Program number in this sense is just like conventional channel number. The table_id field identifies the contents of a TS PSI section as shown in Table 1-7.

**Table 1-7 table_id Assignment Values (MPEG IS: Table 2-26)**

| Value | Description |
|---|---|
| 0x00 | Program_association_section |
| 0x01 | Conditional_access_section |
| 0x02 | TS_program_map_section |
| 0x03 | TS_description_section |
| 0x04 | ISO_IEC_14496_scene_description_section |
| 0x05 | ISO_IEC_14496_object_description_section |
| 0x06~0x37 | ITU-T Rec. H.222.0\| ISO/IEC 13818-6 |
| 0x38~0x3f | Defined in ISO/IEC 13818-6 |
| 0x40~0xfe | User private |
| 0xff | forbidden |

The transport_stream_id is 16-bit field that serves as a label to identify this TS from any other multiplex within a network. The current_next_indicator indicates whether PAT sent is currently applicable or the next table shall become valid. The section_number field gives the number of this section. The first section shall be 0x00 and the value shall be increased by one with each additional section. The program_number field specifies the program to which the program_map_PID is applicable. For 0x0000, the following PID reference shall be the network PID as shown in Figure 1-21 and Figure 1-24.

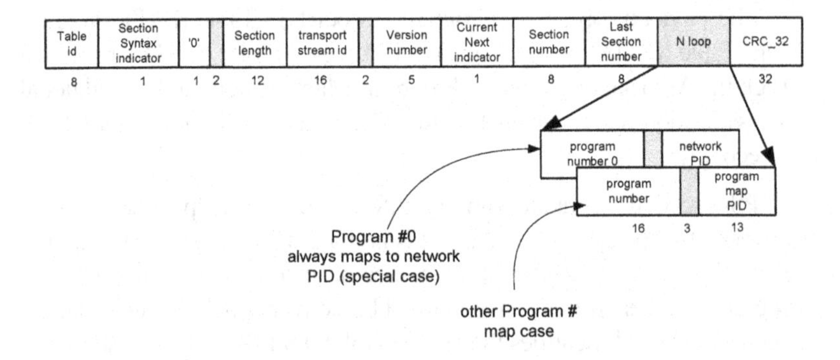


**Figure 1-21 PA Section Syntax Diagram for PAT**

A fundamental function for TS processing is packet filtering based on PID as shown in Figure 1-24. As shown in Figure 1-17, TS packets with PID #0 contain always PAT data. When all necessary data is extracted, PAT is obtained as in Figure 1-24. In the example, PID #15 contains PMT for Program #9, while PID #27 contains PMT for Program #35. The PID #14 is network PID through which Network Information Table (NIT) can be found. NIT is not mandatory to be present, but provides information about physical network parameters such as FDM frequencies and satellite transponder numbers.

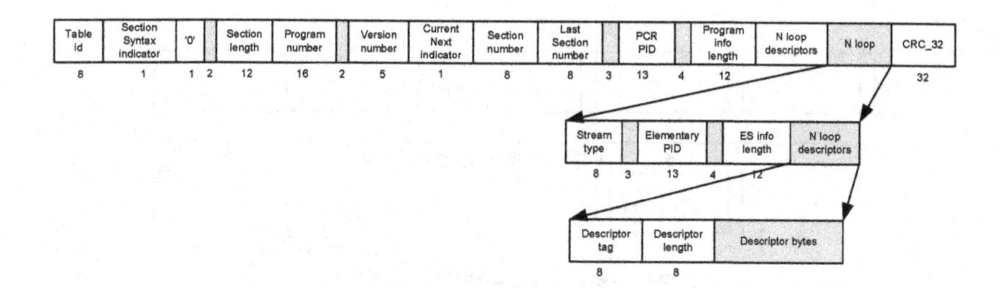


**Figure 1-22 PM Section Syntax Diagram for PMT**

Let's assume that Program #9 is selected to be received by an observer. Then, TS packets with PID #15 will be filtered to be received as shown in Figure 1-25. From PID #15, we can extract PMT as depicted in Figure 1-25. It basically indicates that TS packets with PID #46, PID #47 and PID #48 all belong to Program #9. By looking at stream_type, media type can be found too.

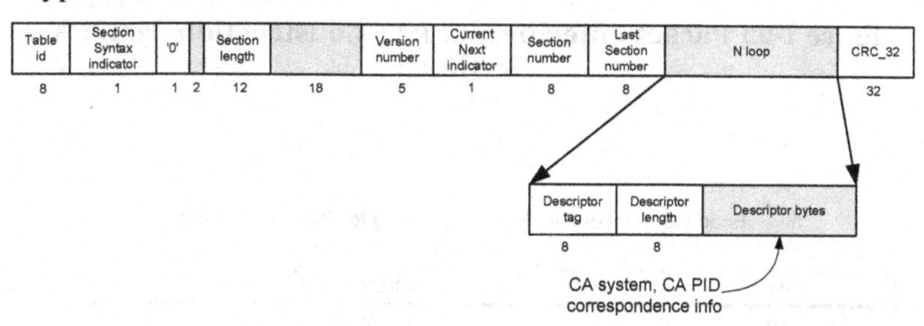


**Figure 1-23 CA Section Syntax Diagram for CAT**

CA is another table with fixed PID number. If TS packets with PID #1 are filtered, those are of CA section in the payload of TSs. CAT provides correspondence between CA systems and their Entitlement Management Message (EMM) streams that are system-wide conditional access management information specifying authorization levels of specific decoders. This may be used in both the TS_program_map_section and the program_stream_map. If any system-wide conditional access management information exists within a TS, a CA descriptor shall be present in the CA section as shown in Figure 1-23, where the CA_PID points to packets containing system-wide EMM. If any elementary stream is scrambled, a CA descriptor shall be present for the program that contains the elementary stream.

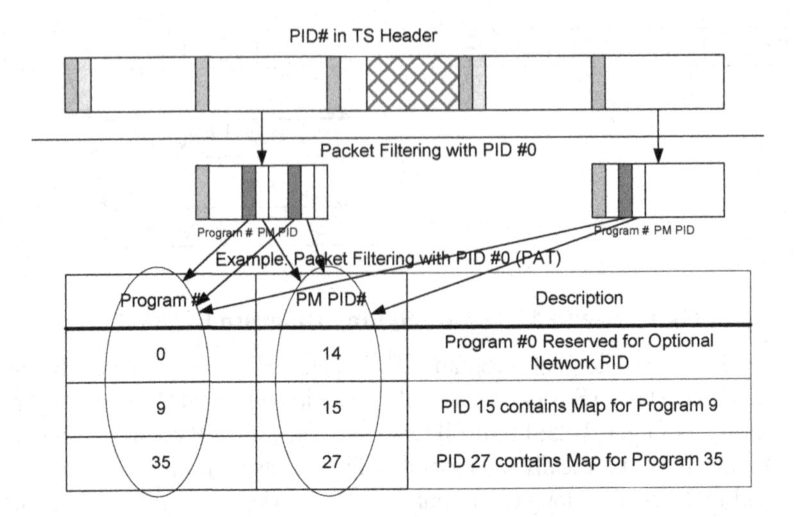


| Program # | PM PID# | Description |
|---|---|---|
| 0 | 14 | Program #0 Reserved for Optional Network PID |
| 9 | 15 | PID 15 contains Map for Program 9 |
| 35 | 27 | PID 27 contains Map for Program 35 |

**Figure 1-24 Packet Filtering and PAT Construction from PA Section**

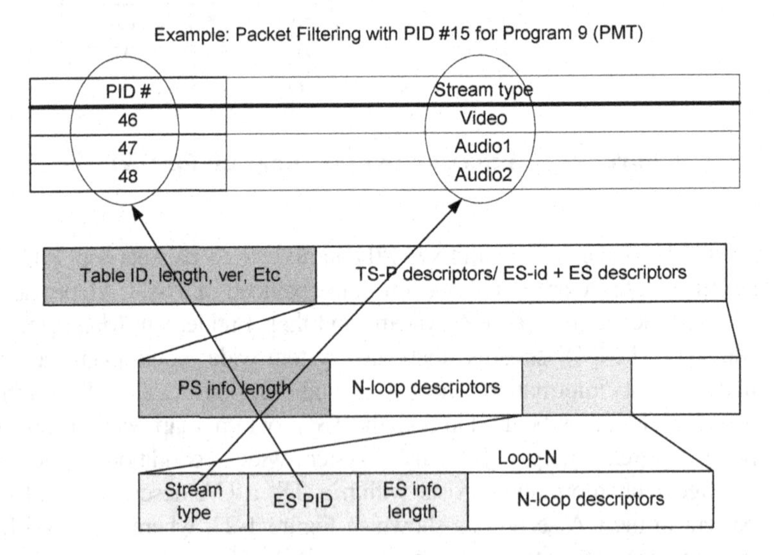


| PID # | Stream type |
|---|---|
| 46 | Video |
| 47 | Audio1 |
| 48 | Audio2 |

**Figure 1-25 PMT Construction from PM Section**

In summary, the tuning process to select a program from broadcast streams can be depicted as Figure 1-26. The PSI data are given as the earlier example used in this subsection.

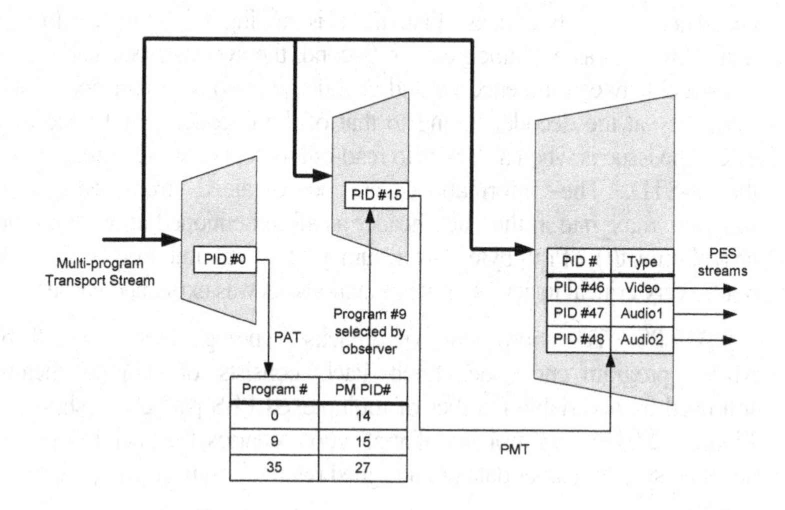


**Figure 1-26 Program Tuning Example summarized**

## Program Stream

PSs are defined for local playback environments with negligible errors. Therefore, an important part of PS is to represent/ store bitstreams in DSM such as DVD in compact and efficient manners. The synchronization issue in PSs is different from that of TSs. In TS packet transmission, the clock speeds of the sender and the receiver were different in general. However, there is no need to synchronize transport clocks in PSs playback mainly because all media data are already available locally at the decoder—no worry about transmission buffer overflow or underflow. Note that System Clock Reference (SCR) in the notion of PSs replaces PCR in TSs. The conceptual difference is that the PCR Time Stamp is constructed based on the reference clock of an encoder, while the SCR Time Stamp is constructed just based on the reference clock of decoders.

Ideally, the local clock at decoders can be adjusted based on SCR values with a VCO as was performed for TSs with PCR. This might provide perfectly synchronized timing with that of the encoder at which the bitstream was encoded. However, TS-like synchronization is not necessary due to two facts: first, there is no input buffer overflow or underflow as was explained earlier. Second, the system clock variance is too small between the encoder and decoder pair, so a correct operational frequency at the decoder tuning to that of the encoder is not necessary. Then, the issue is when a Pack is to read-out to deliver at the input point of the P-STD. The information can be obtained from SCR and program_max_rate in the Pack header as aforementioned about the input arrival time, t(i), of i-th byte. The accuracy of arrival time in P-STD might not be very critical in actual implementations, as was explained earlier.

A PS is composed of Packs ending with a 32-bit MPEG_program_end_code. Each Pack consists of a pack header followed by a variable number of multiplexed PES packets as shown in Figure 1-27. Due to simplicity of application senarios, the Pack header and attached system header data are designed relatively simply compared with headers in TS.

Figure 1-28 depicts the PS syntax diagram in detail. The MPEG_program_end_code is 0x000001b9 and it terminates the PS. The SCR indicates the intended time of arrival of the byte containing the last bit of the system_clock_reference_base at the input of P-STD. The program_mux_rate specifies the rate at which the P-STD receives the PS during the Pack – byte arrival time is computed based on SCR and this value.

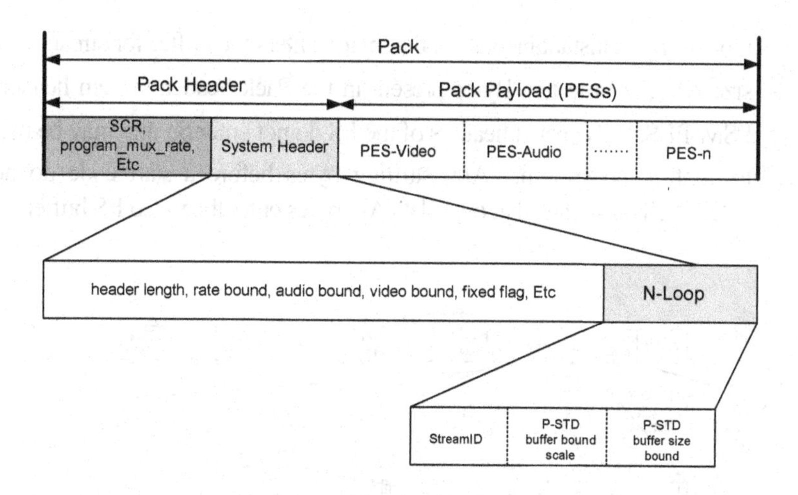


**Figure 1-27 Encapsulation of PES with Pack**

The system header is additionally but optionally defined in the Pack header. Basically any decoding system can just check system header parameters to see if it can handle the worst case scenario on the PS in terms of system resources. The rate_bound is an integer value greater than or equal to the maximum value of the program_mux_rate field coded in any Pack of the PS. The audio_bound means the maximum number of ISO/IEC 13818-3 and ISO/IEC 11172-3 audio streams in the PS for which the decoding processes are simultaneously active. The video_bound is similarly interpreted with ITU-T Rec. H.262| ISO/IEC 13818-2 and ISO/IEC 11172-2 streams. The fixed_flag tells whether either fixed or variable operation is indicated. The system_audio_lock_flag specifies a constant rational relationship between the audio sampling rate and the system_clock_frequency in the STD. The system_video_lock_flag is similarly interpreted between video frame rate and the system clock frequency. The stream_id indicates the coding and elementary stream number of the stream to which the following P-STD_buffer_bound_scale and P-STD_buffer_size_bound fields refer.

Data enters the P-STD at the rate specified by the value of the field program_mux_rate in the Pack header. The PES packet data bytes from video ES n are passed to the input $B_n$. Transfer of byte i from the STD input to $B_n$ is instantaneous, so that byte i enters the buffer for stream n, of size $BS_n$, at time t(i). Bytes present in the Pack header, system headers, PSM-PES, PES packet headers of the PS do not enter $B_n$ and may be used to control the system. Any stuffing bytes before a start code do not enter $B_n$. This means that only data AU bytes enter the video ES buffer.

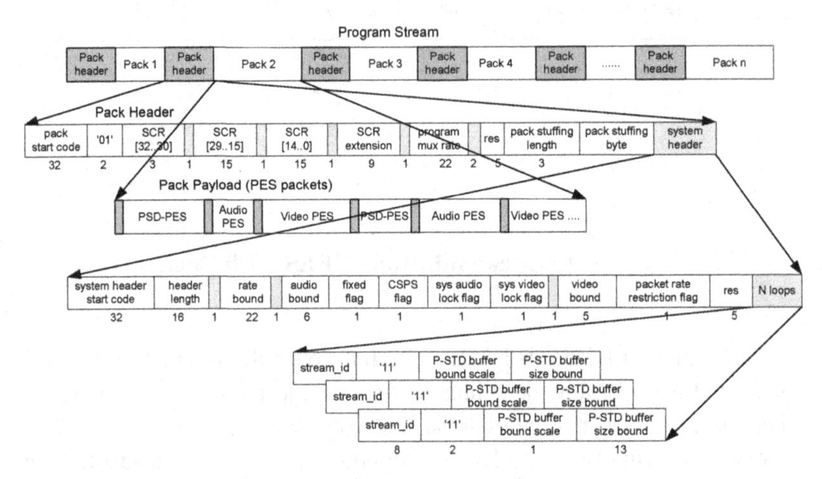


**Figure 1-28 PS Syntax Diagram**

## 1.3 Video Codec Internal and Data Flow

### VC-1 Encoder

Introduction to semantics and syntax of VC-1 and H.264 is covered in a reasonable level in Chapter 2. In this section, only fundamental processing units and data flow/ exchange among them are mentioned in the order of data processing. There are 8 basic functional blocks for VC-1 encoders – Intra-Prediction (Intra-P), Motion Estimation (ME), Motion Compensation (MC), Transform-Quantization (TQ), inverse Quantization-inverse Transform (iQ-iT), Overlap Transform Smoothing (OLT), In Loop Filtering (ILF) and Context Adpative Variable Length Coding (CA-VLC) modules as shown in Figure 1-29. The algorithm blocks shown in the Figure 1-29 can be pretty much mapped to Hardware Accelerators (HWAs). Each block or certain combination of blocks can constitute internal pipelining stages for higher performance in a practical implementation. To ensure good design, hardware/ software partitions can be further examined through profiling computational complexity of each algorithm module. On top of these blocks, rate control unit can be implemented as a separate HWA where a lot of statistics are extracted from a given video to be used as background information for mode decision. If intra mode is used for the MB, the inverse OLT and forward-transform are performed. Intra-P is then performed in the transform domain with DC/ AC transform coefficients. In VC-1, Intra-P happens to be in transform domain with DC/ AC prediction. The residual signal in the transform domain is to be coded after DC/ AC prediction. If inter mode is used for the MB, the ME is to apply to get temporal residual signals in pixel domain. This temporal residual signal is again forward-transformed. Note that the ME unit here contains the subtraction operation. The MC unit performs the inverse operation of ME. Note that MC unit contains the addition operation. The TQ block maps spatial intra or inter residual signals into the transform domain, while the iQ-iT block backward-transforms intra or inter residual signals to the pixel domain. The MC and iQ-iT blocks are necessary in the encoder since the encoder needs to trace down actual operation/ data of the decoder in the loop for correct prediction. If this is not maintained correctly, the encoder and decoder pair drifts from each other. In the course of encoding, the reconstructed signal is smoothed with both the OLT and the ILF functional blocks. Both the OLT and the ILF are all in the loop, so their effects are propagated to future picture coding. The CA-VLC block maps symbols to binary codes.

Some of functional units require data from adjacent blocks. For example, Intra-P needs neighborhood pixel values from adjacent blocks. However, some blocks don't have enough data when they are located on edges. ME/MC blocks generally need data from reference pictures. However, some blocks need data from outside of reference pictures. Then, a pre-defined padding data might be required. OLT and ILF need neighborhood data, too.

In VC-1, slice boundaries are defined at the edge of MBs row. Slices can be more than one row of MBs. Processing that overlaps slices is prohibited in VC-1 from propagating any errors through slices. In VC-1 a slice is not the main body of coded picture. At every frame time, a picture header is delivered and slices are not dependent on the coding types of MBs.

When the ME/MC blocks are needed, a reference data block is fetched/ interpolated to a quarter-pel resolution. When the Intra-P block is needed, the internal or external boundary data of the MB is fetched/ used for Intra-P. When the ILF block is needed, the top or left part of the MB in adjacent MBs are fetched/ used for deblocking filter operations.

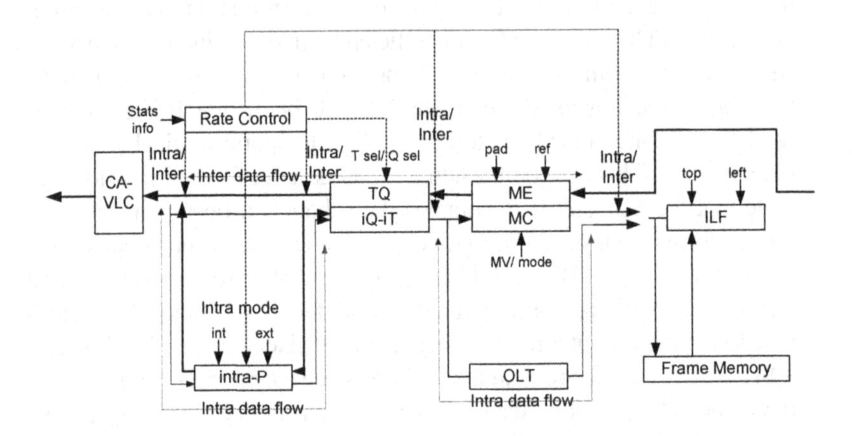


**Figure 1-29 VC-1 Encoder Functional Blocks and Data Flow**

## VC-1 Decoder

There are six basic functional blocks for VC-1 decoders – VLD, iQ-iT, MC, Intra-P, OLT and ILF as shown in Figure 1-30. The VLD block maps binary codes to symbols. Intra-P is then performed in the transform domain with DC/ AC transform coefficients. The iQ-iT block backward-transforms intra or inter residual signal to pixel domain. If intra mode is used for the MB, the OLT is to apply to get reconstructed signal in pixel domain. If inter mode is used for the MB, the MC is to apply to get temporally reconstructed signal temporally in the pixel domain. Note that the MC block contains addition operation in the block. And then, the reconstructed signal is smoothed with the ILF functional blocks. Both the OLT and the ILF are in the loop, so their effects are propagated to future picture coding.

Some functional units require data from adjacent blocks. For example, Intra-P needs DC and AC values from adjacent blocks. However, some blocks don't have enough data when they are located on edges. The MC block generally needs data from reference pictures. However, some blocks need data from outside of reference pictures. Then, a pre-defined padding data might be required. OLT and ILF need neighbor data, too.

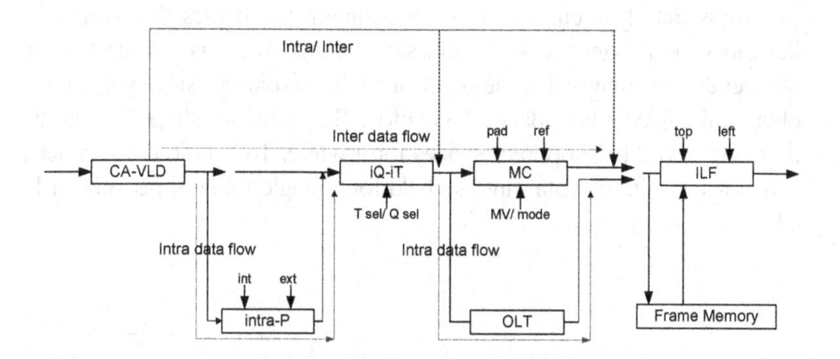


**Figure 1-30 VC-1 Decoder Blocks and Data Flow**

## H.264 Encoder

There are seven basic functional blocks for H.264 encoders – Intra-P, ME, MC, TQ, iQ-iT, ILF and CA-VLC/ CA-BAC modules as shown in Figure 1-31. The algorithm blocks shown in the Figure 1-31 can be pretty much mapped to HWAs. On top of these blocks, a rate control unit can be implemented as a separate HWA where a lot of statistics are extracted from a given video to be used as background information for mode decision. If intra mode is used for the MB, the Intra-P is performed in the spatial domain. The residual signal in the spatial domain is to be transformed after Intra-P. If inter mode is used for the MB, ME is applied to get temporal residual signals in pixel domain. This temporal residual signal is again forward-transformed. Note that the ME unit here contains the subtraction operation. The MC unit performs the inverse operation of ME. Note that the MC unit contains the addition operation. The TQ block maps spatial intra or inter residual signals into the transform domain, while the iQ-iT block backward-transforms intra or inter residual signals to the pixel domain. In the course of encoding, the reconstructed signal is smoothed with the ILF functional block. The ILF is in the loop, so its effect is propagated to future picture coding. The CA-VLC/ CA-BAC block is to map symbols to binary codes. In H.264, CA-BAC is a very computationally intensive unit. HWA implementation for this unit is a good option.

H.264 allows Arbitrary Slice Order (ASO) and Multiple Slice Group (MSG) in Baseline Profile, while it does not allow them in other Profiles. The tools mainly focus on error concealment techniques that would be keys in various patterns of transmission errors. There is a restriction on MB-level pipelining when the order of MBs is arbitrary. Since we need to obtain all adjacent boundary MBs before ILF, arbitrary shaped slices are decoded/ saved to temporary spaces for a while. To handle more generic situations, the frame data must be held for a single frame time to be ILF-ed.

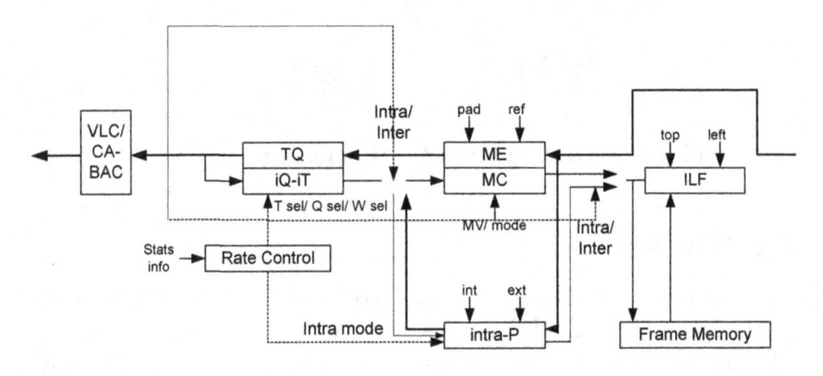


**Figure 1-31 H.264 Encoder Functional Blocks and Data Flow**

## H.264 Decoder

There are five basic functional blocks for H.264 decoders – VLD, iQ-iT, MC, Intra-P and ILF as shown in Figure 1-32. The VLD block maps binary codes to symbols. The iQ-iT block backward-transforms intra or inter residual signals to the pixel domain. If intra mode is used for the MB, the intra-P is applied to get the reconstructed signal. If inter mode is used for the MB, the MC is applied to get the reconstructed signal. Note that the MC contains addition operation in the block. Finally, the reconstructed signal is smoothed with the ILF block.

The restriction about aforementioned MB-level pipelining still exists in the Baseline Profile.

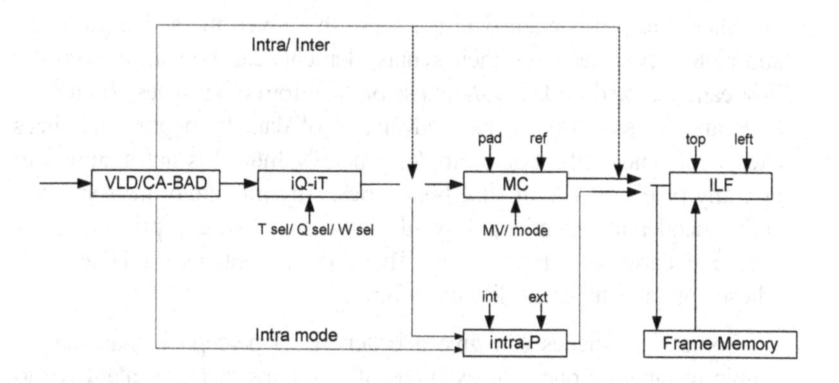


**Figure 1-32 H.264 Decoder Functional Blocks and Data Flow**

# 1.4 Independent Slice Decoder

## Slices and Errors

Conventional slices are defined as a sequence of MBs in raster scan order. The main purpose of slice composition is to provide immunity under transmission errors. If a slice is corrupted, such a decoder can move ahead to the next slice point to continue decoding. Corruption of a bitstream is determined with the VLD unit. When parsing the bitstream fails at a certain point, it is interpreted to contain an error at the point in the bitstream. Typically the entire slice is discarded unless a Reversible VLD (RVLD) mechanism is employed to localize the error point. To provide good visual quality at an error point, error concealment techniques can be used at the decoder.

Since slices are basic units for error immunity, an independent slice decoder is preferred in the implementation. Of course, slice decoding cannot be independent in general. For example, decoding a slice of P-MBs requires reference data. In fact, the term "independent" means to have something to do with errors. Any corrupted input information for slice decoding can initiate error concealment techniques to complete slice decoding for the given error slice. If all inputs are provided, slice decoding proceeds in a regular manner with or without error recovery techniques. If inputs are provided with some missing information, the independent slice decoding is performed with default contexts.

Slices are self-contained in the sense that given the active sequence and picture parameter set, their syntax elements can be parsed from the bitstream and be decoded without use of data from other slices. Therefore, tools are not supposed to take advantage of data from previous slices except reference data. For example, generally Intra-P is not assumed to take any pixels outside of slice boundaries. Any mistake of interpretation at the encoder and decoder pair would cause drift. One exception to cross-slice boundaries is ILF in H.264. There are the options for H.264 ILF whose options filter cross slice boundaries.

Some slice shapes can give a better visual perception than that of simple rectangular one. For example, slice groups that can interleave to compose a frame provide a better visual quality at bitstream errors as shown in Figure 1-36. Recent H.264 defines seven different slice groups to this end. And, these are called Multiple Slice Groups (MSG).

### Slices in MPEG-2

A Slice is a series of an arbitrary number of consecutive MBs. The first and last MBs of a slice shall not be skipped MBs. Every slice shall contain at least one MB and slices shall not overlap. There are two kinds of slice structures – general slice structure and restricted slice structure, as shown in Figure 1-33.

The general slice structure does not need a picture to be entirely covered by slices as shown in Figure 1-33 (a). The area not covered by slices is not coded and no information is needed for such area. The vertical position of an arbitrary slice is basically controlled by the syntax element "slice_vertical_position" in Slice header. This information is described in the last 8-bits of the slice_start_code. It is an unsigned integer giving the vertical position in MB units of the first MB in the slice. The horizontal position of an arbitrary slice is controlled by the syntax element "macroblock_address_increment" in MB header. This is a variable length coded integer that indicates the difference between macroblock_address and previous_macroblock_address. The maximum value of macroblock_address_increment is 33. In addition, predictions shall only be made from those regions of the picture that are enclosed in slices if the picture is subsequently used to form predictions.

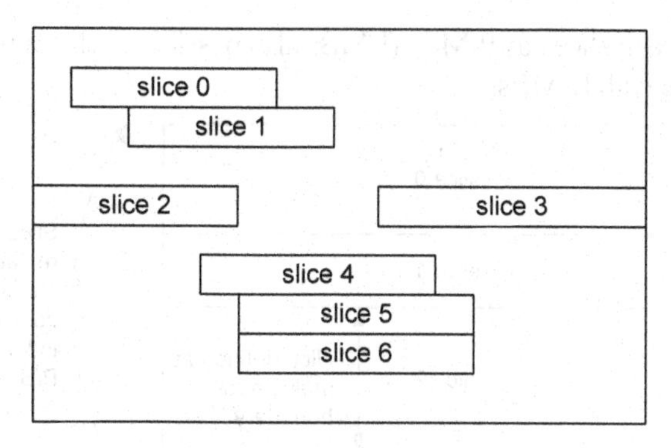


(a) general slice structure

slice 0
slice 1
slice 2 | slice 3
slice 4 | slice 5 | slice 6
slice 7
slice 8
slice 9 | slice 10
slice 11
slice 12
slice 13 | slice 14

(b) restricted slice structure

**Figure 1-33 Slice Shape in MPEG-2**

The restricted slice structure does need a picture to be entirely covered by slices as shown in Figure 1-33 (b). In this case, every MB in the picture shall be enclosed in a slice.

## Slices in VC-1

VC-1 defines a slice as a rectangular shape along with the raster scan order and MB row boundaries as shown in Figure 1-34 (a). The height of the slice can be more than 1 MB-height. VC-1 slices have nothing to do with coding modes – it is just a simple geographical division. If slices are taken from an I frame, each slice has all I-MBs. If slices are taken from a P frame, each slice has P-MBs (I-MBs also possible). The same is true for a B frame with B-MBs.

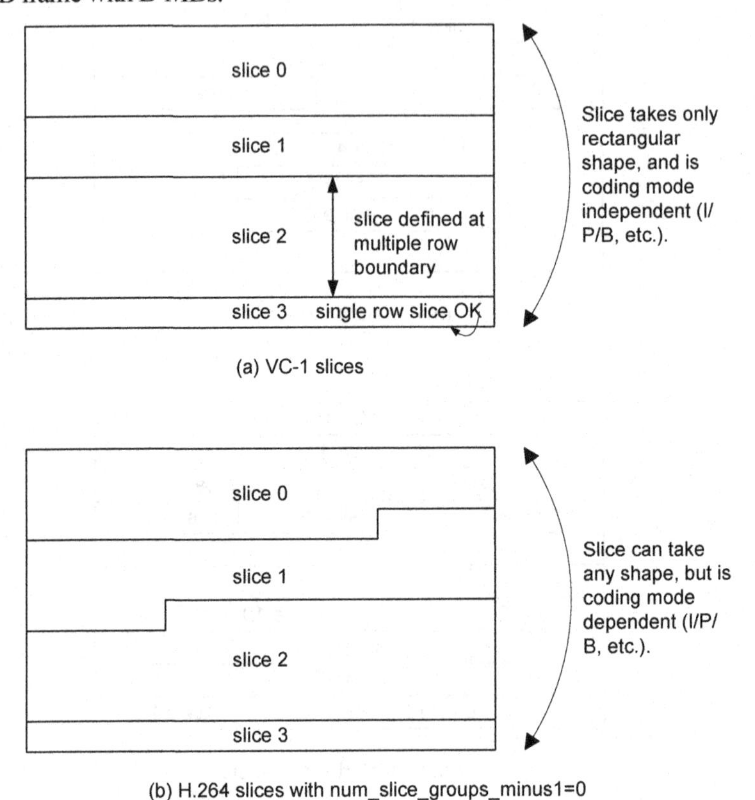


**Figure 1-34 Slice Shape Comparison between VC-1 and H.264**

### Slices in H.264

H.264 defines a slice as a group of MBs with a scanning order. A popular shape of slices in H.264 is the rectangular shape with the raster scanning order, where the slice can start and end at any MB as shown in Figure 1-34 (b). The height of the slice can be more than 1 MB-height. H.264 slices are defined with coding modes – I slice, P slice, B slice, etc. If an I slice is used, the slice has all I-MBs. If a P slice is used, the slice has P-MBs (I-MBs also possible). The same is true for a B slice with B-MBs. The coding modes of slices are called "slice types."

H.264 defines seven slice group types with different slice type to give a better visual perception under different circumstances. The six slice group types are depicted in Figure 1-36, Figure 1-37 and Figure 1-38. Each figure shows the effect of errors when such a slice type is used in the transmission. Slice groups defined in Figure 1-36 have the effect of distributing burst errors in the bitstream to a visually ubiquitous pattern that pleases observers in many cases. Slice groups defined in Figure 1-37 (a) have the effect of selecting segments – foreground or background – to distribute burst errors. Slice groups defined in Figure 1-37 (b) and Figure 1-38 have the effect of increasing the area of visual segments to distribute burst errors. The 7th slice group can be defined explicitly by users, thus making it the most general.

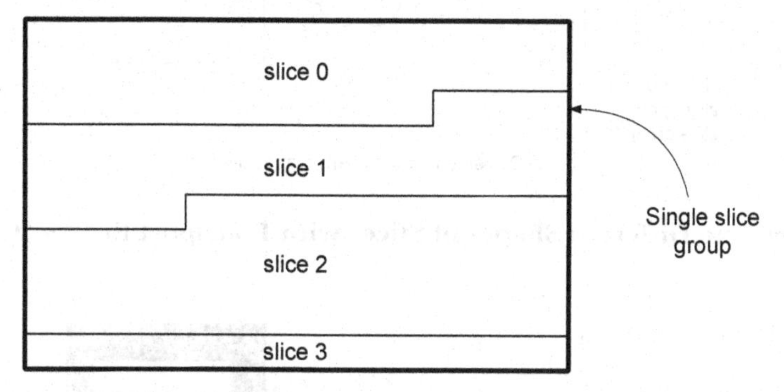


(a) Single slice group with num_slice_groups_minus1=0

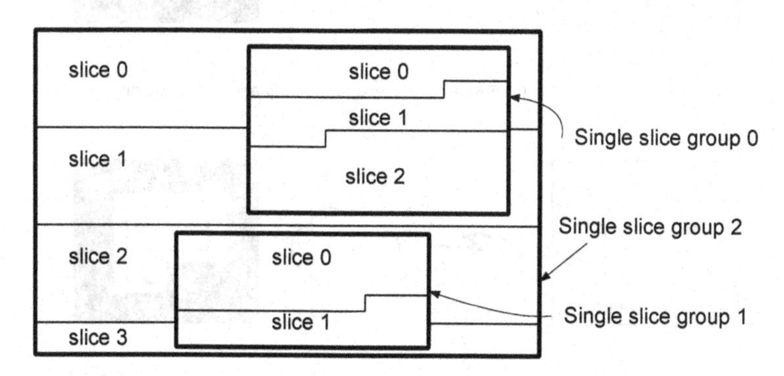


(b) Multiple slice groups with num_slice_groups_minus1=2

**Figure 1-35 Single vs. Multiple Slice Groups in H.264**

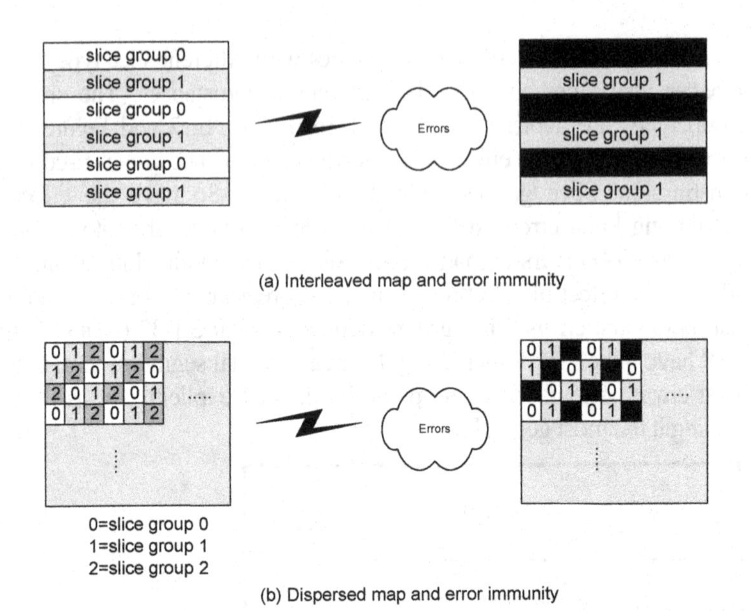


**Figure 1-36 Different Shapes of Slices with Transport Errors (1)**

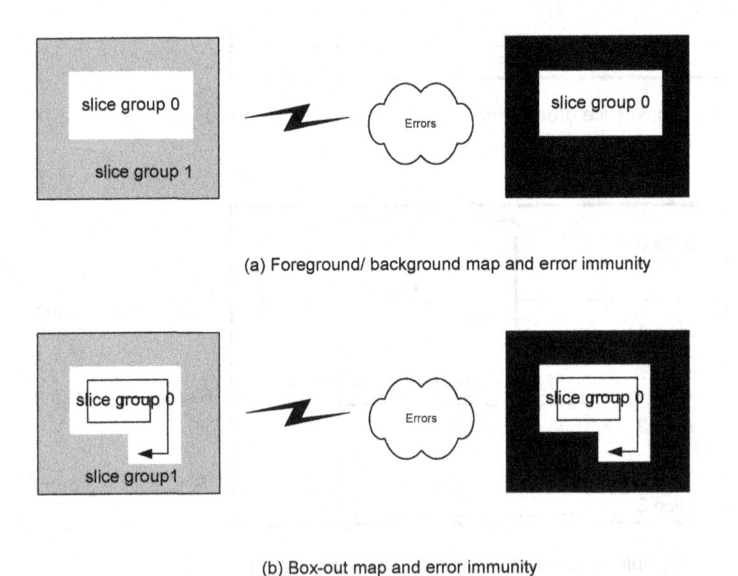


**Figure 1-37 Different Shapes of Slices with Transport Errors (2)**

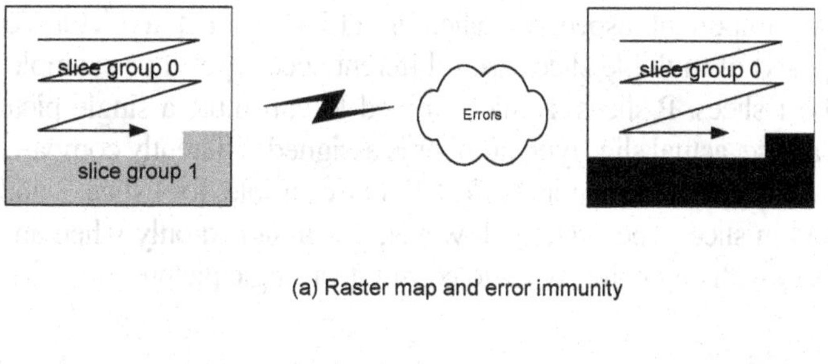


(a) Raster map and error immunity

(b) Wipe map and error immunity

**Figure 1-38 Different Shapes of Slices with Transport Errors (3)**

H.264 allows MSG in Baseline Profile as explained earlier, while it does not allow them in other Profiles. In addition, H.264 allows Arbitrary Slice Order (ASO) as well as shown in Figure 1-39.

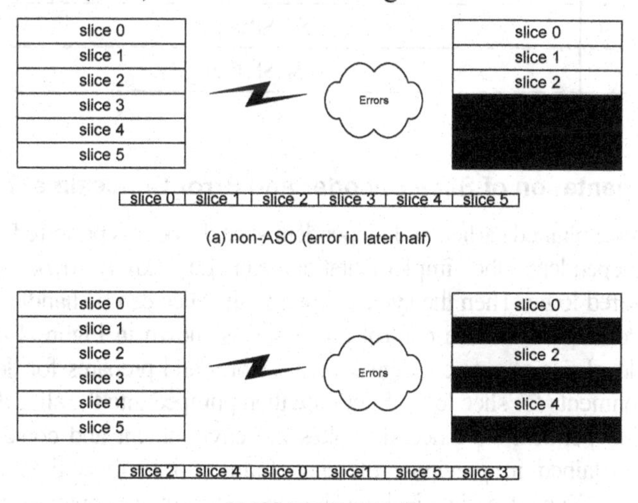


(a) non-ASO (error in later half)

(b) ASO (error in later half)

**Figure 1-39 H.264 Arbitrary Slice Order with Transport Errors**

An important aspect of Slice in H.264 is that a picture can be composed of multiple slices with different slice types. For example, there can be I slices, P slices, B slices mixed to constitute a single picture. In such a case, actual slice type number is assigned differently compared with pure slice type as shown in Table 1-8. For example, for I slice, 2 and 7 are defined in slice_type syntax. However, 2 is assigned only when an I slice is mixed with other slice types to constitute a single picture.

**Table 1-8 Name Association to slice_type**

| Slice_type | Name of slice_type |
|---|---|
| 0 | P (P slice) |
| 1 | B (B slice) |
| 2 | I (I slice) |
| 3 | SP (SP slice) |
| 4 | SI (SI slice) |
| 5 | P (P slice) |
| 6 | B (B slice) |
| 7 | I (I slice) |
| 8 | SP (SP slice) |
| 9 | SI (SI slice) |

### Implementation of Slice Decoder and Error Concealment

As explained earlier, an independent slice decoder is preferred. Since it is independent, the implementation can be heavily dependent on hardwired logic. Then the syntax down to the slice can be handled by the high level parser and a control processor as shown in Figure 1-40. The high level parser extracts high level variables and prepares for decoding environments for slice level. Slices are then pumped into the slice decoder. The independent slice decoder takes the environment and contexts that were obtained in previous reference frames to perform the decoding function. When decoding is done, the output constitutes contexts for future use. If the slice decoding fails due to errors, error concealment techniques can be employed. Note that error concealment techniques are not mandatory, but optional. Therefore, the algorithms are to be proprietary, if any are implemented at all.

There are many known error concealment techniques effective in implementations. One approach is to use a variant of spatio-temporal interpolation that infers the missing data based on guessed correlation. If a previous reference is believed to have more correlation, temporal inference is preferred. If intra neighbor blocks are believed to contain more correlation, spatial inference is performed. Error concealment algorithms are treated as highly proprietary by many companies in the field.

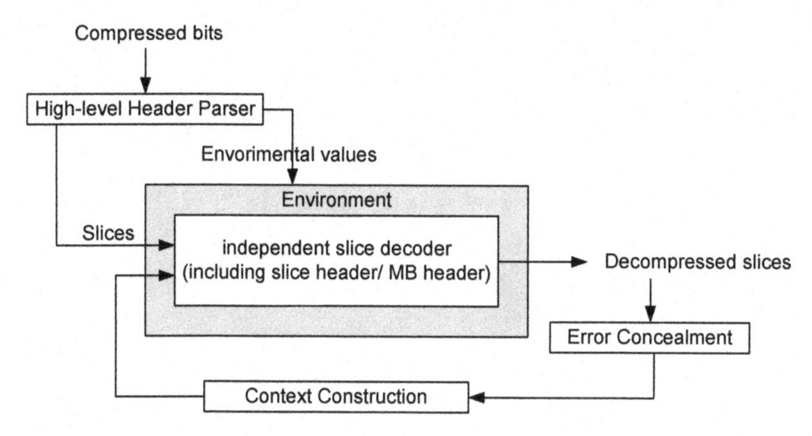


**Figure 1-40 Independent Slice Decoder and Error Concealment**

# 2. Syntax Hierarchies and Encapsulation

## 2.1 VC-1 Syntax Hierarchy in Bitstreams

### WMV-9 and VC-1 Standards

WMV-9 is a video codec developed by Microsoft. It is widely used for streaming media over the Internet due to the popularity of MS Windows Operating Systems. Since WMV-9 is a generic coder, many of its algorithms/tools can be used for a variety of applications under different operating conditions. Originally, three profiles were defined – Simple Profile, Main Profile and Complex Profile. However, Complex Profile was unofficially dropped. Consequently, WMV-9 focuses more on compression technology for progressive video up to Main Profile, while VC-1 has been developed for broadcast interlaced video as well as progressive video [srinivasan:WMV9, SMPTE:VC1].

VC-1 has three profiles – Simple (SP), Main (MP) and Advanced Profiles (AP). Simple and Main Profiles of VC-1 correspond to Simple and Main Profiles of WMV-9, respectively. Advanced Profile is mainly targeted for broadcast applications. Those two technologies are almost identical in philosophy and the tools they provide except where interlaced video is concerned. VC-1 is a pure video compression technology derived from Microsoft's proprietary WMV-9, and is expected to be deployed as a key engine in satellite TV, IP set-tops and HD-DVD/ Bluray-DVD recorders. HD-DVD and Bluray-DVD adopt only the Advanced Profile of VC-1.

### Key Compression Tools for WMV-9 Video

Like all other MPEG standards, WMV-9 is based on motion compensated transform coding. Originally YUV4:2:0 and YUV4:1:1 were defined as input formats for progressive and interlaced video, respectively. Since interlaced video is no longer considered with WMV-9, 8-bit YUV4:2:0 is the only input format.

There is no fixed GOP structure in WMV-9. I, P, B, BI and Skipped P are defined as picture/frame types. I (Intra) frames do not have to be periodic. Rather, any reference can be either an I or P (Predicted) frame. Therefore, if there is no big scene change for a lengthy period of time, there could be only P frames as references.

The number of B frames (Bi-directionally predicted frames) between two reference frames can vary. Maximally, there could be seven B frames. BI frames are almost identical to I frames. If there are major continuous scene changes, some B frames may not capture similarities from two reference frames. In such a case, intra mode performance might be better than prediction mode performance. BI frame compression is a good choice for this case. Since BI frames are not used as reference, dropping those frames is possible under certain conditions such as lack of computation or bandwidth.

The last frame type is the Skipped P frame. If the total length of the data comprising a compressed frame is 8 bits, then this signals that the frame was coded as a non-coded P frame in an encoder.

A key compression tool in WMV-9 is the adaptive block size transform. Transform block size can change adaptively, while the block size for motion compensation is either 16x16 or 8x8. Note that this is quite the opposite to that of H.264. H.264 normally uses fixed size 4x4 (or 8x8 in High Profile) transform with various block sizes for motion compensation. There are four transform sizes – 8x8, 4x8, 8x4 and 4x4. The transforms are 16-bit transforms where both sums and products of all 16-bit values produce results with 16 bits – the inverse transform can be implemented in 16-bit fixed point arithmetic. Note that the transform approximates a DCT, and the norms of the basis function between transforms are identical to enable the same quantization scheme through various transform types.

There are three main options for Motion Compensation (MC): 1. Either half-pel or quarter-pel resolution MC can be used. 2. Either bi-cubic or bi-linear filter can be used for the interpolation. 3. Either 16x16 or 8x8 block size can be used. These are all combined into a single MC mode with MVMODE and MVMODE2 syntax elements to be represented at the Frame level. Note that combinations are allowed in a specific way that clearly prioritizes the performance of MC. There is a tool mode in Sequence layer called FASTUVMC for motion vector computation in

Chroma components. If this is on, computed Chroma MVs are all rounded to the half-pel domain. Thus, interpolation for quarter points is not necessary for Chroma data at decoders – this saves a lot of computation in software-based decoder implementations.

Quantization is generally defined with two parameters in video standards – Qp and Dead-zone. There are two choices for Dead-zone in WMV-9 – 3Qp and 5Qp. There are two levels where this can be described: 1. Sequence header has QUANT field for this description – 3Qp or 5Qp for entire sequence. 2. Explicit option is written in each Picture header, or Implicit option is to describe it through PQindex. In I frames, PQAUNT is applied to entire MBs. However, DQUANT is used to adaptively describe Qp in each MB in P/B frames. In addition, there are other options to use only two Qps for an entire frame depending on either boundary MB or non-boundary MB.

There are two techniques used in WMV-9 to reduce blocky effects around a transform boundary – Overlapped Transform (OLT) smoothing and In Loop deblocking Filtering (ILF). OLT is a unique and interesting technique to reduce blocky effect based on an accurately defined pre-/post-processing pair. The idea is that forward and inverse operations are defined in such a way that original data is recovered perfectly when operations are serially applied (forward and inverse). The forward transform exchanges information across boundary edges in adjacent blocks. The forward operation is performed before the main coding stage. Consider an example where one block preserves relatively good edge data, while the other block loses details of edge data. In this case, the blocky effect is very visible. At the decoder side, the inverse operation is required to exchange the edge data back again to obtain original data. By doing so, good quality and bad quality edges diffuse each other. Therefore, the blocky effect is significantly reduced.

On the other hand, ILF is a more or less heuristic way to reduce blocky effects. Blocky pattern is considered to be high frequency since abrupt value changes are happening around block edges. Considering that original data quality might also contain high frequency, relatively simple non-linear low pass filtering is applied about block edges in ILF. ILF is performed on I and P reference frames. Thus, the result of filtering affects the quality of pictures that use ILFed frames as references.

Entropy coding used in WMV-9 is a kind of Context-Adaptive VLC. Based on Qp, from which the coded quality can be guessed, a new set of VLC tables is introduced. Such examples include mid-rate VLC tables and high-rate VLC tables. In addition, based on MVs, another set of VLC tables is introduced. Such examples include low-motion DC differential tables and high-motion DC differential tables.

## WMV-9 Video Specific Semantics and Syntax

There are five levels of headers inWMV-9 video bitstream syntax – Sequence, Picture, Slice (not clearly defined in WMV-9), MB, and Block. Sequence header contains basic parameters such as profile, interlace, frame rate, bit rate, loop filter, overlap filter and some other global parameters. Picture header contains information about type of picture/ BFRACTION/ PQindex/ MVMODE/ MVMODE2/ LumScale/ LumShift/ DQUANT related/ TTMBF/ TTFRM/ DCTACMBF/ DCTACFRM, etc. BFRACTION data is relative temporal position of B that is factored into the computation of direct mode vectors. Note that the number of B frames inserted with geometrical position of B can be determined from this value. PQindex is interpreted for quantizer scale (QS) and quantizer types (3QP/5QP) in Implicit case, while quantizer types are explicitly defined in Sequence or Picture header in other cases. A combination of MVMODE and MVMODE2 represents a prioritized MC mode adopted for the current frame. LumScale/ LumShift are Intensity Compensation parameters. TTMBF is the flag that tells whether the additional field for Transform Type is in MB level or in Frame level. DCTACMBF is the flag that tells whether DCT AC Huffman table is defined in MB level or Frame level. TTFRM may be used to force a certain transform type in the frame level for P or B frames. DCTACFRM may be used to force a certain DCT AC Huffman table in the frame level. Slices are not clearly defined in WMV-9. When STARTCODE is set in the Sequence header, MB header contains SKIPMBBIT/ MVMODEBIT/ MVDATA/ TTMB, etc. SKIPMBBIT indicates whether the MB is "Skipped"; for MBs in P or B frames (i.e., P-MBs or B-MBs). This representation is extended to take Hybrid mode in WMV-9. MVMODEBIT is present in P-MBs in P frames if the picture is coded in "Mixed-MV" mode. If MVMODEBIT==0, the MB shall be coded in 1MV mode. If MVMODEBIT==1, the MB shall be coded in 4MV mode.

MVDATA tells whether the blocks are coded as Intra or Inter type. If they are coded as Inter 1MV, then MVDATA indicates MV differentials. If they are coded as Inter 4MV, then BLKMVDATA in each Block header indicates MV differentials. Block layer contains all transform coefficients. Sub-block pattern data is included in sub-blocks.

### Simple and Main Profiles for VC-1 Video

The Simple Profile (SP) and Main Profile (MP) of VC-1 are the same as those of WMV-9 in progressive compression. However, there is no Advanced Profile (AP) in WMV-9 to compare with the AP in VC-1. VC-1 AP mainly focuses on interlaced video compression technology. Legacy Complex Profile (CP) in WMV-9 can handle the interlaced video. However, the interlace tools cannot be compared directly in both standards. Even the input interlaced video formats are different -- the input format is YUV4:1:1 for WMV-9, but is YUV4:2:0 for VC-1. Note that only Field-based prediction and Frame-based prediction are selectively applied in each MB in the legacy CP of WMV-9. This tool is applied when the INTERLACE flag is turned on in the Sequence header in the legacy CP. The CP of WMV-9 was not considered for VC-1 interlaced video when the SP and the MP of WMV-9 were adopted in the VC-1 without change.

### Advanced Profile for VC-1 Video

Advanced Profile adds interlace tools to the Main Profile. The number of B frames is not fixed between two references in VC-1. New distance value comes into the Picture Structure at Entry Point Layer with REFDIST_FLAG for interlaced video. REFDIST data indicates the number of pictures between the current picture and the reference one. Progressive-Picture/ Frame-Picture/ Field-Picture can be mixed in VC-1 AP. It is the encoder's job to construct each picture with a Frame Coding Mode (FCM).

The maximum number of references is two for a P Field-Picture. The references are specified in the Picture layer. If both references are used, the selection of reference information is described in MB level and Block level. The number of references is always four for a B Field-Picture — no

Picture layer selection is needed. So, the selection of reference is always in MB level and Block level. Note that one of the reference fields for a bottom field in a B frame is its top field itself.

A P Field-Picture has two MC modes (1MV with 16x16, 4MV with 8x8), while a B Field-Picture has three MC modes (1MV with 16x16 in forward or backward modes, 2MV with 16x16 in interpolative or direct modes, 4MV with 8x8 only in forward or backward modes). A P Frame-Picture has 4 MC modes (1MV Frame-based prediction with 16x16, 4MV Frame-based prediction with 8x8, 2MV Field-based prediction with 16x8 (each field), 4MV Field-based prediction with 8x8 (each field 16x8 divided to left/right 8x8)). A B Frame-Picture has four MC modes (1MV Frame-based prediction with 16x16 in forward or backward modes, 2MV Frame-based prediction with 16x16 in interpolative or direct modes, 2MV Field-based prediction with 16x8 (each field) in forward or backward modes, 4MV Field-based prediction with 16x8 (each field) in interpolative or direct modes). Once residual data is obtained after motion-estimation in encoders, a transform is applied on it.

In Intra MBs or Intra Blocks, a transform is applied on original data. There are two transforms – Frame-transform and Field-transform. Frame-transform applies the transform on Frame-Picture data without any reordering, while Field-transform applies the transform on Frame-Picture data with sorted top/ bottom field data. Note that this option is only available in Frame-Pictures. Encoders decide which transform mode is applied in each MB. In the case of Intra MBs, the mode determined is written in FIELDTX. In Inter MBs, however, the mode is written in MBMODE.

Transform block size can change adaptively, while the size of motion compensation is one of 16x16/ 16x8/ 8x8 in VC-1 interlace video. There are four transform sizes as are in WMV-9 – 8x8, 4x8, 8x4 and 4x4.

The same two techniques are used in VC-1 to reduce blocky effects around transform boundaries – OLT smoothing and ILF. One important difference in the OLT technique between WMV-9 and VC-1 is to have the control even on MB level in I frame with CONDOVER and OVERFLAGS. The 128 level-shift is done on all the Intra MBs and Intra Blocks in VC-1, while the level-shift is performed only on Intra MBs and Intra Blocks that undergo OLT in WMV-9. In interlaced video, the OLT smoothing is applied only for vertical direction in Frame-Pictures, while it is performed for both horizontal and vertical directions in Field-Pictures. Note that horizontal edge filtering might require top and bottom fields together as inputs in Frame-Pictures – this would make potential output filtered data blurry. That is why only vertical direction edges are OLT-filtered for Frame-Pictures in the VC-1 standard.

On the other hand, ILF filters both horizontal and vertical directions in Field-Pictures of interlaced video. In Frame-Pictures of interlaced video, however, horizontal and vertical ILFs are performed differently. ILF in vertical edges is the same as that of Field-Pictures, while ILF in horizontal edges is performed based on Field-based ILF filtering. In other words, only the same polarity data are considered in ILF filtering.

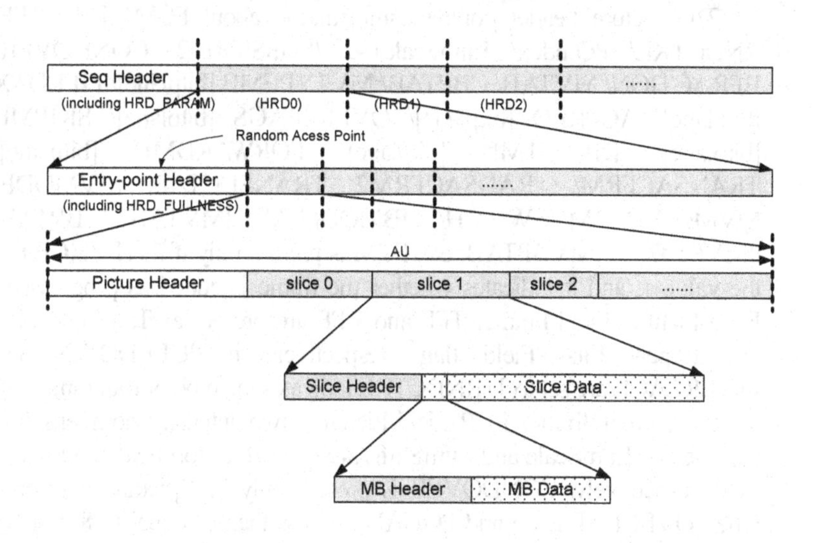


**Figure 2-1 Syntax Hierarchy for VC-1**

## VC-1 Video Specific Semantics and the Syntax

There are six levels of headers inVC-1 video bitstream syntax – Sequence, Entry Point, Picture, Slice, MB and Block. AP has explicit Sequence header, but SP/ MP don't have any Sequence header or Entry Point header in VC-1. The data necessary in the Sequence header should be provided by an external means. Sequence header contains basic parameters such as profile/ level, interlace, loop filter, max_coded_width, max_coded_height and some other global parameters. This includes display related metadata and HRD parameters.

The Entry Point header is the random access point, and is present only in AP. It is used to signal control parameter changes in the decoding. Examples include broken_link, closed_entry, refdist_flag, loopfilter, overlap, coded_width, coded_height and other global parameters until the next Entry Point header. It also contains HRD fullness information.

A key syntax layer is the Picture and it is the access unit (AU) in VC-1. A Picture is composed of many slices as shown in Figure 2-1. A Slice is composed of many MB rows/ MBs as explained in Chapter 1. An MB is composed of six 8x8 blocks as a coding unit.

The Picture header contains information about FCM/ TFF/ RFF/ RNDCTRL/ PQindex/ LumScale1~2 /LumShift1~2/ CONDOVER/ BFRACTION/ MVTAB/ CBPTAB/ MVTYPEMB [bitplane]/ FIELDTX [bitplane]/ ACPRED [bitplane]/ OVERFLAGS [bitplane]/ SKIPMB [bitplane]/ DIRECTMB [bitplane]/ FORWARDMB [bitplane]/ TRANSACFRM/ TRANSACFRM2/ TRANSDCTAB/ MVMODE/ MVMODE2/ 4MVSWITCH/ MBMODETAB/ IMVTAB/ ICBPTAB/ 2MVBPTAB/ 4MVBPTAB, etc. FCM is present only if INTERACE has the value 1, and it indicates whether the frame is coded as progressive/ Field-Picture/ Field-Frame. TFF and RFF are present as Top Field First and Repeat First Field flags respectively if PULLDOWN and INTERLACE are set to 1. RNDCTRL indicates the type of rounding used for the current frame. In P Field-Pictures, two intensity compensation parameters (LumScale and LumShift) are needed for top field and bottom field, respectively. CONDOVER is present only in I pictures and only when OVERLAP is on and PQUANT is less than or equal to 8. For SP and MP, PQUANT is the only condition for the operation of the OLT. Note that the OLT is performed only when PQUANT is larger and equal to 9. For AP, OLT can be even performed when PQUANT is less than or equal to 8 with introduction of CONDOVER syntax element. The syntax elements tagged with "[bitplane]" mean that there are corresponding syntax elements in MB header and the syntax elements in the Picture header are used instead with Raw mode turned off. For example, DIRECTMB syntax is just bitplane coding flags corresponding to DIRECTBBIT in each MB header. Bitplane coding is explained in Chapter 9. TRANSACFRM and TRANSACFRM2 are used to define AC Huffman coding tables selected for CbCr and Y, respectively.

TRANSDCTAB is used to indicate whether the low motion table or the high motion table shall be used for DC value decoding. 4MVSWITCH shall be present in Interlace Frame P. If 4MVSWITCH==0, the MBs in the picture shall have only one or two MVs depending on the MB has been frame-coded or field-coded. If 4MVSWITCH==1, there shall be either one, two or four MVs per MB – this syntax shall not be present in Interlace Frame B. IMVTAB and ICBPTAB are syntax elements to select tables for Intensity Compensation. MBMODETAB is used to select Huffman tables for decoding MBMODE syntax element. The 2MVBPTAB and 4MVBPTAB syntax elements signal which one of four tables are used to decode 2MVBP and 4MVBP syntax elements, respectively.

The Slice header provides information about SLICE_ADDR/ PIC_HEADER_FLAG. Slice Address is from 1 to 511, where the row address of the first MB in the slice is binary encoded. The picture header information is repeated in the slice header if the PIC_HEADER_FLAG is set to 1.

The MB header has TTMB/ OVERFLAGMB/ MVMODEBIT/ SKIPMBBIT/ FIELDTX/ CBPCY/ ACPRED/ MQDIFF/ ABSMQ/ MVDATA/ BLKMVDATA/ HYBRIDPRED/ MBVTYPE/ BMV1/ BMV2/ BMVTYPE/ DIRECTBBIT/ MBMODE/ 2MVBP/ 4MVBP/ MVSW, etc. MBMODE indicates whether Intra, Inter-1MV, Inter-4MV, CBP and MVDATA are present. OVERFLAGMB is present when CONDOVER has the binary value 11. OVERFLAGMB indicates whether to perform OLT within the block and neighboring blocks. MBVTYPE is BMV1 and BMV2 are used for the 1st MVDATA and the 2nd MVDATA in B-MBs, respectively. The decoding procedure for BMV1 and BMV2 shall be identical to the procedure for MVDATA. BMVTYPE indicates whether the MB uses forward, backward or interpolative prediction modes for B-MBs. Note that direct prediction mode is separately described in DIRECTBBIT syntax element for B-MBs. MBMODE is defined for Interlace Field P/B and Interlace Frame P/B MBs to indicates coding modes including FIELDTX and CBPCY. 2MVBP is defined for Interlace Frame P/B MBs to indicates which of two luma blocks contain non-zero MVD – this syntax element shall be present if the MBMODE syntax element indicates that the MB has two field MVs. 4MVBP is defined for Interlace Field P/B and Interlace Frame P/B MBs to indicates which of four luma blocks contain non-zero MVD – this syntax element shall be present if the MBMODE syntax element indicates that the MB has two field MVs under the interpolative mode. MVSW shall be present in B-MBs if the MB is in field mode and BMVTYPE is forward or backward prediction mode. If MVSW==1, it shall indicate that the MV type and prediction type changes from forward to backward (or backward to forward) in going from the top to bottom field. If MVSW==0, the prediction type shall not change in going from the top to the bottom field. Other data can similarly be interpreted as those in WMV-9.

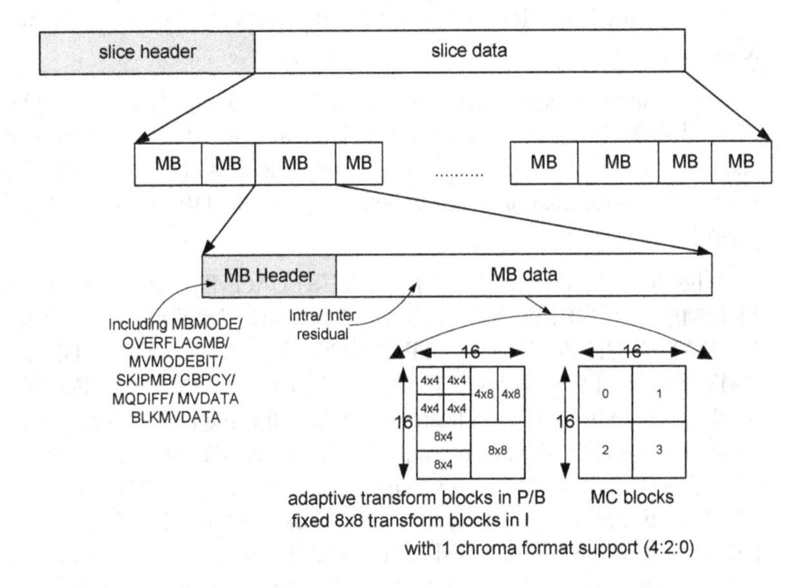


**Figure 2-2 Slice Syntax Hierarchy for VC-1**

There are many coding options for each MB as shown in Figure 2-2. There are four transforms shown to apply in each 8x8 block, where the same transform is specified to apply. For example, an 8x8 block cannot be partitioned into one 8x4 block and two 4x4 blocks. The choice among transform sizes is basically performed by the encoder to take advantage of statistics of input signals. Typically a long size transform is effective when the input signal has high correlation among localized spatial data, while a short size transform is effective when the input signal has relatively low correlation. There are two MC modes in terms of size and the choice among MC sizes is made by the encoder to take advantage of statistics of input signals. Typically a large area is effective when a rigid object is moving (uniform motion in the area), while a small area is effective when tearing object (no object) and/or overlapped objects are moving (random motion in the area).

### VC-1 Profiles/ Tools

The Simple Profile targets low-rate internet streaming and low-complexity applications such as mobile communications, or play back of media in personal digital assistants. There are two levels in the profile – Low and Medium levels.

The Main Profile targets high-rate internet applications such as streaming, movie delivery via IP, or TV/ VOD over IP. This profile has three levels – Low, Medium and High.

The Advanced Profile targets broadcast applications, such as digital TV, HD-DVD for PC play back, or HDTV. It is the only profile that supports interlaced content. In addition, this profile contains the required syntax elements to transmit video bitstreams with encapsulations of generic systems, such as MPEG-2 Transport or Program Streams. This profile contains five levels – L0, L1, L2, L3 and L4.

Each class of bitstreams contains tool sets defined in Table 2-1 and parameters set defined in Table 2-2. Table 2-1 indicates the constraints on the algorithms or compression features for each of the profiles. Note that dynamic resolution change refers to scaling the coded picture size by a factor of two via the RESPIC syntax element in the MP.

Note that range adjustment refers to range reduction by a factor of two via the RANGEREDFRM syntax element in the MP, and range mapping by arbitrary factors via RANGE_MAPY and RANGE_MAPUV syntax elements in the AP.

**Table 2-1 VC-1 Profiles and Tools**

| Tool Options | Simple Profile | Main Profile | Advanced Profile |
|---|---|---|---|
| Baseline Intra Frame Compression | x | x | x |
| Variable-sized Transform | x | x | x |
| 16-bit Transform | x | x | x |
| Overlapped Transform | x | x | x |
| 8x8 and 16x16 Motion Modes | x | x | x |
| Quarter-pixel Motion Compensation Y | x | x | x |
| Quarter-pixel Motion Compensation U, V | | x | x |
| Start Codes | | x | x |
| Extended Motion Vectors | | x | x |
| Loop Filter | | x | x |
| Dynamic Resolution Change | | x | x |
| Adaptive MB Quantization | | x | x |
| Bidirectional (B) Frames | | x | x |
| Intensity Compensation | | x | x |
| Range Adjustment | | x | x |
| Interlace: Field/ Frame Coding Modes | | | x |
| Self Descriptive Fields/ Flags | | | x |
| GOP Layer/ Entry Points | | | x |
| Display Metadata (Pan/ Scan, Colorimetry, Aspect Ratio, Pulldown, Top Field First, Repeat First Field, etc.) | | | x |

**Table 2-2 VC-1 Levels and Limitations**

| Profile @Level | MB/s | MB/f | Examples | B | I | Rmax | Bmax | MV [H]x[V] |
|---|---|---|---|---|---|---|---|---|
| SP@LL | 1,485 | 99 | QCIF(176x144) @15fps | | | 96 | 22 | [-64,63 ¾] x [-32,31 ¾] |
| SP@ML | 5,940 | 396 | CIF(352x288) @15fps, 240x176 @30fps | | | 384 | 77 | [-64,63 ¾] x [-32,31 ¾] |
| MP@LL | 7,200 | 396 | QVGA(320x240) @24fps, CIF(352x288) @15fps | x | | 2,000 | 306 | [-128,127 ¾] x [-64,63 ¾] |
| MP@ML | 40,500 | 1,620 | 480p(720x480) @30fps, 576p(720x576) @25fps | x | | 10,000 | 611 | [-512,511 ¾] x [-128,127 ¾] |
| MP@HL | 245,760 | 8,192 | 1080p(1920x1080) @25/30fps | x | | 20,000 | 2,442 | [-1024,1023 ¾] x [-256,255 ¾] |
| AP@L0 | 11,880 | 396 | CIF(352x288) @25/30fps, SIF(352x240) @30fps | x | | 2,000 | 250 | [-128,127 ¾] x [-64,63 ¾] |
| AP@L1 | 48,600 | 1,620 | 480i-SD, 704x480 @30fps, 576i-SD, 720x576 @25fps | x | x | 10,000 | 1,250 | [-512,511 ¾] x [-128,127 ¾] |
| AP@L2 | 110,400 | 3,680 | 480p, 704x480 @60fps, 720p, 1280x720 @25/30fps | x | x | 20,000 | 2,500 | [-512,511 ¾] x [-128,127 ¾] |
| AP@L3 | 245,760 | 8,192 | 1080i/p, 1920x1080 @25/30fps, 720p, 1280x720 @50/60fps, 2048x1024 @30fps | x | x | 45,000 | 5,500 | [-1024,1023 ¾] x [-256,255 ¾] |
| AP@L4 | 491,520 | 16,384 | 1080p, 1920x1080 | x | x | 135,000 | 16,500 | [-1024,1023 ¾] |

| | | | | | | | | |
|---|---|---|---|---|---|---|---|---|
| | | | @50/60fps, 2048x1536 @24fps, 2048x2048 @30fps | | | | | x [-256,255 ¾] |

There are several levels for each of the profiles. Each level limits the video resolution, frame rate, HRD bit rate, HRD buffer requirements, and the motion vector range. These limitations are defined in Table 2-2. The column marked "B" denotes B frames and loop filter support, and "I" denotes interlace support. For Interlace, picture rate is described in "frames" per second. "Fields" per second is twice that value.

## 2.2 VC-1 Encapsulation in MPEG-2 Systems

### Entry Point and Access Unit in VC-1

Entry Point headers are used in VC-1 Advanced Profile (AP) to support random access. An Access Unit (AU) contains all the data of a picture and padding bits that follow up to the next AU start code. Figure 2-3 shows the structure of the AUs in VC-1 bit stream. Coded picture data represents a video frame, regardless of whether the frame has been encoded as a progressive frame, interlace frame or interlace field mode. AUs start on a byte boundary and begin with either a Sequence start code, an Entry point start code or a Frame start code.

If the frame is not preceded by a Sequence start code, Sequence header or Entry Point header, the AU shall begin with a Frame start code. Otherwise, the AU shall start with the first byte of the first of these structures (excluding any stuffing bytes) before the Frame start code. An AU shall also include any user data start code and user data bytes at the sequence, entry point, frame or field level.

In the case of SP/ MP, an AU shall include all the coded data for a video frame, including the VC-1_SPMP_PESpacket_PayloadFormatHeader( ) bytes that precedes it as well as any flushing bits present to ensure byte alignment that follows it up to, but not including, the start code of the next AU. The start of the next AU shall be the first byte in the next VC-1_SPMP_PESpacket_PayloadFormatHeader( ) which is either a Sequence start code or a Frame start code.

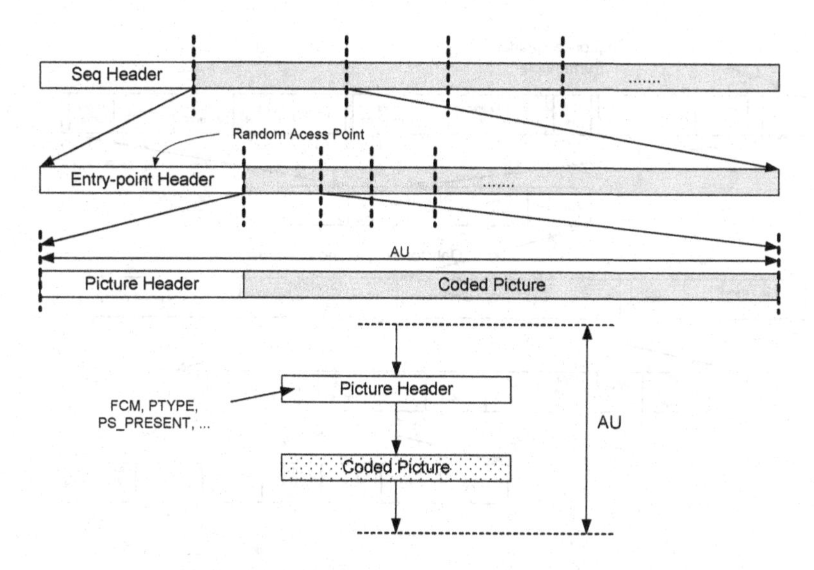


**Figure 2-3 AU for VC-1**

## Encapsulation of VC-1 in PES

The carriage of VC-1 in MPEG-2 Systems is defined in SMPTE RP227 specification. VC-1 streams can be carried in either an MPEG-2 Transport Stream or an MPEG-2 Program stream [SMPTE:VC1systems].

MPEG-2 Systems require PES encapsulation of video data in order to generate PES as described in Chapter 1. The PES structures for MPEG-2 video and VC-1 video are largely similar. The VC-1 specific encapsulation requirements for PES are discussed in the section.

The PES of VC-1 is exactly the same as that of MPEG-2 video with the exception of VC-1 SP/ MP PES packet payload format header as shown in Figure 2-4. PES_packet_length field indicates VC-1 video. A value of "0" means that the semantics shall be extended to VC-1 video ESs. The extended fields contain all necessary parameters set at Sequence level for SP/ MP.

The stream_id for VC-1 ESs shall be set to 0xfd to indicate the use of an extension mechanism mentioned in Chapter 1.

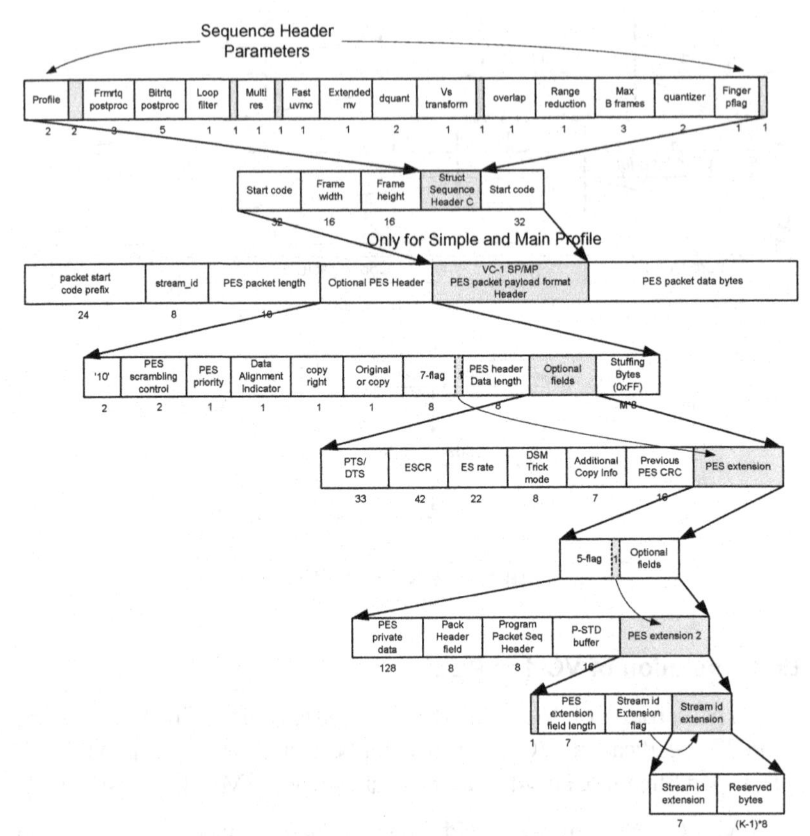


**Figure 2-4 PES Syntax Diagram for VC-1**

The data_alignment_indicator for VC-1 ESs shall follow the same semantics mentioned in Chapter 1, except that it applies to the Data Alignment sub-descriptor values defined in Table 2-5. In particular, the data_alignment_indicator in the PES header shall be set to "1" if there is Data Alignment sub-descriptor associated with the VC-1 ES in the PMT. If the data_alignment_indicator is set to "1" and there is no Data Alignment sub-descriptor associated with the VC-1 ES in the PMT, the default alignment type value shall be equal to 0x02. The data_alignment_indicator value "0" indicates that the alignment is unspecified. For SP/ MP, the value of the data_alignment_indicator field shall always be set to "1" and there shall not be any Data Alignment sub-descriptor associated with the VC-1 ES.

The PTS/ DTS are used in exactly the same manner as in MPEG-2. In particular, the values of the PTS/ DTS fields shall pertain to the first video AU that starts in the payload of the PES packet.

The stream_id_extension field for VC-1 ESs shall have any value in the range between 0x55 and 0x5f. The combination of stream_id and stream_id_extension unambiguously define the stream of PES packets carrying VC-1 video data.

A VC-1 access point is defined as follows:

- The first byte of a VC-1 Sequence header can be an access point if there is no Sequence start code preceding the Sequence header.
- The first byte of the Sequence start code can be an access point if a Sequence start code immediately precedes the Sequence header.

As explained in Chapter 1, the discontinuity_indicator is 1-bit field that indicates whether the discontinuity state is true for the current TS packet. After a continuity counter discontinues in a TS packet with VC-1 data, the first byte of ES data in a TS packet of the same PID shall be the first byte of a VC-1 access point or a VC-1 end-of-sequence start code followed by an access point.

All other data such as random_access_point, elementary_stream_priority_indicator, splice_countdown, seamless_splice_flag, slice_type are interpreted as explained in Chapter 1.

For VC-1 Simple or Main Profile ESs, a VC-1_SPMP_PESpacket_PayloadFormatHeader( ) structure shall be present at the beginning of every AU. A VC-1_SPMP_PESpacket_PayloadFormatHeader( ) shall always start with a start_code presenting a Sequence start code (value of "0x0000010f") or a Frame start code (value of "0x0000010d"). The start code emulation prevention shall be applied to the bytes in the PES packet following the start_code field in the VC-1_SPMP_PESpacket_PayloadFormatHeader( ) structure to protect the start code from occurring in any other locations in the PES packet payload. The start code emulation prevention mechanism is described in Annex E of SMPTE 421M.

The structure shown in Figure 2-5 below corresponds to the Sequence header for the VC-1 SP/ MP. There shall not be any VC-1_SPMP_PESpacket_PayloadFormatHeader( ) structure at the beginning of any PES packet payload.

```
VC-1_SPMP_PESpacket_PayloadFormatHeader( ){
  start_code                    (32 bit)      // bslbf
  if( start_code == 0x0000010f) {
    frame_width                 (16 bit)      // uimsbf
    frame_height                (16 bit)      // uimsbf
    STRUCT_SEQUENCE_HEADER_C( )   (32 bit)//bslbf
    Start_code                  (32 bit)      // 0x0000010d
  } else if (start_code == 0x0000010d) {
  }
}
```

**Figure 2-5 Syntax for VC-1_SPMP_PESpacket_PayloadFormatHeader( ) structure**

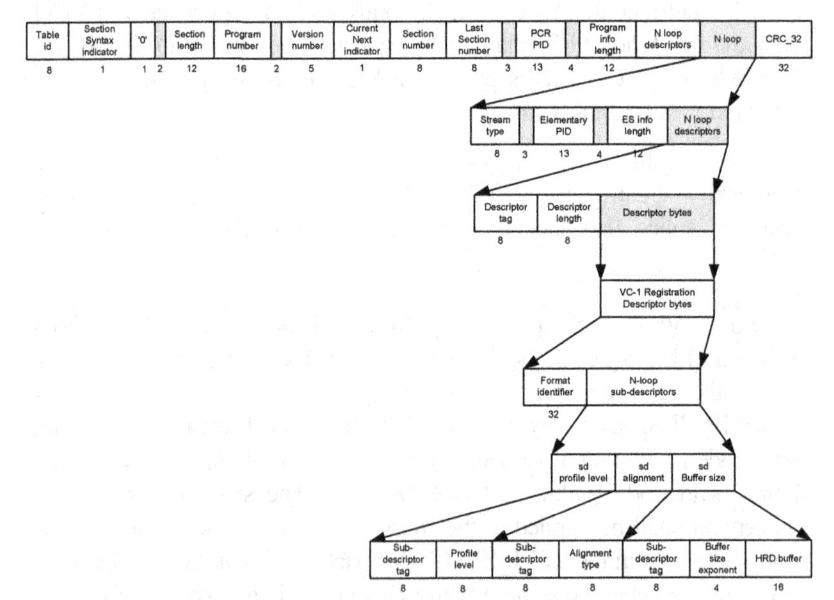


**Figure 2-6 TS PM Section Diagram for VC-1**

### Encapsulation of VC-1 in TS

In a TS, Programs are signaled using a collection of Tables transmitted cyclically and known as PSI (Program Specific Information). This was explained in detail in Chapter 1. Specifically, a Program Map Table (PMT) provides the Program details and specifies necessary information such as PID to find and decode the component ESs. Delivery of VC-1 ESs in MPEG-2 TSs shall be governed by a T-STD in MPEG-2 Systems provisions.

The Registration Descriptor in MPEG-2 Systems is designed to identify formats of "private" data uniquely and unambiguously. The registration_descriptor( ), whose structure is originally defined for MPEG-2 Systems, resides in the inner descriptor loop of the MPEG-2 Program Element (PE) in the TS Program Map (PM) section corresponding to a VC-1 ES. The PM section is depicted in Figure 2-6. Figure 2-7 depicts the registration_descriptor( ). The format_identifier is a number assigned/registered with the SC92 committee to eliminate any confusion for future private data.

```
registration_descriptor ( ){
  descriptor_tag            (8 bit)      // 0x05
  descriptor_length         (8 bit)      // uimsbf
  format_identifier         (32 bit)     // 0x56432d31
  for (i=0; i<K; i++) {
      sub-descriptor( )     (N*8)        //uimsbf
  }
}
```

**Figure 2-7 Syntax for Registration Descriptor**

The sd_profile_level( ) sub-descriptor signals the Profile and Level for the associated VC-1 ES. Syntax and semantics for this sub-descriptor are defined in Figure 2-8.

```
sd_profile_level ( ){
  subdescriptor_tag         (8 bit)      // 0x01
  profile_level             (8 bit)      // uimsbf
}
```

**Figure 2-8 Syntax for Profile and Level Sub-Descriptor**

The sub-descriptor( ) has a data structure as shown in Figure 2-8, Figure 2-9 and Figure 2-10. The first field of a sub-descriptor( ) shall be an 8-bit value known as the subdescriptor_tag which identifies syntax and semantics for a particular sub-descriptor. Table 2-3 defines values of the subdescriptor_tag.

**Table 2-3 Value for subdescriptor_tag**

| Assigned value | Description |
|---|---|
| 0x00 | Indicates a null sub-descriptor. A null sub-descriptor consists only of an 8-bit sub-descriptor_tag field |
| 0x01 | Profile and Level sub-descriptor defined for VC-1 |
| 0x02 | Alignment sub-descriptor defined for VC-1 |
| 0x03 | Buffer size sub-descriptor defines for VC-1 |
| 0x04~0xfe | SMPTE reserved for future applications |
| 0xff | Indicates a null sub-descriptor. A null sub-descriptor consists only of an 8-bit sub-descriptor_tag field |

An alignment sub-descriptor sd_alignment( ) is used to define explicitly which type of alignment exists between the coded byte sequence and a PES packet. Syntax and semantics for this sub-descriptor are defined in Figure 2-9.

```
sd_alignment ( ){
  subdescriptor_tag       (8 bit)      // 0x02
  alignment_type          (8 bit)      // uimsbf
}
```

**Figure 2-9 Syntax for Profile and Level Sub-Descriptor**

```
sd_buffer_size( ){
  subdescriptor_tag        (8 bit)      // 0x02
  reserved                 (8 bit)      // uimsbf
  buffer_size_exponent     (4 bit)      // uimsbf
  hrd_buffer               (16 bit)     // uimsbf
}
```

**Figure 2-10 Syntax for Sd_buffer_size Sub-Descriptor**

**Table 2-4 Value for profile_level field**

| Assigned value | Description |
|---|---|
| 0x00 | SMPTE reserved |
| 0x11 | Simple Profile, Low Level |
| 0x12 | Simple Profile, Medium Level |
| 0x13~0x50 | SMPTE reserved |
| 0x51 | Main Profile, Low Level |
| 0x52 | Main Profile, Medium Level |
| 0x53 | Main Profile, High Level |
| 0x54~0x90 | SMPTE reserved |
| 0x91 | Advanced Profile, Level L0 |
| 0x92 | Advanced Profile, Level L1 |
| 0x93 | Advanced Profile, Level L2 |
| 0x94 | Advanced Profile, Level L3 |
| 0x95 | Advanced Profile, Level L4 |
| 0x96~0xff | SMPTE reserved |

**Table 2-5 Syntax for Alignment Sub-Descriptor**

| Assigned value | Description |
|---|---|
| 0x00 | SMPTE reserved |
| 0x01 | Slice or Video AU |
| 0x02 | Video AU |
| 0x03 | Entry Point or Sequence |
| 0x04 | Sequence |
| 0x05 | Frame |
| 0x06~0xff | SMPTE reserved |

A buffer size sub-descriptor specifies the minimum size of the video ES buffer $EB_n$ in TSs or $B_n$ in PSs needed in the decoder to decode the ES conveyed in the MPEG-2 Program Element associated with this sub-descriptor. This descriptor is provided to allow receivers to check compatibility of their decoding capability against the decoding requirements for the ES. The buffer associated with the VC-1 Profile and Level shall be assumed if this sub-descriptor is not present and if no HRD parameters are specified in the Sequence header.

Conventionally defined descriptors in MPEG-2 Systems shall be used with VC-1 ES in certain cases. Such descriptors include Target Background Grid Descriptor, Video Window Descriptor, CA Descriptor, ISO 639 Language Descriptor, Multiplex Buffer Utilization Descriptor, Smoothing Buffer Descriptor, Copyright Descriptor, Maximum Bitrate Descriptor, Private Data Indicator Descriptor, IBP Descriptor, and STD Descriptor.

The size of buffer $TB_n$, known as $TBS_n$, shall be equal to 512 bytes, which is exactly the same as in MPEG-2 Systems. In addition, the buffer size MBS and rate Rx are similar as those of MPEG-2 Systems.

When there is no data in $TB_n$:

$$Rx_n = 0. \tag{2-1}$$

Otherwise,

$$Rx_n = 1.2 \times R_{\max}[profile, level]. \tag{2-2}$$

The multiplexing buffer size $MBS_n$ is defined as follows:

$$MBS_n = BS_{mux} + BS_{oh} \tag{2-3}$$

where PES packet overhead buffering

$$BS_{oh} = (1/750)\sec \times R_{\max}[profile, level], \tag{2-4}$$

and additional multiplex buffering

$$BS_{mux} = (0.004\sec) \times R_{\max}[profile, level]. \tag{2-5}$$

Note that the transfer data from MB to EB shall be governed by a leak method only. Use of a leak method shall be signaled via one of the following two methods: 1. There is no MPEG-2 STD Descriptor present in the inner descriptor loop of the MPEG-2 PE in the TS_program_map_section corresponding to the VC-1 ES. 2. An MPEG-2 STD Descriptor is present in the inner descriptor loop of the MPEG-2 PE in the TS_program_map_section corresponding to the VC-1 ES and the leak_valid flag has the value of "1."

The leak method transfers data from MB to EB using the leak rate Rbx as:

$$Rbx_n = R_{\max}[profile, level]. \tag{2-6}$$

The default size $EBS_n$ of the ES buffer shall be the $VBV_n[profile, level]$ associated with the profile and level of the VC-1 ES. The profile and level may be defined by the field profile_level in the sd_profile_level( ) sub-descriptor. With an assumption that the incoming service delivery rate R is known through ES_rate field in PES header, a receiver may opt to use a smaller ES buffer for its internal representation of the HRD. However, the size of ES buffer shall always be equal to the minimum buffer value $B_{\min}$ specified by the Generalized Hypothetical Reference Decoder for rate R as follows (in units of bits):

$$EBS_n[k] = (hrd\_buffer[k]+1) \times 2^{(buffer\_size\_exponent+4)} \quad (2\text{-}7)$$

and the associated rate R[k] may be computed from the hrd_rate[k] and the bit_rate_exponent fields as follows (bits/sec):

$$R[k] = (hrd\_rate[k]+1) \times 2^{(bit\_rate\_exponent+6)}. \quad (2\text{-}8)$$

## Encapsulation of VC-1 in PS

Delivery of VC-1 Elementary Streams in MPEG-2 PSs shall be governed by a P-STD in MPEG-2 Systems provisions. The stream type value 0xea and the use of registration descriptor and sub-descriptors defined in the previous section shall also be applicable to the carriage of a VC-1 ES in an MPEG-2 PS. The only difference is that in the case of an MPEG-2 PS, the structures where these fields are used is the PSM-PES as opposed to the PM section in TSs. Headers and payload formats shall be identical to the format described in MPEG-2 Systems. PSM-PES for VC-1 is depicted in Figure 2-11.

The input buffers $B_n$ in the P-STD shall not overflow. Furthermore, they shall not underflow except where the value of HRD_PARAM_FLAG field in the Sequence header of the VC-1 ES is equal to "0", in which case the codec operates in variable delay mode as described in Chapter 3. Data enters the P-STD at the rate specified by the value of the field program_mux_rate in the Pack header. The PES packet data bytes from VC-1 ES number n are passed to the input $B_n$. Unless specified by the P-STD_buffer_scale and P-STD_buffer_size field in the PES header, the input buffer sizes $B_n$ are equal to the sum of $vbv\_\max[profile, level]$ value of the VC-1 ES and the value $BS_{add}$ defined in ISO 13818-1 as:

$$BS_{add} \le Max[6 \times 1024, R_{v\max} \times 0.001] \text{ bytes} \quad (2\text{-}9)$$

With an assumption that the incoming service delivery rate R is known through ES_rate field in PES header, a receiver may opt to use a smaller ES buffer for its internal representation of the HRD. However, the size of ES buffer shall always be equal to the minimum buffer value $B_{\min}$ specified by the Generalized Hypothetical Reference Decoder for rate R as follows (in units of bits):

$$BS_n[k] = (hrd\_buffer[k]+1) \times 2^{(buffer\_size\_exponent+4)} \qquad (2\text{-}10)$$

and the associated rate R[k] may be computed from the hrd_rate[k] and the bit_rate_exponent fields as follows (bits/sec):

$$R[k] = (hrd\_rate[k]+1) \times 2^{(bit\_rate\_exponent+6)}. \qquad (2\text{-}11)$$

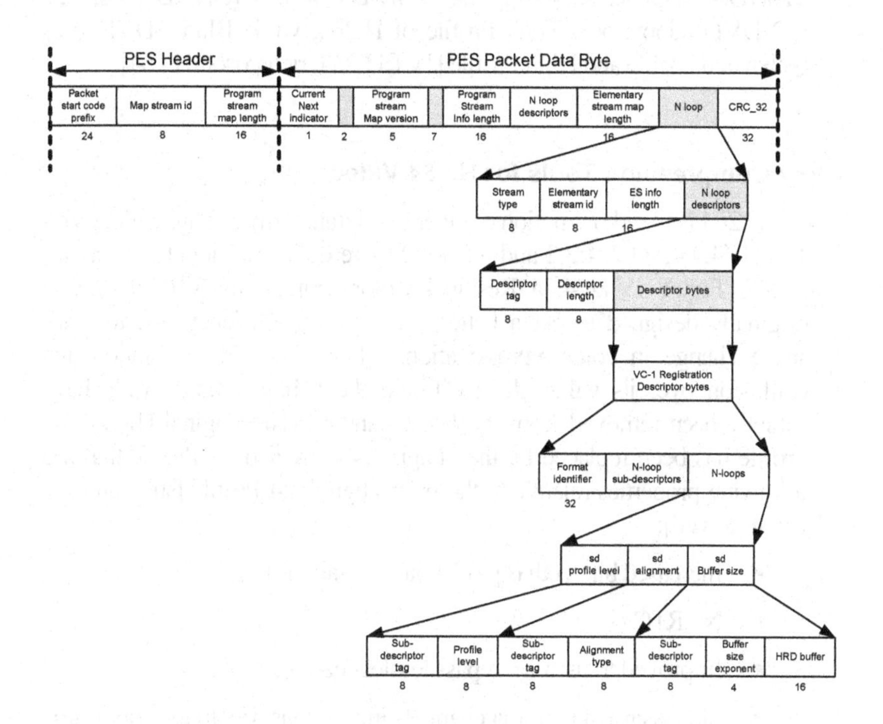


**Figure 2-11 PSM-PES Diagram for VC-1**

## 2.3 H.264 Syntax Hierarchy in Bitstreams

### H.264 Standard

MPEG-4 Part 10 or H.264 video compression standard was developed to enhance compression performance over the current de facto standard MPEG-2 that was developed about 10 years ago primarily for digital TV systems with interlaced video coding. H.264 is known to achieve a significant improvement in rate-distortion efficiency compared with existing standards, and is designed for broadcast TV over cable/DSL/satellite, IP set-tops and HD-DVD/ Bluray-DVD recorders. HD-DVD adopts only High Profile of H.264, while Bluray-DVD uses both Main Profile and High Profile [JVT:H.264, richardson:H.264].

### Key Compression Tools for H.264 Video

H.264 is based on motion compensated transform coding. Unlike VC-1, YUV4:4:4, YUV4:2:2 and YUV4:2:0 are defined as input formats for H.264. For YUV4:4:4, a Residual Color Transform (RTC) tool was originally designed to get a better compression efficiency due to color space change in color representation. After some reconsideration and confusion over its value, the RCT and the "High 4:4:4" Profile have actually been removed from the latest standard. The original High 4:4:4 Profile has been replaced by the "High 4:4:4 Predictive" Profile that has following properties relative to the prior High 4:4:4 Profile [sullivan:new, lee:improved]:

- Increased bit depth (upto 14 bits per sample)
- No RTC
- Improved transform-bypass lossless coding
- Processing of Chroma channels in a similar way to the way Luma channels are processed (i.e., in terms of MC, interpolation and Intra prediction)
- Two modes of operation – the "Common" and "Separate" modes control

The key concept in H.264 is the Slice. The information of two higher layers over Slice is Sequence parameters set and Picture parameters set. Note that those are parameters set to be used either directly or indirectly by each Slice as shown in Figure 2-12. Sequence parameter ID ranges from 0 to 31 while Picture parameter ID ranges from 0 to 255.

There is no GOP structure since referencing order is decoupled from coding order in H.264. But there is an imaginary Picture structure that is composed of one or more Slices. Here, the term "imaginary" means that there is no Picture layer in the bitstream structure, but a picture is generated through the Slice decoding process. The characteristic of the picture is declared with primary_pic_type, where the value indicates what kind of slices are in the primary coded picture. Note that many different types of Slices can constitute a single picture. For example, a primary coded picture can be composed of I, P and B Slices. There are 10 Slice types – two I (Intra), two P (Predicted), two B (Bi-predictive), two SP (Switching P) and two SI (Switching I) as shown in Table 1-8. Each coding mode slice has two different numbers assigned to it since a picture in H.264 can be composed of multiple slices with different slice types. An example includes I slices, P slices, B slices mixed to constitute a single picture.

In H.264, the geometrical structure of Slices is very generic. There are six pre-defined Slice geometrical structures—Interleaved, Dispersed, Foreground and Background, Box-out, Raster scan and Wipe as discussed in Chapter 1. There is a 7th option to define any kind of Slice shape explicitly. Note that the geometrical structure of a Slice can help visual recovery of lost data. For example, interleaving of two slices improves visual perceptual when one of them is lost as shown in Chapter 1.

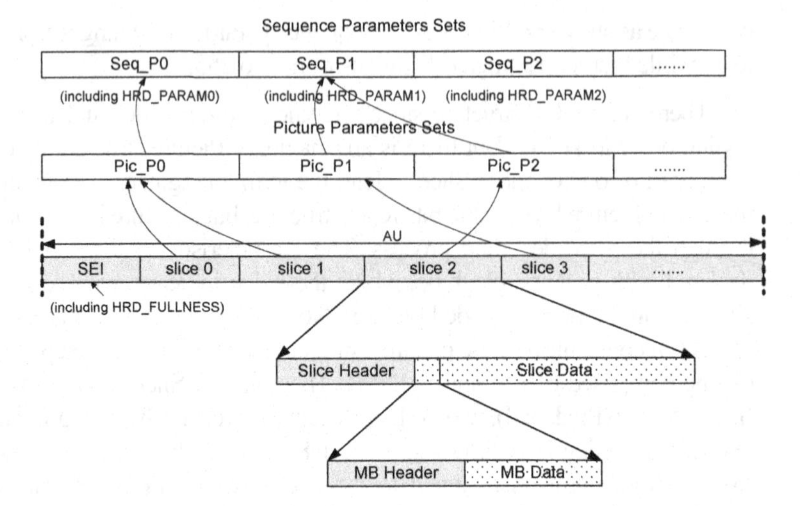


**Figure 2-12 Syntax Hierarchy for H.264**

Multiple reference pictures can be used in motion compensation. The syntax element, num_ref_frames, specifies the maximum total number of short-term and long-term reference frames, and it ranges from 0 to 16. In addition, B frames can be a reference in H.264. References are stored in a DPB (Decoded Picture Buffer) in both the encoder and the decoder. The encoder and the decoder maintain a list of previously coded pictures -- reference picture list0 for P Slices and reference picture list0 and list1 for B Slices. These two lists can contain short-term and long-term pictures, where either list0 or list1 can be used for past coded pictures, while the other one can be used for future coded pictures. There is pre-defined index management called "sliding window" memory control. For example, a coded picture is reconstructed by the encoder and marked as a short-term picture that is identified with its PicNum for list0. However, the reference list0 can be also controlled with "adaptive" memory control in a customized manner by an encoder. For example, if a certain pattern is continuously used as a background video pattern during 10 minutes in a video, an encoder may want to keep it as a long-term picture reference. That kind of encoder decision about list management can be written in the bitstreams with two syntax elements – reference picture list reordering syntax and decoded reference picture marking syntax. These could be thought of as on the fly memory control commands in the bitstreams.

Also, an encoder can send an IDR (Instantaneous Decoder Refresh) coded picture made up of I- or SI-Slices to clear all DPB contents. This means that all subsequent transmitted slices can be decoded without reference to any frame decoded prior to the IDR picture.

A key compression tool in H.264 is adaptive size MC (motion compensation) with a small 4x4 transform. Note that the 8x8 transform has been recently introduced. The size of MC is described in two levels – MB partition and MB sub-partition. The MB partition size can be broken down into sizes of 16x16, 16x8, 8x16 and 8x8. If 8x8 size is chosen in MB partition, each MB sub-partition (8x8 block) can be broken into sizes of 8x8, 8x4, 4x8 or 4x4. The color components, Cb and Cr, have the same, half, or quarter the size of luma components based on video formats such as YUV4:4:4, YUV4:2:2 or YUV4:2:0, but each chroma block is partitioned in the same way as the luma component. Note that the resolution of MVs in luma is quarter-pel, while that of derived MVs in chroma is 1/8-pel. A specific 6 tap-FIR filter and bi-linear filter are used for interpolation. Also, MVs can point past picture boundaries with conceptually extended padding.

The 4x4 transform is a novel integer transform with the following two aspects: 1. It is an integer transform, where all operations can be carried out with integer arithmetic. 2. Mistmatch between encoders and decoders doesn't occur. The 4x4 transform de-correlates well for high frequency patterns. As a fact, small size transform performs well in random-looking areas such as residual data. Consequently, it reduces the "ringing" artifact. Note that directional spatial prediction for Intra coding is introduced in H.264 to eliminate redundancy resulting in random-looking areas, even in I Slices. Directional spatial prediction has a couple of options to best de-correlate intra block data based on previously-decoded parts of the current pictures. Hierarchical transform is applied on DC values of 4x4 transforms in 16x16 Intra mode MB and chroma blocks, since R-D characteristics with longer basis transform are better than that of shorter basis transform over a smooth area. There are improvements in skipped mode and direct mode in motion compensation over existing standards in syntax representation to suppress space for MV data.

The weighted prediction is extended to handle fading scene scenarios by providing weights and offsets. The direct mode uses bi-directional prediction with derived MVs based on the MV of the co-located MB in the subsequent picture reference one as shown in Figure 6-4. Note that a default weight based on geometrical division can be overridden with explicit weights to improve prediction performance. One important difference between H.264 and VC-1 in terms of direct mode technique is that actual blending for a predictor happens based on weights in H.264, while pixel-wise average for a predictor is obtained in VC-1 with weights only being considered for MV computation. Multi-hypothesis mode is worthwhile to note since a joint estimation for a predictor is possible through it. While bi-directional prediction type only allows a linear combination of a forward/ backward prediction pair, any combination of references is possible (i.e., backward/ backward combination or forward/ forward combination).

ILF is used to reduce blocking artifacts. H.264 ILF dynamically adapts the length of FIR filter tap. The decision of filter type and taps is dependent on how serious the blocky effect is; the standard provides ways to measure it. Based on the measurement, 5/4/3 tap filters are applied on horizontal and vertical block boundaries. The handling of 4x4 transform block coefficients is quite different compared with other standards. 3D RLC or 2D RLC is not used, but Token, Sign, Level, run_before groups are coded in the inverse scanning order (the last to the first coefficient).

H.264 standard suggests two different types of entropy coding – CA-VLC and CA-BAC. In CA-VLC, Huffman (VLC) tables are chosen based on neighboring contexts or historical contexts. Examples are following: the choice of Token VLC tables is dependent on the number of non-zero coefficients in upper and left-hand previously coded blocks. The choice of Level VLC computation rule is triggered by the present Level value. If a present Level value is over a certain threshold, new VLC computation rules for Level are used to compress Level symbols.

When CA-BAC is selected for entropy coding, the following four consecutive stages follow: 1. Binarization, 2. Context-model selection, 3. Arithmetic encoding, 4. Probability update. Generally speaking, CA-BAC outperforms CA-VLC. The reason for this is that arithmetic coding is block entropy coding (the block size is the size of overall input bits until another initialization), while Huffman (VLC) coding is scalar entropy coding. An additional benefit for using arithmetic coding is to dynamically adapt the probability model as the probability of symbols change w. r. t. time. There can be many adaptation methods. One is to use accumulation of Symbols up to present for the model probability. However, this method doesn't capture local statistics, but only considers the long-term average.

An excellent method to capture local statistics is to use neighboring contexts as is done in CA-BAC. Note that the context models are initialized depending on the initial value of the Qp. The Binarization method and Context-model selection for each Syntactic element are specified in the standard. A multiplication-free method for the arithmetic coding is used in H.264 based on a small set of representative values of range and probability of symbols. The range is quantized into four distinct values and the probability of symbols is quantized into 64 distinct values. This allows a pre-computed table approach in the arithmetic coding, thus making it multiplication-free.

There are many coding options for each MB as shown in Figure 2-13. There are two transforms choices in each 8x8 block, where the same transform is required to apply – either 4x4 and 8x8. Note that 8x8 transform is only defined for High Profile. The choice between transform sizes is basically determined by the encoder to take advantage of statistics of input signals in High Profile. There are many MC modes in terms of size where the partition is a combination of sub-pattern of a 16x16 block and a 8x8 block as shown in the Figure 2-13. The size can vary from 4x4 to 16x16, thus meaning that the number of MVs can vary from 32 to 1 for a MB. The choice among MC sizes is basically made by the encoder to take advantage of statistics of the input signals. Typically, a large area is effective when a rigid object is moving (uniform motion in the area), while a small area is so when tearing object (no object) and/or overlapped objects are moving (random motion in the area).

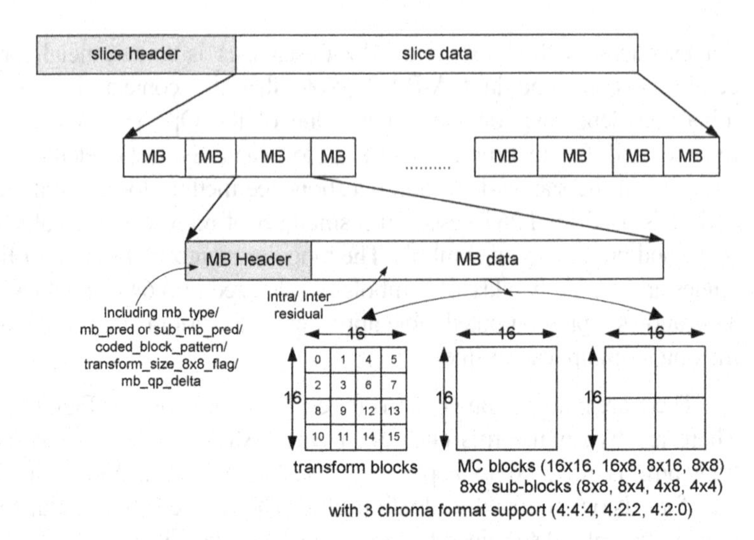


**Figure 2-13 Slice Hierarchy for H.264**

## H.264 Video Specific Semantics and the Syntax

There are five levels of information in the H.264 video bitstream syntax – Sequence Parameter Set (SPS), Picture Parameter Set (PPS), Slice level, MB and Block. SPS contains basic parameters such as profile and level data, seq_parameter_set_id (identifies sequence parameter set that is referred to), MaxFrameNum, picture order related parameters, num_ref_frames (maximum total number of short-term and long-term reference frames), frame_mbs_only_flag, direct_8x8_inference_flag (specifies the method used in the derivation process of luma MVs for B_Skip, B_Direct_16x16, B_Direct_8x8), etc.

PPS contains information about pic_paramter_set_id (identifies the picture parameter set that is referred to in the Slice header), seq_parameter_set_id, entropy_coding_mode_flag (either CA-BAC or CA-VLC), slice definition related parameters, maximum reference index data for reference list0 or list1, weighted_pred_flag (whether weighted prediction is applied to P and SP Slices), weighted_bipred_idc (weighted bi-prediction mode applied to B Slices), Qp related data, deblocking_filter_control_present_flag (syntax element controlling the characteristics of the deblocking filter is present in the Slice header), etc .

In the Slice header, there is information about slice attributes, extra display order related parameters, direct_spatial_mv_pred_flag, override information about reference index for list0 and list1, deblocking filter related data, etc. Slice data contains mb_skip related data and MB layer information. The MB layer contains mb_type, coded_block_pattern, mb_qp_delta. Based on mb_type information, mb_pred or sub_mb_pred data are put into the bitstream. Then, residual data is appended. The mb_pred and sub_mb_pred information contain reference picture indices and MV data. Note that each 8x8 block area must refer to a single reference picture even though different MVs can be used for different 4x4 blocks in a single 8x8 block. There are two residual data syntax flows defined – one for CA-VLC and the other for CA- BAC.

### H.264 Profiles/ Tools

Baseline Profile targets low-rate Internet video conferencing and low-complexity applications such as mobile communications or play back of media in personal digital assistants.

The Main Profile targets broadcast applications such as digital TV and local video storage for PC play back. Bluray-DVD can contain Main Profile contents.

The Streaming Profile targets Internet applications such as streaming, movie delivery via IP, or TV/ VOD over IP.

The High Profile targets high quality applications such as studio editing, HD-DVD, Bluray-DVD, or DVB (digital video broadcast for European TV).

The High 10 Profile targets future Consumer Electronics products supporting for up to 10 bits per sample of decoded picture precision.

The High 4:2:2 Profile targets professional applications that use interlaced video supporting for the 4:2:2 Chroma sub-sampling format and 10 bits per sample of decoded picture precision.

The High 4:4:4 Predictive Profile targets professional applications supporting for the 4:4:4 Chroma sub-sampling format and 14 bits per sample of decoded picture precision.

Each class of bitstreams contains tool sets defined in Table 2-6 and Table 2-7 and parameters set defined in Table 2-8.

**Table 2-6 H.264 Profiles and Tools in Original H.264**

| Tool Options | Baseline Profile | Main Profile | Extended Profile |
|---|---|---|---|
| I and P Slices | x | x | x |
| CA-VLC | x | x | x |
| CA-BAC | | x | |
| B Slices | | x | x |
| Interlaced Coding (PAFF, MBAFF) | | x | x |
| Enh. Err. Resil. (FMO, ASO, RS) | x | | x |
| Further Enh. Err. Resil. (DP) | | | x |
| SI and SP Slices | | | x |

**Table 2-7 H.264 Profiles and Tools in New H.264 Profiles/Amendament**

| Tool Options | High | High 10 | High 4:2:2 | High 4:4:4 Predictive |
|---|---|---|---|---|
| Main Profile Tools | x | x | x | x |
| 4:2:0 Chroma Format | x | x | x | x |
| 8 bit Sample Bit Depth | x | x | x | x |
| 8x8 vs. 4x4 Transform Adaptivity | x | x | x | x |
| Quantization Scaling Matrices | x | x | x | x |
| Separate Cb and Cr QP Control | x | x | x | x |
| Monochrome Video Format | x | x | x | x |
| 9 and 10 bit Sample Bit Depth | | x | x | x |
| 4:2:2 Chroma Format | | | x | x |
| 14 bit Sample Bit Depth | | | | x |

| | | | | |
|---|---|---|---|---|
| 4:4:4 Chroma Format | | | | x |
| Improved Transform-bypass Lossless coding | | | | x |
| 3 Separate Color Plan coding | | | | x |

Additionally, the standard now contains four all-Intra Profiles, which are defined as simple subsets of other corresponding Profiles. These are for professional applications. The reason to define these profiles is to reduce decoder complexity by restricting the coding mode and the entropy coding method. This direction was taken based on the fact that H.264 Intra coding many times outperforms the state of the art in still image coding schemes.

The High 10 Intra Profile is defined when the High 10 Profile is constrained to all-Intra use.

The High 4:2:2 Intra Profile is defined when the High 4:2:2 Profile is constrained to all-Intra use.

The High 4:4:4 Intra Profile is defined when the High 4:4:4 Profile is constrained to all-Intra use.

The CA-VLC 4:4:4 Intra Profile is defined when the High 4:4:4 Predictive Profile is constrained to all-Intra use and to CA-VLC.

**Table 2-8 H.264 Levels and Limitations**

| Level Number | MB/s | MB/f | DPBmax (1024 bytes for 4:2:0) | Rmax (1000 bps-VCL, 1200 bps-NAL) | CPBmax (1000 bps-VCL, 1200 bps-NAL) | MV [Vertical] | CRmin | Max number of MVs per 2 consecutive MBs (MaxMVsPer2Mb) |
|---|---|---|---|---|---|---|---|---|
| 1 | 1,485 | 99 | 148.5 | 64 | 175 | [-64,63 ¾] | 2 | - |
| 1b | 1,485 | 99 | 148.5 | 128 | 350 | [-64,63 ¾] | 2 | - |
| 1.1 | 3,000 | 396 | 337.5 | 192 | 500 | [-128,127 ¾] | 2 | - |
| 1.2 | 6,000 | 396 | 891.0 | 384 | 1,000 | [-128,127 ¾] | 2 | - |

| | | | | | | | | |
|---|---|---|---|---|---|---|---|---|
| 1.3 | 11,880 | 396 | 891.0 | 768 | 2,000 | [-128,127 ¾] | 2 | - |
| 2 | 11,880 | 396 | 891.0 | 2,000 | 2,000 | [-128,127 ¾] | 2 | - |
| 2.1 | 19,800 | 792 | 1,782.0 | 4,000 | 4,000 | [-256,255 ¾] | 2 | - |
| 2.2 | 20,250 | 1,620 | 3,037.5 | 4,000 | 4,000 | [-256,255 ¾] | 2 | - |
| 3 | 40,500 | 1,620 | 3,037.5 | 10,000 | 10,000 | [-256,255 ¾] | 2 | 32 |
| 3.1 | 108,000 | 3,600 | 6,750.0 | 14,000 | 14,000 | [-512,511 ¾] | 4 | 16 |
| 3.2 | 216,000 | 5,120 | 7,680.0 | 20,000 | 20,000 | [-512,511 ¾] | 4 | 16 |
| 4 | 245,760 | 8,192 | 12,288.0 | 20,000 | 25,000 | [-512,511 ¾] | 4 | 16 |
| 4.1 | 245,760 | 8,192 | 12,288.0 | 50,000 | 62,500 | [-512,511 ¾] | 2 | 16 |
| 4.2 | 522,240 | 8,704 | 13,056.0 | 50,000 | 62,500 | [-512,511 ¾] | 2 | 16 |
| 5 | 589,824 | 22,080 | 41,400.0 | 135,000 | 135,000 | [-512,511 ¾] | 2 | 16 |
| 5.1 | 983,040 | 36,864 | 69,120.0 | 240,000 | 240,000 | [-512,511 ¾] | 2 | 16 |

There are several levels for each of the profiles. Each level limits the video resolution, frame rate, HRD bit rate, HRD buffer requirements, and the motion vector range. These limitations are defined in Table 2-8.

## 2.4 H.264 Encapsulation in MPEG-2 Systems

Amendment 3 of ITU-T Recommendation H.222.0| ISO/IEC 13818-1:2000/Amd.3:2004 recommends H.264 bitstream encoding provisions that define a minimum set of rules for the carriage of an H.264 elementary stream in an MPEG-2 Transport Stream with additional intention to provide a generic means of carrying an H.264/ AVC video elementary stream in an MPEG-2 Program Stream as used by the DVD Forum [ISO:MPEG2systems.amd]. This section discusses H.264 bitstream encapsulation in MPEG-2 Systems.

### NAL and VCL

In H.264, the Network Abstraction Layer (NAL) is designed to be self-contained for extensive encapsulation methods so that the coding layer can be separated from delivery/ system encapsulation mechanism as shown in Figure 2-14.

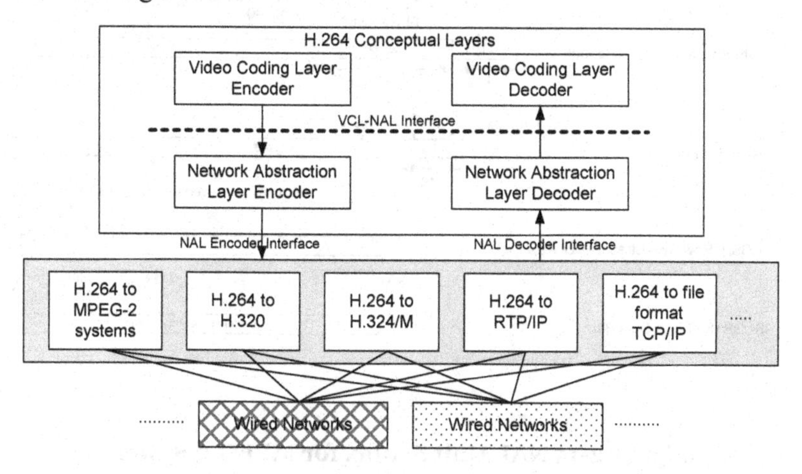


**Figure 2-14 Network Abstraction Layer and Interface**

To this end, NAL and Video Coding Layer (VCL) layers are devised. VCL is designed to efficiently represent the video content, while NAL encapsulates the VCL representation of video with header information in such a way that a variety of transport layers and storage media can easily adopt compressed contents. A NAL unit specifies both byte-stream and packet-based formats. Byte-stream format as shown in Figure 2-15 is used for bitstream-like representation as in MPEG-2, while packet-based format targets applications with coded data carried in an internet-like transport protocol. NAL unit header 1B is composed of 1-bit "forbidden_zero_bit," 2-bit "nal_ref_idc" and 5-bit "nal_unit_type." The forbidden_zero_bit indicates whether the NAL unit has errors – "1" means to be in error. The nal_ref_idc implies whether the NAL unit is disposable. The nal_unit_type provides a peeking function into the payload about NAL unit types such as SEI message, Parameter Sets, VCL data, etc. Packet-based systems can employ NAL units directly.

NAL units are classified into VCL NAL and non-VCL NAL units. The VCL NAL units contain the data that represents the values of the samples in the video pictures, and the non-VCL NAL units contain additional information such as timing information.

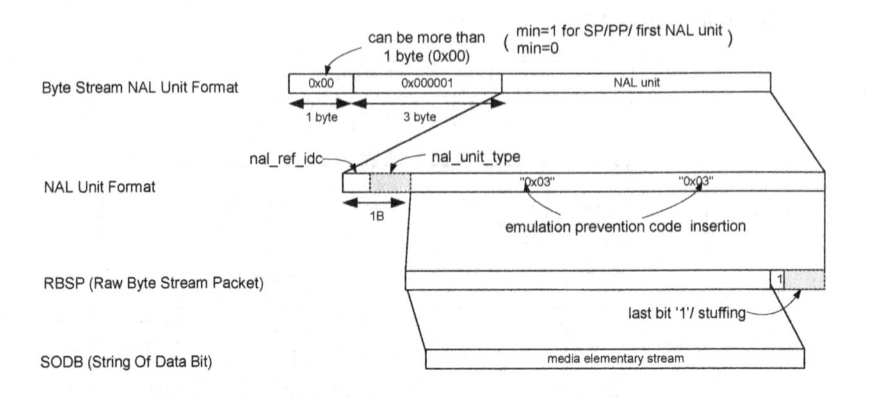


**Figure 2-15 NAL Unit Syntax for MPEG-2 Systems**

## Access Unit and SEI in H.264

A set of NAL units in a specified form as in Figure 2-16 is referred to as an AU in H.264. The decoding of each AU results in one decoded picture. Each AU contains a set of VCL NAL units that together compose a Primary Coded Picture (PCP). It is mandatory to be pre-fixed with an AU delimiter to aid in locating the start of the AU.

Supplemental Enhancement Information (SEI) containing data such as picture timing information may also precede the PCP. The PCP consists of a set of VCL NAL units containing slices or slice data partitions that represent the samples of the video picture. The Redundant Coded Picture (RCP) is composed of additional VCL NAL units that may contain redundant representations of areas of the same video picture. Decoders are not required to decode redundant coded pictures when they are present. Decoders can decode them when any problems occur at PCP decoding.

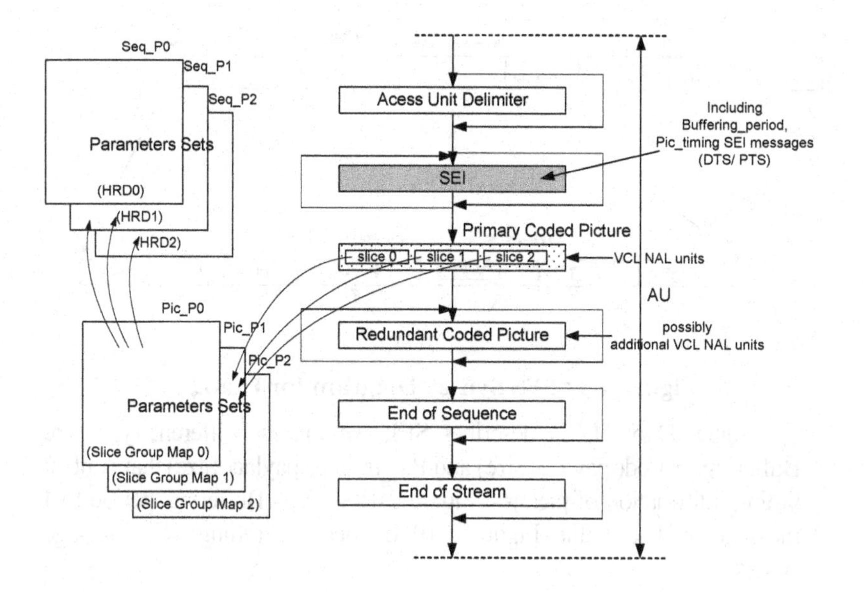


**Figure 2-16 Access Unit in H.264**

If the coded picture is the last picture of the coded video sequence (a sequence of pictures that is independently decodable with one sequence parameter set), an End of Sequence NAL unit may be present to indicate the end of the sequence. Note that in order to activate a different SPS, one must start a new coded video sequence. In other words, one movie can be broken down to a couple of segments each of which is defined as a separate coded sequence. If the coded picture is the last coded picture in the entire NAL unit stream, and End of Stream NAL unit may be present to indicate that the stream is ending. Figure 2-17 depicts PES payload format for H.264.

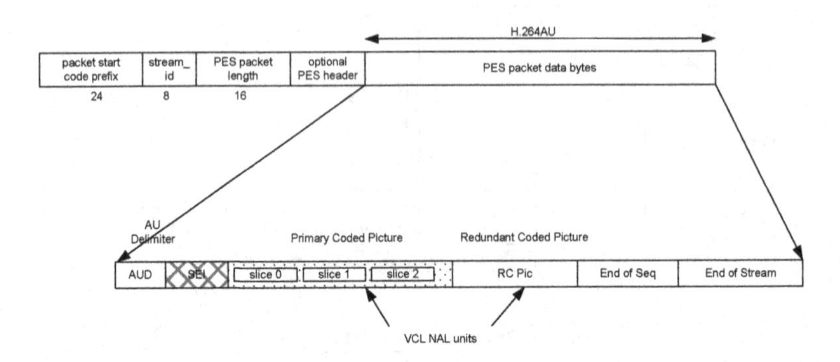


**Figure 2-17 PES Syntax Diagram for H.264**

Annex D of H.264 describes SEI. Among the different types are Buffering_period( payloadSize) and Pic_timing( payloadSize) that contain timing information of pictures. Figure 2-18 defines Buffering_period SEI message syntax, while Figure 2-19 defines Pic_timing SEI message syntax.

The explanation about Buffering_period SEI message syntax as shown in Figure 2-18 is as follows: Seq_parameter_set_id specifies the SPS that contains the sequence HRD attributes. The value of Seq_parameter_set_id is equal to the value of Seq_parameter_set_id in the PPS referenced by the PCP associated with the Buffering period SEI message.

Initial_cpb_removal_delay[ SchedSelIdx] specifies the delay for the SchedSelIdx-th Coded Picture Buffer (CPB) between the time of arrival in the CPB of the first bit of the coded data associated with the AU of the Buffering period SEI message and the time of removal from the CPB of the coded data associated with the same AU. It is for the first buffering period after HRD initialization. The syntax element has a length in bits given by initial_cpb_removal_delay_length_minus1 + 1. It is in units of a 90 kHz clock (a.k.a., counter tick).

The initial_cpb_removal_delay_offset[ SchedSelIdx] is used for the SchedSelIdx-th CPB in combination with the Cpb_removal_delay to specify the initial delivery time of coded AU to the CPB. The syntax element has a length in bits given by initial_cpb_removal_delay_length_minus1 + 1. It is in units of a 90 kHz clock. This syntax element is not used by decoders and is needed only for delivery scheduling at Hypothetical Stream Scheduler (HSS) specified in Annex C.

```
Buffering_period( payloadSize ){
  Seq_parameter_set_id
  If( NalHrdBpPresentFlag){
   For(SchedSelIdx=0;        SchedSelIdx       <=
   Cpb_cnt_minus1; SchedSelIdx ++){
      Initial_cpb_removal_delay[SchedSelIdx]
   Initial_cpb_removal_delay_offset[SchedSelIdx]
      }
  }
  If( VclHrdBpPresentFlag){
   For(SchedSelIdx=0;        SchedSelIdx       <=
   Cpb_cnt_minus1; SchedSelIdx ++){
      Initial_cpb_removal_delay[SchedSelIdx]
   Initial_cpb_removal_delay_offset[SchedSelIdx]
   }
  }
}
```

**Figure 2-18 Buffering_period SEI Message Syntax**

The explanation about Pic_timing SEI message syntax as shown in Figure 2-19 is as follows: Cpb_removal_delay specifies how many clock ticks to wait after removal from the CPB of the AU associated with the most recent buffering period SEI message before removing from the buffer the AU data associated with the picture timing SEI message. The syntax element has a length in bits given by cpb_removal_delay_length_minus1 + 1. This value is also used to calculate the earliest possible time of arrival of AU data into the CPB for the HSS specified in Annex C. The clock tick is defined in VUI as:

$$t_c = \frac{num\_units\_in\_tick}{time\_scale} \tag{2-12}$$

For example, $t_c = 1/29.97$ (sec) when num_units_in_tick=1001 and time_scale = 30000. During the period of $t_c$, the Time Stamp counter increases 3003 based on a 90kHz clock. This setting is most likely used with fixed_frame_rate_flag=0.

```
Pic_timing( payloadSize ){
  If(CpbDpbDelaysPresentFlag){
    Cpb_removal_delay
    Dpb_output_delay
  }
  If( pic_struct_present_flag){
    Pic_struct
    For(I=0; I<NumClockTS; I++){
      Clock_timestamp_flag[I]
      If( clock_timestamp_flag[I]){
        Ct_type
        Nuit_field_based_flag
        Counting_type
        Full_timestamp_flag
        Discontinuity_flag
        Cnt_dropped_flag
        N_frames
        If( full_timestamp_flag){
          Seconds_value
          Minutes_value
          Hours_value
        } else {
          seconds_flag
          if( seconds_flag){
            seconds_value
            minutes_flag
            if( minutes_flag){
              minutes_value
              hours_flag
              if( hours_flag)
                hours_value
            }
          }
        }
        If( time_offset_length>0)
          Time_offset
      }
    }
  }
}
```

**Figure 2-19 Pic_timing SEI Message Syntax**

For example, $t_c = 1/59.94$ (sec) when num_units_in_tick=1001 and time_scale = 60000. This setting is most likely used with fixed_frame_rate_flag=1 that covers certain broadcast video (i.e., Pic_struct=0, 3, 4, 5, 6 etc. in Table 2-9). Note that fixed_frame_rate_flag=1 scenarios are built on the assumption that $1/t_c$ is twice the frame rate.

Dpb_output_delay specifies how many clock ticks to wait after removal of an AU from CPB before the decoded picture can be output from the DPB. This value is used to compute the DPB output time of the picture.

Pic_struct indicates whether a picture should be displayed as a frame or one or more fields according to Table 2-9 below. Frame doubling (7) indicates that the frame should be displayed two times consecutively, and frame tripling (8) indicates that the frame should be displayed three times consecutively.

**Table 2-9 Interpretation of Pic_struct**

| Value | Indicated display of picture | Restrictions | NumClockTS |
|---|---|---|---|
| 0 | Frame | Field_pic_flag shall be 0 | 1 |
| 1 | Top field | Field_pic_flag shall be 1,<br>Bottom_pic_flag shall be 0 | 1 |
| 2 | Bottom field | Field_pic_flag shall be 1,<br>Bottom_pic_flag shall be 1 | 1 |
| 3 | Top field, bottom field, in that order | Field_pic_flag shall be 0 | 2 |
| 4 | Bottom field, top field, in that order | Field_pic_flag shall be 0 | 2 |
| 5 | Top field, bottom field, top field repeat, in that order | Field_pic_flag shall be 0 | 3 |

| | | | |
|---|---|---|---|
| 6 | Bottom field, top field, bottom field repeat, in that order | Field_pic_flag shall be 0 | 3 |
| 7 | Frame doubling | Field_pic_flag shall be 0, Fixed_frame_rate_flag shall be 1 | 2 |
| 8 | Frame tripling | Field_pic_flag shall be 0, Fixed_frame_rate_flag shall be 1 | 3 |
| 9..15 | reserved | | |

**Table 2-10 Definition of counting_type Values**

| Counting_type value | Interpretation |
|---|---|
| 0 | No dropping of N_frames count values and no use of time_offset |
| 1 | No dropping of N_frames count values |
| 2 | Dropping of individual zero values of N_frames count |
| 3 | Dropping of individual MaxFPS-1 values of N_frames count |
| 4 | Dropping of two lowest (value 0 and 1) N_frames counts when Seconds_value is equal to 0 and Minutes_value is not an integer multiple of 10. |
| 5 | Dropping of unspecified individual N_frames count values |
| 6 | Dropping of unspecified numbers of unspecified N_frames count values |
| 7..31 | reserved |

NumClockTS is determined by Pic_struct and it specifies that there are up to NumClockTS sets of Time Stamp information for a picture, as specified by clock_timestamp_flag[I] for each set. The sets of Time Stamp information apply to the field(s) or the frame(s) associated with the picture by Pic_struct. The contents of the Time Stamp syntax elements indicate a time of origin, capture, or alternative ideal display. This indicated time is computed as:

$$\text{ClockTimeStamp} = ((\text{Hours\_value} \times 60 + \text{Minutes\_value}) \times 60 + \text{Second\_value}) \times \text{time\_scale} + \text{N\_frames} \times (\text{num\_units\_in\_tick} \times (1 + \text{Nuit\_field\_based\_flag})) + \text{Time\_offset} \quad (2\text{-}13)$$

in units of clock ticks of a clock with clock frequency equal to time_scale Hz, relative to some unspecified point in time for which ClockTimeStamp is equal to 0. For example, with num_units_in_tick=1001 and time_scale=30000:

$$\text{ClockTimeStamp} = ((\text{Hours\_value} \times 60 + \text{Minutes\_value}) \times 60 + \text{Second\_value}) \times 30000 + \text{N\_frames} \times (1001 \times (1 + \text{Nuit\_field\_based\_flag})) + \text{Time\_offset}. \quad (2\text{-}14)$$

Output order and DPB output timing are not affected by the value of ClockTimeStamp.

Ct_type indicates the scan type (interlaced or progressive) of the source material. Counting_type specifies the method of dropping values of the N_frames as specified in Table 2-10, while Cnt_dropped_flag specifies the skipping of one or more values of N_frames using the counting method specified by Counting_type.

Discontinuity_flag indicates whether the difference between the current value of ClockTimeStamp and the value of ClockTimeStamp counted from the previous ClockTimeStamp in output order should or should not be interpreted as the time difference between the times of origin or capture of the associated frames or fields.

Seconds_value/ Minutes_value/ Hours_value/ Time_offset/ N_frames, etc. are used to compute ClockTimeStamp.

### HRD Parameters in H.264

Most of important parameters used to parse SEI messages are in HRD parameters as shown in Figure 2-20.

```
Hrd_parameters( ){
  Cpb_cnt_minus1
  Bit_rate_scale
  Cpb_size_scale
  For(SchedSelIdx=0;     SchedSelIdx     <=
Cpb_cnt_minus1; SchedSelIdx ++){
    Bit_rate_value_minus1[SchedSelIdx]
    Cpb_size_value_minus1[SchedSelIdx]
    Cbr_flag[SchedSelIdx]
  }
  Initial_cpb_removal_delay_length_minus1
  Cpb_removal_delay_length_minus1
  Dpb_output_delay_length_minus1
  Time_offset_length
}
```

**Figure 2-20 HRD Parameters Syntax**

Cpb_cnt_minus1+1 specifies the number of alternative CPB specifications in the bitstream. When low_delay_hrd_flag is equal to 1, cpb_cnt_minus1 shall be equal to 0. Bit_rate_scale and Bit_rate_value_minus1[SchedSelIdx] specify the maximum input bit rate (the bit rate in bits per sec.) for the SchedSelIdx-th CPB with:

BitRate[SchedSelIdx]=

$$(bit\_rate\_value\_minus1[SchedSelIdx]+1)\times 2^{(6+bit\_rate\_scale)} \quad (2\text{-}15)$$

When the bit_rate_minus1[SchedSelIdx] is not present, BitRate[SchedSelIdx] shall be inferred to be equal to $1000\times$ MaxBR for VCL HRD parameters and to $1200\times$ MaxBR for NAL HRD parameters, respectively.

The CpbSize in bits is given by:

CpbSize[SchedSelIdx]=

$$(cpb\_size\_value\_minus1[SchedSelIdx]+1)\times 2^{(4+cpb\_size\_scale)} \quad (2\text{-}16)$$

When the cpb_size_value_minus1 [SchedSelIdx] is not present, CpbSize [SchedSelIdx] shall be inferred to be equal to $1000 \times$ MaxCPB for VCL HRD and to $1200 \times$ MaxCPB for NAL HRD parameters, respectively.

## Derivation of DTS/ PTS in H.264

In MPEG-2 video, certain important information such as DTS/ PTS is not indicated in the ES. Utilizing ESs without explicit timing information might cause potential problems at random access instances in certain systems. Therefore, MPEG-2 PES becomes a kind of inevitable encapsulator for all applications. Once the time point of decoding and displaying the AUs in the PES is explicitly written at the Time Stamps in MPEG-2 Systems, there is no confusion for decoder and display processor to initiate actions.

Unfortunately, the same argument can be applied for H.264 ESs. Potential problems at random access instances can occur for H.264 ESs since there is no absolute timing information in the NAL units of H.264. To resolve this issue, H.264 can take advantage of already-well-established MPEG-2 Systems as an encapsulator as defined in the ITU-T H.222.0 | ISO/IEC 13818-1:2000/Amd.3 document. In such cases, an encoder's packetizer should be able to extract DTS/ PTS information from the NAL layer to generate the PES packet of MPEG-2 Systems with timing information. Once a PES encapsulated H.264 bitstream is generated, any decoder can process it with correct synchronization actions.

When H.264 bitstreams are not encapsulated in MPEG-2 Systems, DTS/ PTS information might be or might not be explicitly described in any header of the encapsulator. When the timing data is not explicitly written in forms of DTS/ PTS in such an encapsulator, the DTS/ PTS can be derived from timing related information in NAL layer of H.264.

Two key items of information to extract DTS/ PTS are Buffering_period (payloadSize) and Pic_timing (payloadSize) SEI messages . The AU with a buffering period SEI message that initializes the CPB is referred to as AU-0.

**DTS Derivation**

The nominal removal time $t_{r,nom}(0)$ and its DTS(0) of the AU-0 from the CPB are specified by:

$$t_{r,nom}(0) = Initial\_cpb\_removal\_delay[SchedSelIdx]/90000 \text{ and}$$
$$DTS(0) = Initial\_cpb\_removal\_delay[SchedSelIdx] \quad (2\text{-}17)$$

Typically, the HRD is initialized at the beginning of a buffering period in the stream when a random access occurs. For the first AU of a buffering period that does not initialize the HRD, the nominal removal time $t_{r,nom}(n)$ and its DTS(n) of the AU from the CPB are specified by:

$$t_{r,nom}(n) = t_{r,nom}(n_b) + t_c \times cpb\_removal\_delay(n) \text{ and}$$
$$DTS(n) = DTS(n_b) + \frac{num\_units\_in\_tick}{time\_scale} \times 90000 \times cpb\_removal\_delay(n) \quad (2\text{-}18)$$

where $t_{r,nom}(n_b)$ is the nominal removal time of the first picture of the previous buffering period and cpb_removal_delay(n) is specified in the picture timing SEI message associated with AU-n.

When an AU-n is the first AU of a buffering period, $n_b$ and $DTS(n_b)$ are set equal to n and DTS(n) at the removal time of AU-n.

The nominal removal time $t_{r,nom}(n)$ and its DTS(n) of an AU-n that is not the first AU of a buffering period are given by:

$$t_{r,nom}(n) = t_{r,nom}(n_b) + t_c \times cpb\_removal\_delay(n) \text{ and}$$
$$DTS(n) = DTS(n_b) + \frac{num\_units\_in\_tick}{time\_scale} \times 90000 \times cpb\_removal\_delay(n) \quad (2\text{-}19)$$

where $t_{r,nom}(n_b)$ is the nominal removal time of the first picture of the current buffering period and cpb_removal_delay(n) is specified in the picture timing SEI message associated with AU-n.

Finally, the removal time and its DTS(n) of AU-n are specified as follows:

- If low_delay_hrd_flag is equal to 0 or nominal removal time $t_{r,nom}(n) \geq$ final arrival time $t_{af}(n)$, the removal time $t_r(n)$ and its DTS(n) of AU-n are specified by:

$t_r(n) = t_{r,nom}(n)$ and

$$DTS(n) = DTS(n_b) + \frac{num\_units\_in\_tick}{time\_scale} \times 90000 \times cpb\_removal\_delay(n)$$
$$= DTS(0) + \frac{num\_units\_in\_tick}{time\_scale} \times 90000 \times (n_b \cdot \Delta_{cpb} + cpb\_removal\_delay(n))$$
$$= Initial\_cpb\_removal\_delay[SchedSelIdx] + \frac{num\_units\_in\_tick}{time\_scale} \times 90000 \times (n_b \cdot \Delta_{cpb} + cpb\_removal\_delay(n)) \quad (2\text{-}20)$$

Note that DTS(n) in Equation (2-20) shall be rounded to the closest integer value prior to its insertion in a PES header since it is a Time Stamp.

For example, when $t_c = 1/29.97$ (sec) (i.e., $\Delta_{cpb} = 1$ ) with num_units_in_tick=1001 and time_scale = 30000, DTS(n) is as follows:

$$DTS(n) = DTS(n_b) + 3003 \times cpb\_removal\_delay(n)$$
$$= DTS(0) + 3003 \times n_b + 3003 \times cpb\_removal\_delay(n)$$
$$= Initial\_cpb\_removal\_delay[SchedSelIdx] + 3003 \times (n_b + cpb\_removal\_delay(n)) \quad (2\text{-}21)$$

- Otherwise (low_delay_hrd_flag is equal to 1 and $t_{r,nom}(n) < t_{af}(n)$ ), the removal time and its DTS(n) of AU-n are specified by:

$$t_r(n) = t_{r,nom}(n) + t_c \times Ceil((t_{af}(n) - t_{r,nom}(n)) \div t_c) \text{ and}$$

$$DTS(n) = DTS(n_b) + \frac{num\_units\_in\_tick}{time\_scale} \times 90000 \times$$
$$(cpb\_removal\_delay(n) + Ceil((t_{af}(n) - t_{r,nom}(n)) \div t_c))$$
$$= DTS(0) + \frac{num\_units\_in\_tick}{time\_scale} \times 90000 \times$$
$$(n_b \cdot \Delta_{cpb} + cpb\_removal\_delay(n) + Ceil((t_{af}(n) - t_{r,nom}(n)) \div t_c))$$
$$= Initial\_cpb\_removal\_delay[SchedSelIdx] +$$
$$\frac{num\_units\_in\_tick}{time\_scale} \times 90000 \times (n_b \cdot \Delta_{cpb} + cpb\_removal\_delay(n) +$$
$$Ceil((t_{af}(n) - t_{r,nom}(n)) \div t_c))$$

(2-22)

For example, when $t_c = 1/29.97$ (sec) with num_units_in_tick=1001 and time_scale = 30000, DTS(n) is as follows:

$$DTS(n) = DTS(n_b) + 3003 \times (cpb\_removal\_delay(n) +$$
$$Ceil((t_{af}(n) - t_{r,nom}(n)) \div t_c))$$
$$= DTS(0) + 3003 \times (n_b + cpb\_removal\_delay(n) +$$
$$Ceil((t_{af}(n) - t_{r,nom}(n)) \div t_c))$$
$$= Initial\_cpb\_removal\_delay[SchedSelIdx] + 3003 \times$$
$$(n_b + cpb\_removal\_delay(n) + Ceil((t_{af}(n) - t_{r,nom}(n)) \div t_c))$$

(2-23)

This case indicates that the size of AU-n is so large that it prevents removal at the nominal removal time.

### PTS Derivation

Picture n is decoded and its DPB output time $t_{o,dpb}(n)$ and PTS(n) are derived by:

$$t_{o,dpb}(n) = t_r(n) + t_c \times dpb\_output\_delay(n) \text{ and}$$

$$PTS(n) = DTS(n) + \frac{num\_units\_in\_tick}{time\_scale} \times 90000 \times dpb\_output\_delay(n) \qquad (2\text{-}24)$$

The output time of the current picture and its PTS(n) are specified as follows:

- If $t_{o,dpb}(n) = t_r(n)$, the current picture is output.

$t_r(n) = t_{r,nom}(n)$ and

$$PTS(n) = DTS(n) + \frac{num\_units\_in\_tick}{time\_scale} \times 90000 \times dpb\_output\_delay(n) \qquad (2\text{-}25)$$

Note that PTS(n) in Equation (2-25) shall be rounded to the closest integer value prior to its insertion in a PES header since it is a Time Stamp.

For example, when $t_c = 1/29.97$ (sec) with num_units_in_tick=1001 and time_scale = 30000, PTS(n) is as follows:

$$PTS(n) = DTS(n) + 3003 \times dpb\_output\_delay(n) \qquad (2\text{-}26)$$

- Otherwise ($t_{o,dpb}(n) > t_r(n)$), the current picture is output later and will be stored in the DPB. And, the stored picture is output at time $t_{o,dpb}(n)$ unless indicated not to be output by the decoding or inference of no_output_of_prior_pics_flag equal to 1 at a time that precedes $t_{o,dpb}(n)$.

### Artificial Generation of PTS For Special Pic_struct Type

If a picture has a special structure such as frame doubling, the same content of decoded frame would be used again at a later time. If necessary, the PTSs can be artificially generated. ClockTimeStamp formula in a previous subsection may be used to compute multiple PTS for a specific Pic_struct.

### Constraints of Byte-Stream NAL Unit Format for MPEG-2 Systems

An H.264 stream is an element of an ISO/IEC 13818-1 program as defined by the PMT in a TS and the PSM in a PS. The stream_id and stream_type are defined as shown in Table 2-11 and Table 2-12 for H.264. H.264 is also called "AVC video stream" in this context. The carriage and buffer management of AVC video streams is defined using existing parameters from international standards such as PTS and DTS as well as information present within a H.264 video stream.

Carriage of H.264 streams in MPEG-2 Systems defines accurate mapping between STD parameters and HRD parameters that may be present in an AVC video stream. When an H.264 stream is carried in MPEG-2 Systems, coded H.264 bitstreams shall be contained in PES packets. The coded data shall comply with the byte-stream NAL unit format as shown in Figure 2-15.

Extra constraints on byte-stream NAL unit format are as follows:

- Each AVC AU shall contain an AU delimiter NAL Unit.
- Each byte-stream NAL unit that carries the AU delimiter shall contain exactly one zero-byte syntax element.
- All SPS and PPS necessary for decoding the AVC video stream shall be present within the AVC video stream.
- Each AVC video sequence that contains hrd_parameters( ) with the low_delay_hrd_flag set to "1" shall carry VUI parameters where the timing_info_present_flag shall be set to "1."

### Encapsulation of H.264 in MPEG-2 Systems

H.264 video is carried in PES packets in the payload using one of 16 stream_id values assigned to video as shown in Table 2-11, while signaling an H.264 video stream by means of the assigned stream_type value in the PMT or PSM as shown in Table 2-12.

**Table 2-11 Stream_id Assignment (MPEG IS-Amd3: Table 2-18)**

| Stream_id | Stream coding |
|---|---|
| 1011 1100 | Program_stream_map |
| 110x xxxx | ISO/IEC 13818-3 or ISO/IEC 11172-3 or ISO/IEC 13818-7 or ISO/IEC 14496-3 audio stream number x xxxx |
| 1110 xxxx | ITU-T Rec. H.262\| ISO/IEC 13818-2, ISO/IEC 11172-2, ISO/IEC 14496-2 or ITU-T Rec. H.264 \| ISO/IEC 14496-10 video stream number xxxx |
| ………… | ………… |
| 1111 1111 | Program_stream_directory |

**Table 2-12 Stream_type Assignment (MPEG IS-Amd3: Table 2-29)**

| Value | Description |
|---|---|
| 0x00 | ITU-T ISO/IEC Reserved |
| 0x01 | ISO/IEC 11172-2 Video |
| 0x02 | ITU-T Rec. H.262\| ISO/IEC 13818-2 or ISO/IEC 11172-2 video stream |
| 0x03 | ISO/IEC 11172-3 Audio |
| ………. | ………. |
| 0x1b | AVC video stream as defined in ITU-T Rec. H.264\| ISO/IEC 14496-10 video |
| 0x1C~0x7e | ITU-T Rec. H.222.0\| ISO/IEC 13818-1 Reserved |
| 0x7f | IPMP stream |
| 0x80~0ff | User Private |

**Table 2-13 PD and PED Examples (MPEG IS-Amd3: Table 2-39)**

| Descriptor_tag | TS | PS | Identification |
|---|---|---|---|
| .......... | | | .......... |
| 2 | x | x | Video_stream_descriptor |
| 3 | x | x | Audio_stream_descriptor |
| 9 | x | x | CA_descriptor |
| 10 | x | x | ISO_639_language_descriptor |
| 35 | x | | MultiplexeBuffer_descriptor |
| 40 | x | x | AVC video descriptor |
| 42 | x | x | AVC timing and HRD descriptor |
| .......... | | | .......... |

The highest level that may occur in an H.264 video stream as well as a profile that the entire stream conforms to should be signaled using the AVC video descriptor as shown in Table 2-13. If an AVC video descriptor is associated with an H.264 video stream, then this descriptor shall be conveyed in the descriptor loop for the respective elementary stream entry in the PMT or PSM.

The AVC video descriptor provides basic information for identifying coding parameters of the associated H.264 stream, such as on profile and level parameters included in the SPS of a H.264 stream. The AVC video descriptor also signals the presence of AVC still pictures and the presence of AVC 24-hour pictures in the H.264 video stream. If the descriptor is not included in the PMT or PSM for a H.264 video stream, such a H.264 video stream shall not contain AVC still pictures or AVC 24-hour pictures. The syntax for AVC video descriptor is provided in Figure 2-21.

```
AVC_video_descriptor( ){
    descriptor_tag              (8 bit)      // uimsbf
    descriptor_length           (8 bit)      // uimsbf
    profile_idc                 (8 bit)      // uimsbf
    constraint_set0_flag        (1 bit)      // bslbf
    constraint_set1_flag        (1 bit)      // bslbf
    constraint_set2_flag        (1 bit)      // bslbf
    AVC_compatible_flags        (5 bit)      // bslbf
    level_idc                   (8 bit)      // uimsbf
    AVC_still_present           (1 bit)      // bslbf
    AVC_24_hour_picture_flag(1 bit)          // bslbf
    reserved                    (6 bit)      // bslbf
}
```

**Figure 2-21 Syntax for AVC Video Descriptor**

Most of the semantics are exactly same to the those in SPS. Here, AVC_still_present indicates that the AVC video stream may include AVC still pictures, while AVC_24_hour_picture_flag indicates that the associated AVC video stream may contain AVC 24-hour pictures. Note that the definition of AVC still picture is to be an AVC still picture that consists of an AVC AU containing an IDR picture, preceded by SPS and PPS NAL units that carry sufficient information to correctly decode the IDR picture. Preceding an AVC still picture, there shall be another AVC still picture or an End of Sequence NAL unit terminating a preceding coded video sequence. An AVC still picture is repeatedly displayed until the PTS of the next AU. And, the definition of AVC 24-hour picture is an AVC AU with a presentation time that is more than 24 hours in the future. The AVC AU-n has a presentation time that is more than 24 hours in the future if the difference between the initial arrival time $t_{ai}(n)$ and the DPB output time $t_{o,dpb}(n)$ is more than 24 hours.

The AVC timing and HRD descriptor provides timing and HRD parameters of the associated H.264 video stream. For each AVC video stream carried in MPEG-2 Systems, the AVC timing and HRD descriptor shall be included in the PMT or PSM. The H.264 bitstream can carry VUI parameters with the timing_info_present_flag set to "1"

- for each IDR picture.
- and for each picture that is associated with a recovery point SEI message.

Absence of the AVC timing and HRD descriptor in the PMT for a H.264 video stream signals usage of the leak method in the T-STD, but such usage can also be signaled by the hrd_management_valid_flag set to "0" in the AVC timing and HRD descriptor. If this transfer rate is used in the T-STD for the transfer between $MB_n$ to $EB_n$, the AVC timing and HRD descriptor with the hrd_management_valid_flag set to "1" shall be included in the PMT for the H.264 video stream.

```
AVC_timing_and_HRD_descriptor( ){
   descriptor_tag              (8 bit)      // uimsbf
   descriptor_length           (8 bit)      // uimsbf
   hrd_management_valid_flag(1 bit)         // bslbf
   reserved                    (6 bit)      // bslbf
   picture_and_timing_info_present (1 bit)        // bslbf
   if( picture_and_timing_info_present) {
   90kHz_flag                  (1 bit)      // bslbf
   reserved                    (7 bit)      // bslbf
      if( 90kHz_flag=='0'){
            N                  (32 bit)     // uimsbf
            K                  (32 bit)     // uimsbf
      }
   num_units_in_tick           (32 bit)     // uimsbf
   }
   fixed_frame_rate_flag       (1 bit)      // bslbf
   temporal_poc_flag           (1 bit)      // bslbf
   picture_to_display_convesion_flag(1 bit)       // bslbf
   reserved                    (5 bit)      // bslbf
}
```

**Figure 2-22 Syntax for AVC Timing and HRD Descriptor**

When the AVC timing and HRD descriptor is associated to a H.264 video stream carried in a TS, the following applies: If the hrd_management_valid_flag is set to "1," Buffering Period SEI and Picture Timing SEI messages shall be present in the associated H.264 video stream. These Buffering Period SEI messages shall carry coded initial_cpb_removal_delay and initial_cpb_removal_delay_offset values for the NAL HRD. If the hrd_management_valid_flag is set to "1," the transfer of each byte from $MB_n$ to $EB_n$ in the T-STD shall be according to the delivery schedule for that byte into the CPB in the NAL HRD.

When the hrd_management_valid_flag is set to “0,” the leak method shall be used for the transfer from $MB_n$ to $EB_n$ in the T-STD.

The 90kHz_flag, when set to “1,” indicates that the frequency of the AVC time base is 90 kHz. For a H.264 video stream the frequency of the AVC time base is defined by the AVC parameter time_scale in VUI parameters. The relationship between the AVC time_scale and the STC shall be defined by the parameters N and K in this descriptor as follows:

$$time\_scale = \frac{(N \times system\_clock\_frequency)}{K} \qquad (2\text{-}27)$$

where time_scale denotes the exact frequency of the AVC time base with K larger than or equal to N.

When the temporal_poc_flag is set to “1” and the fixed_frame_rate_flag is set to “1,” the associated H.264 video stream shall carry Picture Order Count (POC) information (PicOrderCnt). When the temporal_poc_flag is set to “0,” no information is conveyed regarding any potential relationship between the POC information in the H.264 video stream and time.

For PES packetization, no specific data alignment constraints apply. For synchronization and STD management, PTSs and DTSs are encoded in the header of the PES packet that carries the H.264 video ESs.

**Extended T-STD**

The T-STD model is the same as that of MPEG-2 T-STD except extended DPB as depicted in Figure 2-23. Carriage of a H.264 bitstream over MPEG-2 Systems does not impact the size of buffer $DPB_n$. The size of $DPB_n$ is actually defined in H.264 standard for decoding a H.264 bitstream in the STD. A decoded H.264 AU enters $DPB_n$ instantaneously upon decoding of the H.264 AU, hence at the CPB removal time of the H.264 AU. A decoded H.264 AU is presented at the DPB output time.

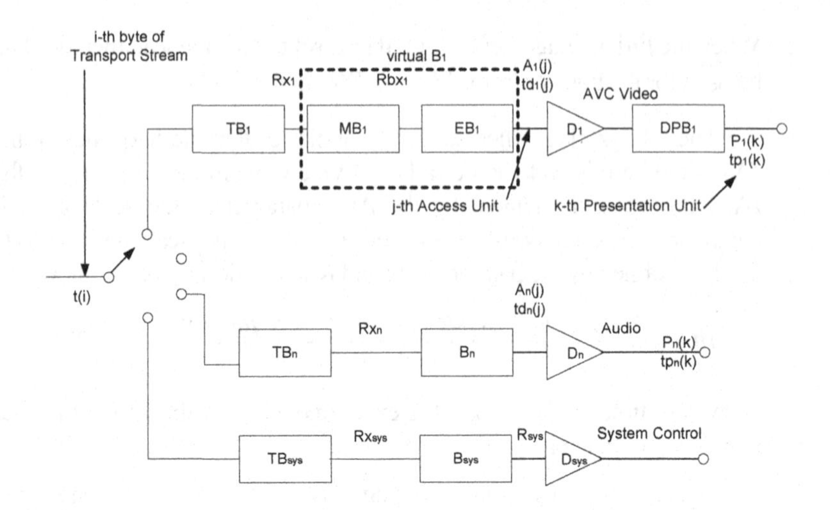


**Figure 2-23 T-STD Model Extension for H.264**

If the H.264 video stream provides insufficient information to determine the CPB removal time and the DPB output time of H.264 AU, then these time instances shall be determined in the STD model from PTS and DTS as follows:

- The CPB removal time of H.264 AU-n is the instant in time indicated by DTS(n) where DTS(n) is the DTS value of the AU-n.
- The DPB output time of H.264 AU-n is the instant in time indicated by PTS(n) where PTS(n) is the PTS value of the AU-n.

The output rate Rx is defined for T-STD based on data types as follows:

- $Rx_n = bit\_rate$ for video data, (2-28)

where bit_rate is the bit rate BitRate[ cpb_cnt_minus1] of data flow into the CPB for the byte-stream format signaled in the NAL hrd_parameters( ) carried in VUI parameters in the H.264 bitstream. If

NAL hrd_parameters( ) are not present, the bit_rate shall be the bitrate $1200 \times$ maxBR[ level] defined in Annex A of H.264 standard.

The purpose of "B" or "virtual B" (video case) is to eliminate packet multiplexer jitter. However, virtual B is broken down into two units of buffers for video – MB and EB. Resource usage schedule should be synchronized between an encoder and a decoder in terms of HRD for video. Since EB is nothing but a CPB in the HRD buffer, the size of $EB_n$ (a.k.a., $EBS_n$) is *cpb_size*. In other words,

- $EBS_n = cpb\_size$ . (2-29)

If NAL hrd_parameters( ) are not present, then the *cpb_size* shall be the size $1200 \times$ maxCPB defined in Annex A of H.264 standard.

The size $MBS_n$ of Buffer $MB_n$ is defined as follows:

- $$MBS_n = BS_{mux} + BS_{oh} + 1200 \times MaxCPB[level] - cpb\_size \quad (2\text{-}30)$$

where PES packet overhead buffering $BS_{oh} = (1/750)\sec\times \max\{1200 \times MaxBR[level],\ 2000000 bits/\sec\}$ (2-31)

and additional multiplex buffering $BS_{mux} = 0.004\sec\times \max\{1200 \times MaxBR[level],\ 2000000 bits/\sec\}$, (2-32)

where MaxCPB[ level] and MaxBR[ level] are defined for the byte-stream format in H.264 standard.

If the AVC_timing_and_HRD_descriptor is present with the hrd_management_valid_flag set to "1," the transfer of data from $MB_n$ to buffer $EB_n$ shall follow the HRD defined scheme for data arrival in the CPB as defined in Annex C in H.264 standard. Otherwise, the leak method shall be used to transfer data from $MB_n$ to $EB_n$ as follows:

- $Rbx_n = 1200 \times \max BR[level]$ . (2-33)

### Extended P-STD

The P-STD model is the same as that of MPEG-2 P-STD except with extended DPB as depicted in Figure 2-24. Carriage of a H.264 bitstream over MPEG-2 Systems does not impact the size of buffer $DPB_n$. For each H.264 video stream n, the size $BS_n$ of buffer $B_n$ in the P-STD is defined by the P-STD_buffer_size field in the PES packet header. Buffer $DPB_n$ shall be managed in exactly the same way as in the extended T-STD aforementioned. The H.264 AU enters buffer $B_n$. At the time $td_n(j)$, H.264 AU $A_n(j)$ is decoded and instantaneously removed from $B_n$. The decoding time $td_n(j)$ is specified by the DTS or by the CPB removal time, derived from information in the H.264 video stream.

PS shall be constructed so that the following conditions for buffer management are satisfied:

- $B_n$ shall not overflow.
- $B_n$ shall not underflow, except when VUI parameters are present for the H.264 video sequence with the low_delay_hrd_flag set to "1" or when trick_mode status is true. Underflow of $B_n$ occurs for H.264 AU $A_n(j)$ when one or more bytes of $A_n(j)$ are not present in $B_n$ at the decoding time $td_n(j)$.

### DTS/ PTS Carriage in PES Packets for AVC Pictures

ITU-T H.222.0 | ISO/IEC 13818-1:2000/Amd.3 describes the extension of MPEG-2 Systems to encapsulate H.264.

When H.264 video bitstreams are encapsulated in MPEG-2 Systems, DTS/ PTS information are explicitly described in PES headers. However, there is no explicit DTS/ PTS value in the PES header for AVC 24-hour pictures or AVC still pictures. For such H.264 AU, decoders shall infer the PTS through HRD parameters in H.264 video streams, as mentioned in previous subsections. Therefore, each H.264 video stream that contains one or more AVC 24-hour pictures

- shall either carry picture timing SEI messages with coded values of Cpb_removal_delay and Dpb_output_delay.
- or shall carry VUI parameters with the fixed_frame_rate_flag set to "1" and shall carry POC information.

If a DTS is present in the PES packet header, it shall refer to the first H.264 AU that commences in this PES packet. An H.264 AU commences in a PES packet if the first byte of the H.264 AU is present in the PES packet.

If a PTS is present in the PES packet header, it shall refer to the first H.264 AU that commences in this PES packet. An H.264 AU commences in a PES packet if the first byte of the H.264 AU is present in the PES packet.

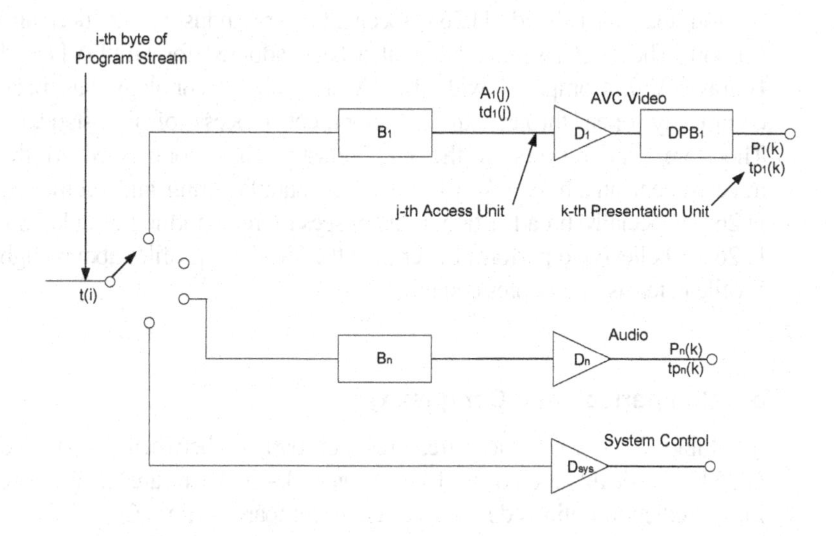


**Figure 2-24 P-STD Model Extension for H.264**

## 2.5 Comparisons between VC-1 and H.264

VC-1 is an emerging video standard primarily developed by Microsoft, while H.264 is an emerging video standard developed by the MPEG community under the auspice of the ISO [srinivasan:WMV9]. The VC-1 standard has been adopted for HD-DVD and Bluray-DVD. Its strength is its computational efficiency – currently low powered Pentium can decode HD video without any SOC support. If power consumption were not a concern, dedicated SOC might not be needed for VC-1 decoding. In fact, VC-1 (i.e., just another name of WMV-9) decoders are widely used as they are part of the Windows Media Player. A key weakness, however, is access to VC-1 encoder technology. There is limited activity on open VC-1 encoder development since no public reference implementation is available for encoding algorithms.

On the other hand, H.264 received more industry attention and support. The H.264 standard has also been adopted for HD-DVD and Bluray-DVD. Compared with the VC-1, H.264 technology has been completely open through the development process of the standard. Therefore, key players in the CE industry have contributed to the development and have taken the lead in manufacturing and promoting H.264. Especially, broadcasting industry sees more opportunities in H.264. H.264 is believed to perform better than the VC-1 for profiles above High Profile in terms of compression ratio.

### Tool Comparison and Complexity

Table 2-14 shows tools/features comparison between VC-1 and H.264. The details are covered in Chapters 4 - 9. There are many more Intra Prediction options devised in H.264 compared with VC-1. For MV accuracy, similar interpolation and resolutions are used. There are many more MC options devised in H.264 compared with VC-1. However, there are many more Transform options devised in H.264 compared with VC-1. In summary, the focus of the VC-1 tools is on simplicity with emphasis on compression ratio, while the focus of the H.264 tools is on rate-distortion performance with many viable options in terms of compression ratio.

**Table 2-14 Tools/ Features Comparison**

| Feature | WMV-9/ VC-1 Main | VC-1 Advanced | H.264 Main | H.264 High |
|---|---|---|---|---|
| Picture type | I, P, B, BI, Skipped P | I, P, B, BI, Skipped P | N/A | N/A |
| Intra prediction | 3 DC/ AC modes in frequency domain | 3 DC/ AC modes in frequency domain | 9(4x4) + 4(16x16) in pixel domain | 9(4x4) + 9(8x8) + 4(16x16) in pixel domain |
| MV accuracy | ¼ pel (4 tap bi-cubic or bi-linear) | ¼ pel (4 tap bi-cubic or bi-linear) | ¼ pel (6+2 tap) | ¼ pel (6+2 tap) |
| MC block size | 16x16, 8x8 | 16x16, 8x8 | 16x16, 16x8, 8x16, 8x8, 8x4, 4x8, 4x4 | 16x16, 16x8, 8x16, 8x8, 8x4, 4x8, 4x4 |
| Num reference frames | 2 | 2 | 16 (limited by DPBmax) | 16 (limited by DPBmax) |
| Trasform (I – Integer TX ) | 8x8 (I), 8x4(I), 4x8(I), 4x4(I) | 8x8 (I), 8x4(I), 4x8(I), 4x4(I) | 4x4(I) | 4x4(I), 8x8(I) |
| Transform scaling matrices | N/A | N/A | N/A | 8 adjustable (per Seq or Pic) |
| Entropy coding | VLC, kind of CA-VLC | VLC, kind of CA-VLC | CA-VLC, CA-BAC | CA-VLC, CA-BAC |
| Interlace | N/A | MPEG-2 like | PAFF, MBAFF | PAFF, MBAFF |
| In-loop deblocking filter (# taps, pels modified, pel compared) | 6, 2, 8 (non-linear filtering) | 6, 2, 8 (non-linear filtering) | 5 (max), 6, 8 (linear filtering) | 5 (max), 6, 8 (linear filtering) |

VC-1 is more complex to decode than MPEG-2 and H.264 is more complex to decode than VC-1. The comparison based on computation measure is shown in Table 2-15. Even VC-1 Main Profile (MP) decoding takes much lower computation than that of H.264 Baseline Profile (BP) as shown in Table 2-15. This implies that low power/ low cost implementation with minimal efforts is possible with VC-1.

**Table 2-15 Complexity Comparison**

| Sequence | Millions of ARM cycles/ second | |
|---|---|---|
| | VC-1 Main | H.264/ AVC Baseline (Optimized code) |
| Foreman | 27 | 38 |
| News | 17 | 22 |
| Container | 19 | 24 |
| Slient | 18 | 25 |
| Glasgow | 25 | 30 |
| Average | 21.2 | 27.8 |

## Objective Tests

In VC-1, low motion video and high motion video distinctively use different Huffman tables based on Qp. Therefore, tests are divided into two categories to compare VC-1 MP and MPEG-2. Figure 2-25 shows general characteristics of PSNR and bitrate for slow moving video contents. In contrast, Figure 2-26 shows general characteristics of PSNR and bitrate for fast moving video contents. Given a bitrate, the quality of VC-1 coded bitstreams is significantly better than that of MPEG-2 coded bitstreams. The performance of VC-1 for slow moving scenes is actually much better than that for fast moving scenes.

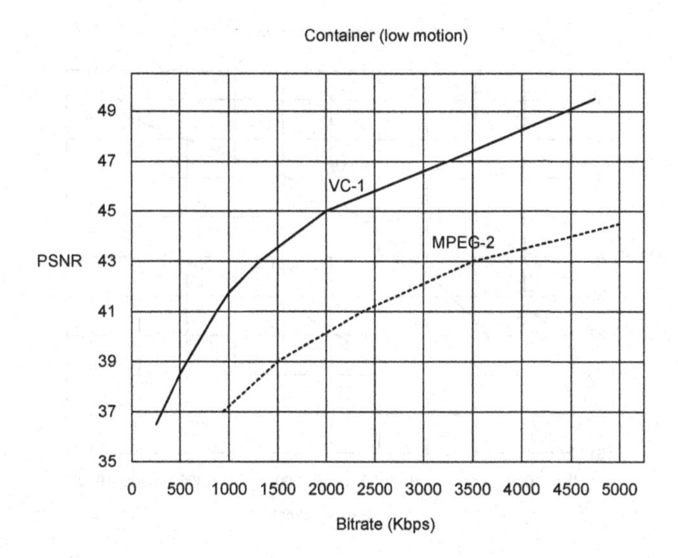


**Figure 2-25 MPEG-2 vs. VC-1 MP (WMV-9) for Low Motion Sequence**

For example, for slow moving scenes as shown in Figure 2-25, the quality of MPEG-2 coded at 2500 Kbps is same as the quality of VC-1 video coded at about 800 Kbps. In other words, the performance of VC-1 is 3 times better than that of MPEG-2 in terms of compression performance. For fast moving scenes as shown in Figure 2-26, the quality of MPEG-2 video at 6300 Kbps is same as the quality of VC-1 video at 2500 Kbps. In other words, the performance of VC-1 is 2.5 times better than that of MPEG-2 in terms of compression performance. Even though compression performance varies based on characteristics of input video and profiles/tools, it is safe to say that the performance of VC-1 is more than 2 times better than that of MPEG-2.

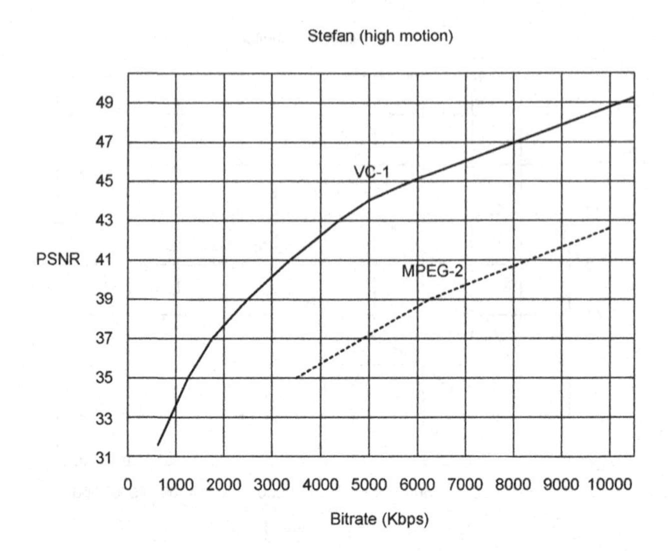


**Figure 2-26 MPEG-2 vs. VC-1 MP (WMV-9) for High Motion Sequence**

In VC-1 and H.264, inter prediction options are significantly different. Apart from various coding modes in MC block size and interpolation options, H.264 can utilize many more reference pictures compared to a maximum of two allowed by VC-1. To compare coding modes fairly, the same number of reference pictures should be at least allocated for both standards. Figure 2-27 illustrates results of VC-1 MP and H.264 BP. Since B slices are not allowed in H.264 BP, only one reference picture (with P slices) is used to compare coding options in Figure 2-27. For VC-1 MP, only one reference picture is used with the picture type order of I, P, P, P,... The implication of the result shown in Figure 2-27 is that the performance of H.264 is not significantly better than that of VC-1 without multi-picture MCP option. In fact, the performance of VC-1 shows better than that of H.264 without multi-picture MCP option as shown in Figure 2-27. For example, the quality at 1500 Kbps with H.264 is achieved at about 1300 Kbps with VC-1. In other words, the performance of VC-1 is 1.2 times better than that of H.264 in terms of compression performance without multi-picture MCP option on. The compression performance varies based on characteristics of input video and profiles/tools, Figure 2-27 shows only one case of results with a specific input video.

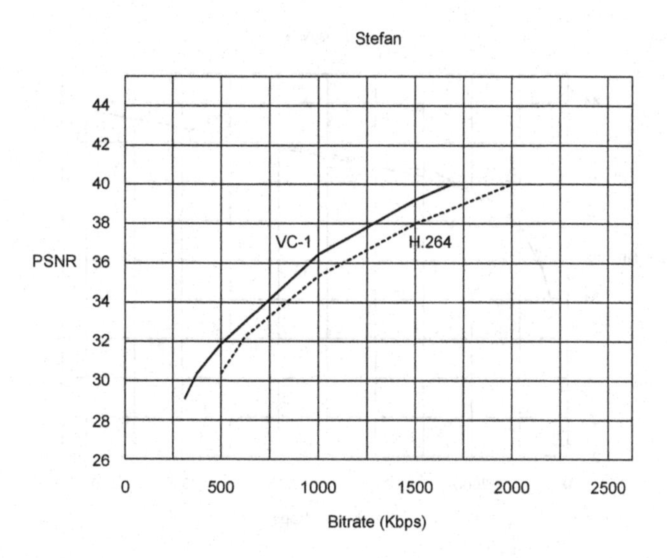


**Figure 2-27 VC-1 MP (WMV-9) vs. H.264 BP with 1 Reference Picture**

Figure 2-28 illustrates results of VC-1 AP and H.264 High Profile (HP). Compared with Figure 2-27, the results of Figure 2-28 are obtained with multi-picture MCP tool on. The performance of H.264 is better than that of VC-1 as shown in Figure 2-28. However, the performance of H.264 is not significantly better than that of VC-1 even with multi-picture MCP option. For example, the quality at 6000 Kbps with VC-1 is achieved at about 5000 Kbps with H.264. In other words, the performance of H.264 is 1.2 times better than that of VC-1 in terms of compression performance with all reasonable options turned on. The compression performance varies based on characteristics of input video and profiles/tools, Figure 2-28 shows only one case of results with a specific input video.

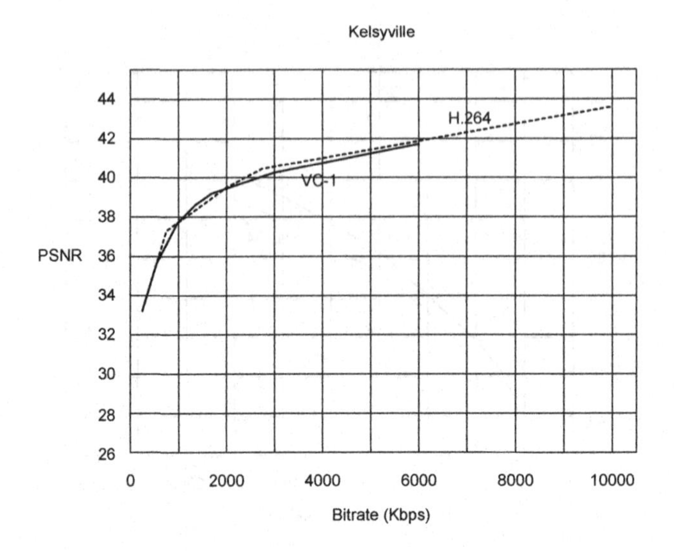


**Figure 2-28 VC-1 Advanced vs. H.264 High**

PSNR is generally taken as a reasonably reliable measurement for distortion. Based on PSNR, presented results imply that the performance of VC-1 is in the similar rate distortion ballpark as that of H.264, if not the same. The results are shown similarly in the MP and the HP of H.264 compared to the MP and the AP of VC-1.

## Subjective Tests

Objective measures provide just one kind of measurement. In many cases, subjective measures provide more reliable results than objective measures since objective measures do not properly represent the characteristics of human visual systems.

Figure 2-29 shows results of the DVD Forum Codec comparison tests. The Forum tested the performance of multiple video codecs (e.g., MPEG-2, MPEG-4 ASP, H.264, VC-1 (WMV-9), etc.) in six film clips of time length 90s and resolution 1920x1080. Industry reference D-VHS (MPEG-2 at 24 Mbps) and original D5 master (nearly lossless compression at 235 Mbps) are included to compare with VC-1 as shown in the Figure 2-29. The subjective test scores for VC-1 with D-VHS and D5 are presented in Figure 2-29. VC-1 shows strong subjective test results as presented in Figure 2-29.

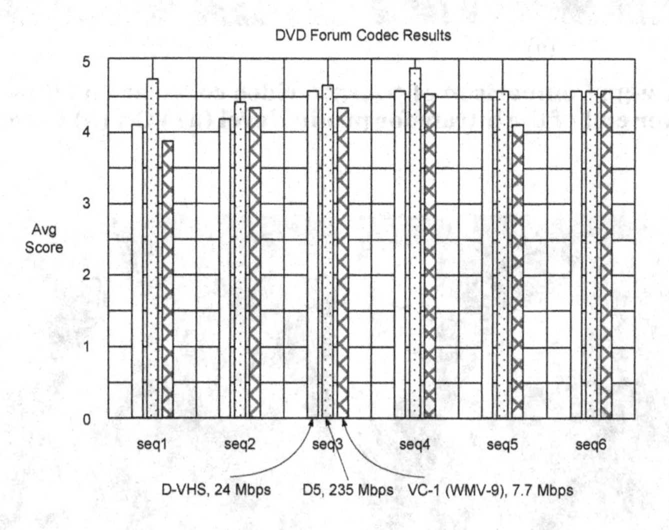


**Figure 2-29 DVD Forum Codec Results**

Figure 2-30 and 2-31 illustrate results of subjective comparison tests for VC-1 and H.264. Generally, they look equally good to bare human eyes with similar target bitrates when they are set at a higher operational bitrate in the allowed range in a specific profile/level combination. On the other hand, H.264 looks better with similar target bitrates when they are set at a medium or lower operational bitrate in the allowed range in a specific profile/level. Microsoft claims that generally VC-1 and H.264 show measures in the similar performance ballpark, but VC-1 has the advantage of being simpler in terms of implementation. For example, H.264 multi-picture MCP tool imposes harsh conditions for memory accesses on SOC implementation, while the VC-1 memory accesses patterns are similar to that of MPEG-2. Both VC-1 and MPEG-2 have a maximum of two reference pictures.

(a) (b)

**Figure 2-30 Comparison of 704x576 video coded at 1.4 Mbps, the lower end of the bitrate for profile/level (a) VC-1 (b) H.264**

(a) (b)

**Figure 2-31 Comparison of 352x28 video coded at 917 Kbps, the higher end of the bitrate for profile/level (a) VC-1 (b) H.264**

# 3. HRD Models and Rate Control

## 3.1 Video Buffer Verifier (VBV) Model

### VBV Model in MPEG-2

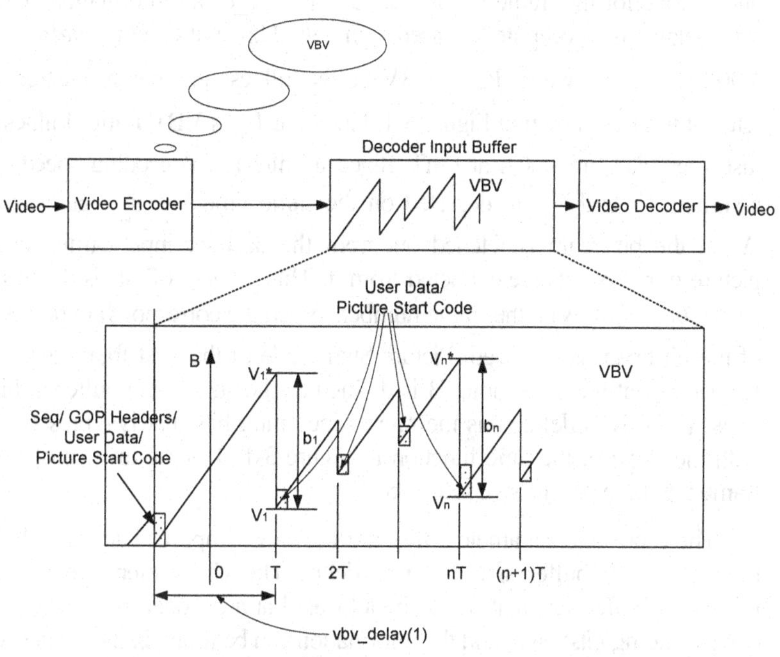


**Figure 3-1 VBV Model with Constant Bit Rate**

The Video Buffer Verifier (VBV) is the MPEG-2 hypothetical buffer model for a video decoder [yu:MPEG2systems, ITU:MPEG2systems]. VBV is meant to connect to the output of an encoder, in concept, while encoding through a specific rate control algorithm. This is a virtual buffer system through which the encoder can emulate the actual decoder input buffer behavior during the encoding process, thus checking available buffer resources for a rate control algorithm. At the decoder, actual buffer behavior follows through VBV accurately so that any guess made at the encoder is perfectly matched at the decoder. In the course of bitstream creation, the VBV fullness must be checked to ensure that it does not overflow or underflow.

A dummy decoder is assumed with certain predefined behavior such as infinite processing speed at decoding as shown as in Figure 3-1. As sequence header/ GOP headers/ user data and the first Picture Start Code are parsed, buffering of the bitstream is initiated. The waiting time to fetch and decode for the frame is considered with vbv_delay information. The vbv_delay is computed based on 90kHz with $vbv\_delay = 90000 \times V_n^* / R$, where $V_n^*$ is VBV buffer fullness just before bitstream fetch at t=nT as shown in Figure 3-1. Note that $V_n$ is VBV buffer fullness just after bitstream fetch at t=nT. Since infinite fetch/decoding speed is assumed, $V_n^*$ and $V_n$ are depicted on the same time line in Figure 3-2. When the bitstream is fetched out from the decoder input buffer, one picture worth of bits are extracted from it. This amount of bits is denoted as $b_1$, $b_2$, …, $b_n$ w.r.t. the frame number. When the corresponding portion of the bitstream is removed, Picture Start Code of the next frame is also fetched out at the same time. B is defined as the size of the buffer in bit units, while R is defined as the rate of incoming bits that is represented with the slope of the time function in Figure 3-1. T is the inverse of the frame rate for a progressive sequence.

There are two parameters that are the most important in the VBV model— vbv_buffer_size and vbv_delay. The vbv_buffer_size is the minimum buffer size that has to be allocated at a decoder to decode the corresponding bitstream, and the information can be found in the sequence header. The vbv_delay is the time to fill up the VBV buffer to be able to decode without buffer underflow and the information can be found in each picture header as shown in Figure 3-2. To manage the buffer resource of the decoder for any picture at a random access instance, vbv_delay information is required in each picture header. Note that the Picture Start Code for the current picture is depicted to be included in the bitstream portion fetched out for the previous picture in Figure 3-2.

The low delay mode of MPEG-2 VBV is useful for video conferencing/ monitoring applications, where low bi-directional delay for communication is recommended. In such applications, only I and P pictures are used since B pictures cause reordering delays due to display handling. Since P pictures are not efficient in terms of Motion Compensation (MC) compared with B pictures, sometimes pictures with a large amount of bits are (a.k.a., big pictures) created in the encoder. When a big picture is created/ transmitted, the next frame typically suffers buffer overflow. To handle this situation, variable frame rate encoding is applied to skip a picture right after a big picture to maintain bandwidth and the buffer occupancy. It is the encoder's job to decide whether the next picture is coded with very poor quality or skipped.

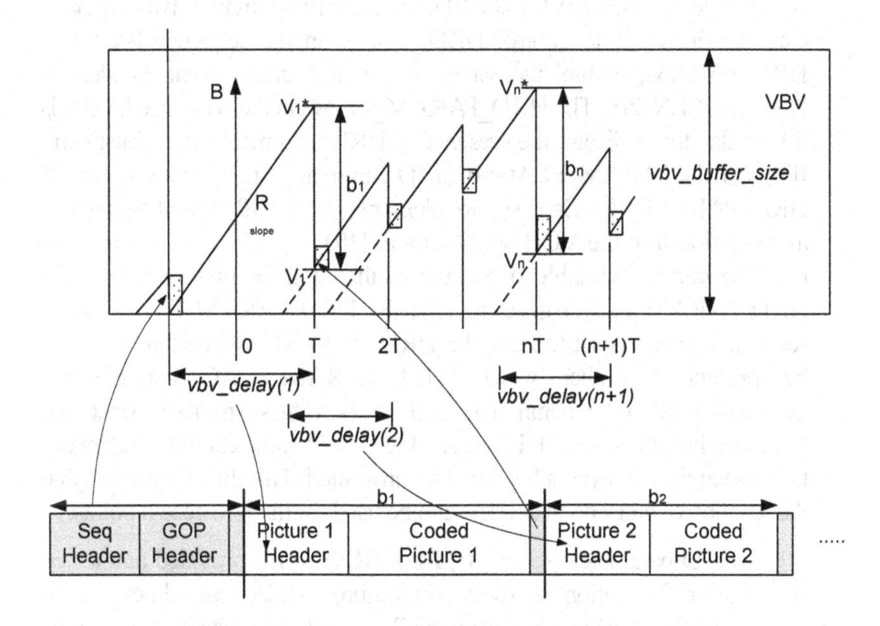


**Figure 3-2 Parameters in VBV model**

There are only two reference pictures used for ME/ MC in MPEG-2 video. The reference pictures can be I or P pictures. The B pictures in between are not used as reference pictures. Since the referencing structure is regular, the decoder output buffer management becomes relatively simple. Therefore, there is no need to design commands to allocate/ release the output display buffer. Since the maximum number of references is two, how to handle the decoder output buffer is an implementation issue. The actual implementation might require more picture buffers based on the internal pipeline of the decoder implemented.

## 3.2 HRD Model in VC-1 Video

### Constant Delay CBR HRD in VC-1

The notion of VBV model in MPEG-2 is adopted as Hypothetical Reference Decoder (HRD) model in VC-1 and H.264. The constant delay HRD parameters with CBR can be adaptively changed in VC-1 compression [SMPTE:VC1, JVT:H.264]. There are mainly two syntactic elements in VC-1 to control HRD. One is in the sequence header as HRD_PARAM, while the other is in the entry point header as HRD_FULLNESS. The HRD_PARAM_FLAG in the sequence header is a 1-bit flag that indicates the presence of HRD parameters in the bitstream. If this HRD_PARAM_FLAG=0, HRD parameters shall not be present. If HRD_PARAM_FLAG=1, syntax elements of the HRD shall be present as in section 6 of the VC-1 specification. HRD_FULLNESS in the entry point header is a variable size structure that shall be present only if the HRD_PARAM_FLAG is set to 1. If the HRD_PARAM_FLAG in the sequence header is set to zero, the HRD_FULLNESS structure shall not be present. In other words, HRD_PARAM_FLAG controls the availability of HRD parameters and R, B values are fixed once the sequence header is parsed. However, the F or D value can be given every time when the entry point header is retransmitted. This directly implies that the location of entry point headers can be used for random access points.

A HRD parameters set (R, B, F) or (R, B, D) is provided and tested with buffer fluctuation as shown in Figure 3-3. A leaky bucket with parameters (R, B, F) is said to "contain" a coded video bitstream if there is no underflow of the decoder buffer (i.e., $V_i \geq 0$ for all i). A leaky bucket with parameters (R, B, F) shall "contain" a coded CBR video bitstream if the following constraints hold:

$$V_1 = F - b_1 \quad (3\text{-}1)$$

$$V_i = V_{i-1} + R_i(t_i - t_{i-1}) - b_i \qquad \text{for all i} \quad (3\text{-}2)$$

$$R_i = R \text{ for all i} \quad (3\text{-}3)$$

$$V_i + b_i \leq B \qquad \text{for all i.} \quad (3\text{-}4)$$

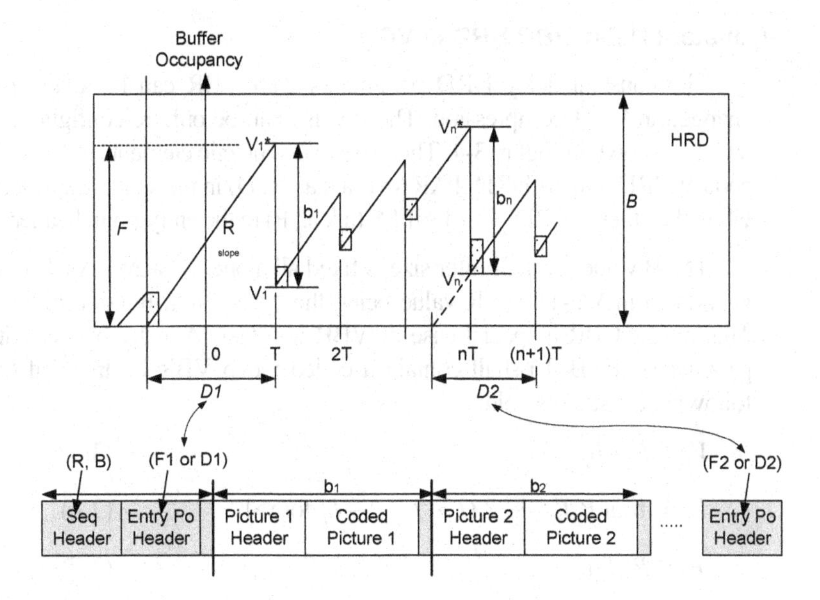


**Figure 3-3 Parameters in HRD Model in VC-1**

The decoder buffer fullness $V_i$ after removing frame i, with i>1, shall meet above conditions, where $t_i$ is the decoding time for frame i, and $b_i$ is the number of bits for frame i. CBR here means that $R_i = R$ for all i. In the leaky bucket model defined for VC-1 HRD, the decoder buffer may fill up, but shall not overflow. The buffer fullness at any time instance shall be less than or equal to B. When the decoder buffer is full, the encoder shall not send any more bits until there is room in the HRD buffer.

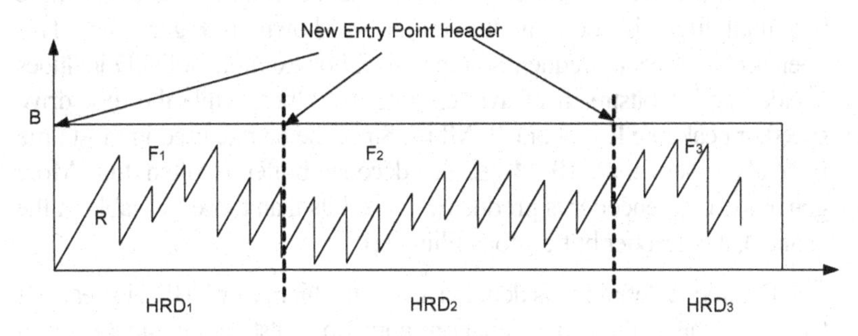


**Figure 3-4 Parameters Insertion in HRD Model**

### Constant Delay VBR HRD in VC-1

The constant delay HRD parameters with VBR can be adaptively changed in VC-1 compression. The F value can be only re-configured in VC-1 as shown in Figure 3-4. There are two syntactic elements in VC-1 to control HRD. One is HRD_PARAM (a.k.a., R, D) in the sequence header, while the other is HRD_FULLNESS (a.k.a., F) in the entry point header.

The B value for the buffer size is fixed all alone. However, VBR case is handled in VC-1 with R value being the "peak" transmission bit rate. This makes CBR a special case of VBR in VC-1. A leaky bucket with parameters (R, B, F) shall contain a coded video VBR bitstream if the following constraints hold:

$$V_1 = F - b_1 \tag{3-5}$$

$$V_i = \min(B, V_{i-1} + R_i(t_i - t_{i-1})) - b_i \text{ for i>1} \tag{3-6}$$

$$R_i \leq R \text{ all i} \tag{3-7}$$

$$V_i \geq 0 \text{ all i.} \tag{3-8}$$

Here, $R_i$ is the average bit rate that enters the buffer during the time interval $(t_i, t_{i-1})$ and shall be such that $R_i \leq R$ for all i. In the leaky bucket model defined for VC-1 HRD, the decoder buffer may fill up, but shall not overflow. The buffer fullness at any time instance shall be less than or equal to B. As a result, the min (B,x) operator implies that $V_i \leq B$ for all i. An example of a decoder buffer that fills up in several periods of time is shown in Figure 3-5. Some of the intervals indicate that the average bit rate is a useful measure.

When the decoder buffer is full, the encoder shall not send any more bits until there is room in the buffer as shown in Figure 3-5. This phenomenon occurs frequently in practice. For example, a DVD includes a video coded bitstream of average rate 4~6Mbps, while the disk drive speed or peak rate R is about 10 Mbps. Since the bit rate used in most time intervals is less than 10 Mbps, the decoder buffer is often full. More generally, if an encoder is producing fewer bits than those available in the channel, the decoder buffer stops filling up.

The above definition is denoted a variable bit rate or VBR bitstream. If the constraints in the above equations apply to a bitstream without the min

(B, x) operator (a.k.a., $V_i = V_{i-1} + R_i(t_i - t_{i-1}) - b_i$ for all i), but with $R_i = R$ for all i, and if there is no buffer overflow (a.k.a., $V_i + b_i \le B$ for all i), the bitstream is a CBR bitstream.

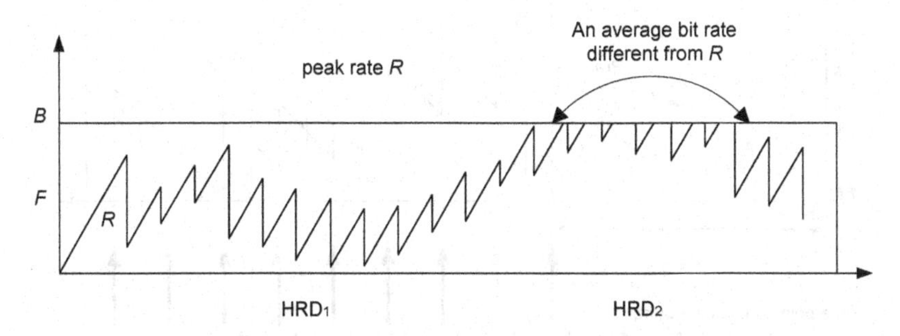


**Figure 3-5 Peak R VBR Buffer Model**

### Variable Delay HRD in VC-1

The notion of a low delay VBV model in MPEG-2 is adopted as variable delay HRD model in VC-1. The variable delay mode of the HRD is useful for video conferencing applications. This mode of operation is signaled when HRD parameters are not signaled in the sequence header (i.e., HRD_PARAM_FLAG is equal to value 0).

In this mode:

The leaky bucket shall be ($R_1$, $B_1$, $F_1$), where $R_1$ and $B_1$ correspond to the values $R_{max}$ and $B_{max}$ for the given profile and level of the bitstream as shown in Table 2-2.

The initial buffer fullness shall be equal to the buffer size (i.e., $F_1$=$B_1$) in the formal VC-1 provision. In practice, $F_1$ may be equal to the number bits for the first frame (i.e., $F_1$=$b_1$) as shown in Figure 3-6, which is similar to that of MPEG-2. Note that the formal provision of VC-1 can be used for random access or channel hopping scenarios. When the next entry point header is reached, HRD parameters (in the case of HRD_PARAM_FLAG ==1) may or may not be used based on specific implementation architectures. Implementations could take a trade-off condition between initial buffer fullness and buffer examination time point.

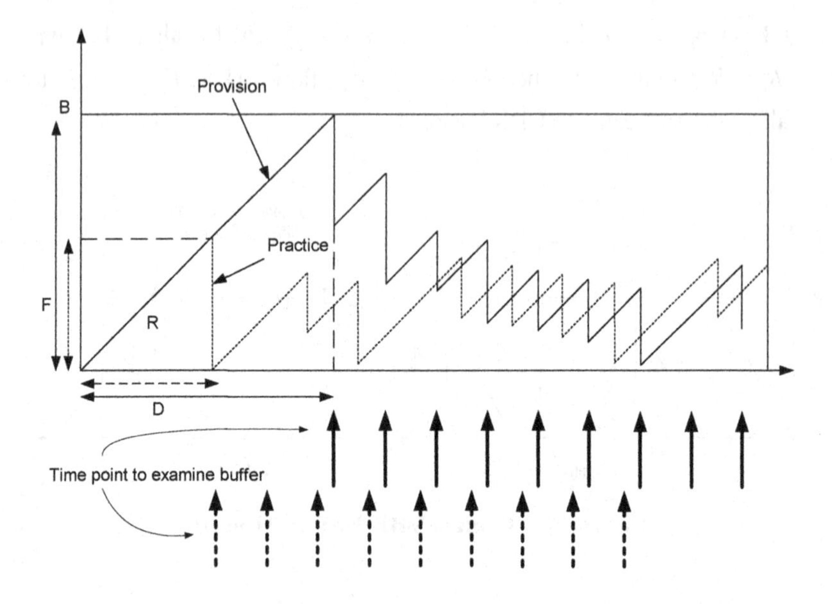


**Figure 3-6 Variable Delay HRD**

The decoder buffer in the HRD shall be examined every T seconds, where T is the inverse of the maximum frame rate in Table 2-2 for the respective profile and level of the given bitstream. If at least one complete coded picture is in the buffer, then all the data for the earliest picture in the bitstream order shall be instantaneously removed. Immediately before removing the picture, the buffer fullness shall be less than $B_1$.

In this mode, the HRD shall wait until a complete video frame has arrived at the buffer before decoding the frame. As a result, the delay is minimized for a given frame, but the end-to-end delay is not constant. The mode enables the encoder to send big pictures to the decoder while preventing buffer overflow.

### Multiple HRD in VC-1

A bitstream may be contained in multiple leaky buckets.

A video bitstream can be transmitted at any peak transmission bit rate (regardless of the average bit rate of the sequence) without suffering decoder buffer underflow, as long as the buffer size and initial delay are large enough. For example, if R approaches 0, the buffer size and initial buffer fullness can be as large as the bitstream itself. Figure 3-7 illustrates the peak bit rate $R_{min}$ and buffer size $B_{min}$ values for a given video bitstream. This curve indicates that in order to transmit the stream at a peak bit rate R, the decoder needs to buffer at least $B_{min}(R)$ bits. It is a fact that higher peak rates require smaller buffer sizes. Alternatively, if the size of the decoder buffer is B, the minimum peak rate required for transmitting the bitstream is the associated $R_{min}(B)$.

A key observation proven in some research is that the curve of $(R_{\min}, B_{\min})$ pairs, or that of $(R_{\min}, F_{\min})$, for any bitstream is piecewise linear and convex. Due to the property of the convexity with the observation, the decoder can linearly interpolate the values to arrive at some points $(R_{interp}, B_{interp}, F_{interp})$ that are slightly but safely larger than $(R_{\min}, B_{\min}, F_{\min})$.

When operational $(R_n, B_n)$ pairs are given as shown in Figure 3-7, the interpolated buffer size B between points n and n+1 follows the straight line:

$$B = \frac{R_{n+1} - R}{R_{n+1} - R_n} B_n + \frac{R - R_n}{R_{n+1} - R_n} B_{n+1}, \quad R_n < R < R_{n+1}. \quad (3\text{-}9)$$

And the initial decoder buffer fullness F can be linearly interpolated:

$$F = \frac{R_{n+1} - R}{R_{n+1} - R_n} F_n + \frac{R - R_n}{R_{n+1} - R_n} F_{n+1}, \quad R_n < R < R_{n+1}. \quad (3\text{-}10)$$

The derived leaky bucket with parameters $(R, B, F)$ is guaranteed to contain the bitstream since the minimum buffer size $B_{\min}$ is convex in both R and F.

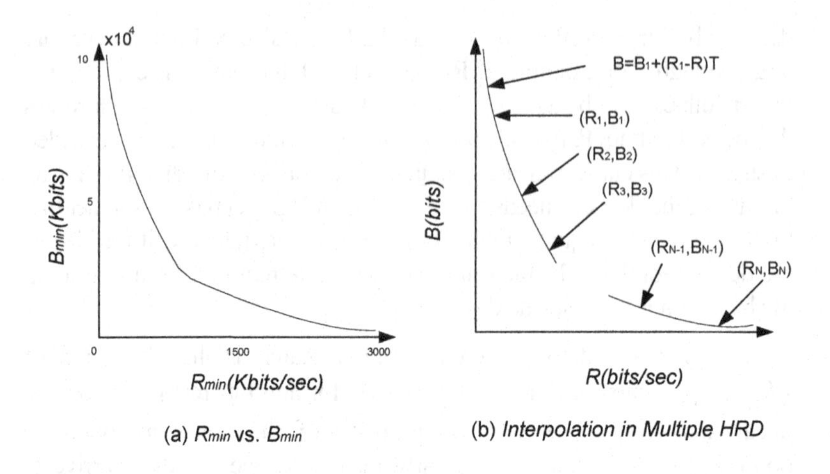


**Figure 3-7 Multiple Leaky Buckets**

If $R$ is larger than $R_N$, the leaky bucket $(R, B_N, F_N)$ also contains the bitstream. If $R$ is smaller than $R_1$, the upper bound $B = B_1 + (R_1 - R)T$ may be used (and one may set $F = B$), where $T$ is the time length of the video sequence in seconds.

## Display Order and Buffer Management in VC-1

There are only two reference pictures used for ME/ MC in VC-1. The reference pictures can be I or P pictures. The B pictures in between are not used as reference pictures, just as in MPEG-2 video. Figure 3-8 depicts the relationship between coding order and display order. Since the referencing structure is regular, the decoder output buffer management becomes relatively simple like in MPEG-2. Because the number of references is two, at least two pictures worth of decoder output buffer must be allocated. Actual implementation might require a different number of picture buffers based on the internal pipeline of the decoder.

In practical implementation, the display re-ordering procedure for ESs can be typically taken as follows: First, the first reference (#0) is to delay for one picture time after decoding to display. Second, B pictures can be displayed immediately after decoding. Third, any reference can be displayed when very next reference picture is parsed. However, it is not de-allocated until a 2nd reference picture is parsed. This procedure would cause only one picture time delay from decoding to display and require two reference picture buffers.

For PES packets, display re-ordering is not considered systematically since PTS in the PES header can be directly used. Note that efforts for display re-ordering and buffer management can be significantly reduced when the decoder interface is PES packets. The allocation/release timing of the decoder input buffer and the allocation/release timing of the decoder output buffer are directly commanded through DTS/ PTS in the PES headers. As is the case for MPEG-2, DTS/ PTS can be interpolated if they are missing.

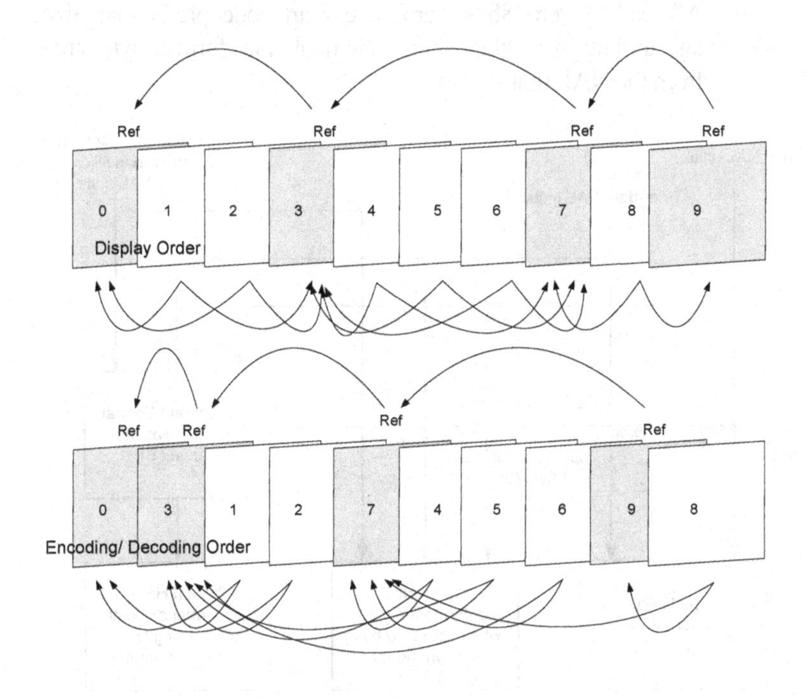


**Figure 3-8 Display Order and Coding Order in VC-1**

## 3.3 HRD Model in H.264 Video

### HRD Buffer Model in H.264

Two types of bitstreams are subject to HRD conformance checking for H.264 [JVT:H.264]. The first type of conformance bitstream is Type I bitstream that is a NAL unit stream containing only the VCL NAL units and filler data NAL units for all AUs as shown in Figure 3-9. The second type of bitstream is Type II bitstream that contains, in addition to the VCL NAL units and filler data NAL units for all AUs, at least one of the following:

- Additional non-VCL NAL units other than filler data NAL unit
- All leading_zero_8bits, zero_byte, start_code_prefix_one_3byte, and trailing_zero_8bits syntax elements that form a byte stream from the NAL unit stream.

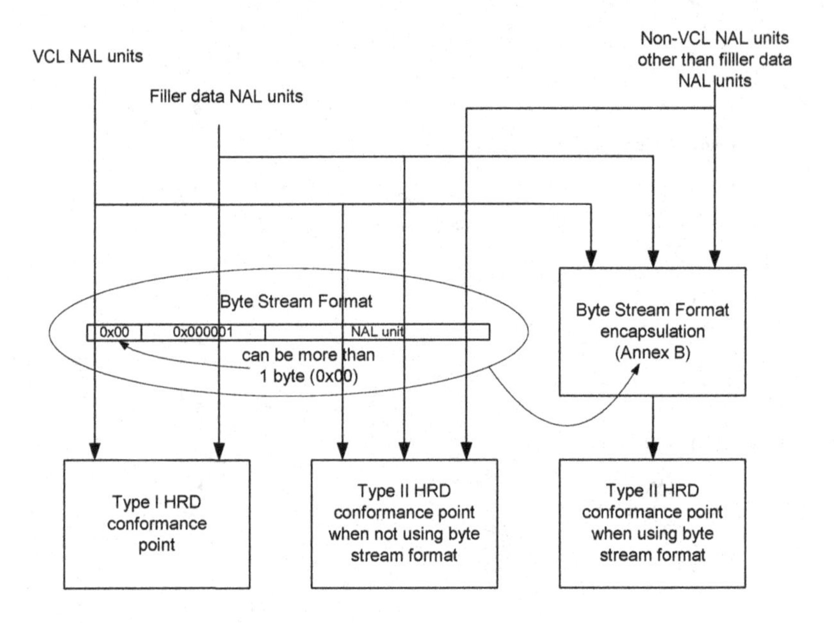


**Figure 3-9 Structure of Byte Streams and NAL Unit Streams**

In order to check conformance of a bitstream using the HRD, all SPSs and PPSs referred to in the VCL NAL units, along with corresponding buffer period and picture timing SEI messages, shall be conveyed to the HRD, in a timely manner, either in the bitstream (by non-VCL NAL units), or by other means not specified in the standard.

The HRD contains a Coded Picture Buffer (CPB), a video decoder that performs an instantaneous decoding process, a Decoded Picture Buffer (DPB), and output cropping, as shown in Figure 3-10. Note that DPB is part of the HRD buffer model since multiple reference pictures are allowed with allocation or release of them being controlled by only the encoder. This situation requires both the encoder and the decoder to synchronize reference picture buffer resources on both sides.

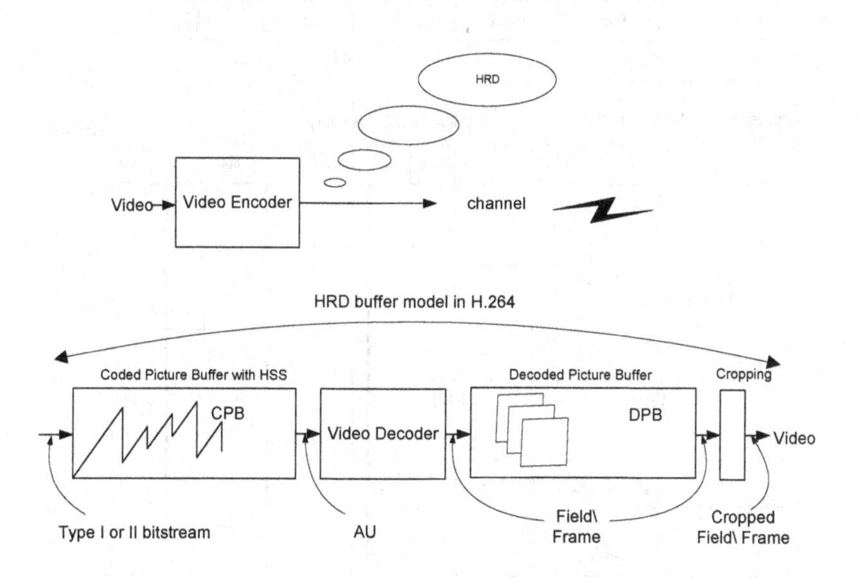


**Figure 3-10 HRD Buffer Model in H.264**

The CPB size (number of bits) is *CpbSize[ SchedSelIdx]*. The DPB size (number of frame buffers) is *Max( 1, max_dec_frame_buffering)*. The HRD operates as follows: data associated with AUs that flow into the CPB according to a specified arrival schedule are delivered by the Hypothetical Stream Scheduler (HSS) as shown in Figure 3-11. The associated data with each AU are removed and decoded instantaneously by the instantaneous decoding process at CPB removal times. This is depicted in Figure 3-11. The bits are accumulated linearly (slope R) in the CPB of the

HRD buffer and removed immediately at each picture decoding time. The notation R for the leaky bucket model for H.264 is again "peak" transmission rate (bits/sec) like that of VC-1. Each decoded picture is placed in the DPB at its CPB removal time unless it is output at its CPB removal time and is a non-reference picture. If a picture is placed in the DPB, it is removed from the DPB at the later of the DPB output time or the time that it is marked as "unused for reference." Figure 3-12 depicts the behavior of the DPB.

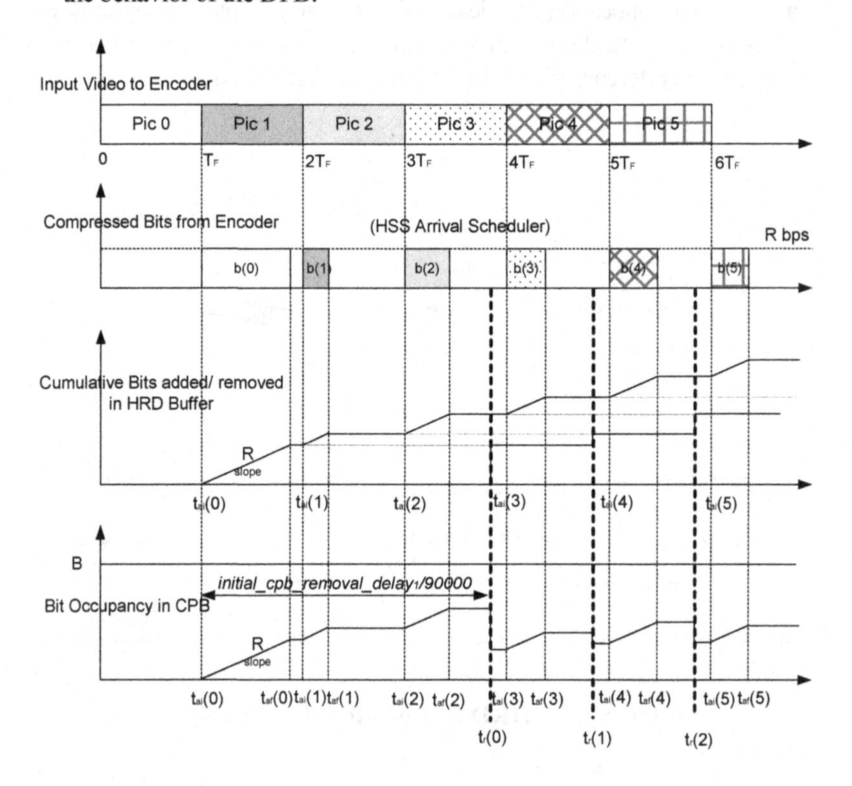


**Figure 3-11 Behavior Model of CPB in H.264 HRD**

The bitstream portion for the first picture in a buffering period is output from the CPB *initial_cpb_removal_delay* counter ticks after $t_{ai}(0)$. To be more general, the first picture removal delay in the buffering period $m$ is given with $initial\_cpb\_removal\_delay_m / 90000$ and its offset is provided with $initial\_cpb\_removal\_delay\_offset_m$ $/90000$. Note that *initial_cpb_removal_delay* and *initial_cpb_removal_delay_ offset* are only applied to the very first buffering period processed when a random access occurs. It then is all based on *cpb_removal_delay* and *dpb_output_delay* in clock tick units. The $t_{ai}(n)$ is the initial arrival time of picture *n*, while the $t_{af}(n)$ is the final arrival time of picture *n*. During the interval $[t_{ai}(n), t_{af}(n)]$, *b(n)* bits arrive at a rate of R bps where R is the peak transmission rate. Therefore, the final arrival time $t_{af}(n)$ can be derived as $t_{af}(n) = t_{ai}(n) + b(n)/R$. Note that $t_{ai}(n) = t_{af}(n-1)$ for CBR case, while $t_{ai}(n) = \max\{t_{af}(n-1), t_{ai,earliest}(n)\}$ for VBR case.

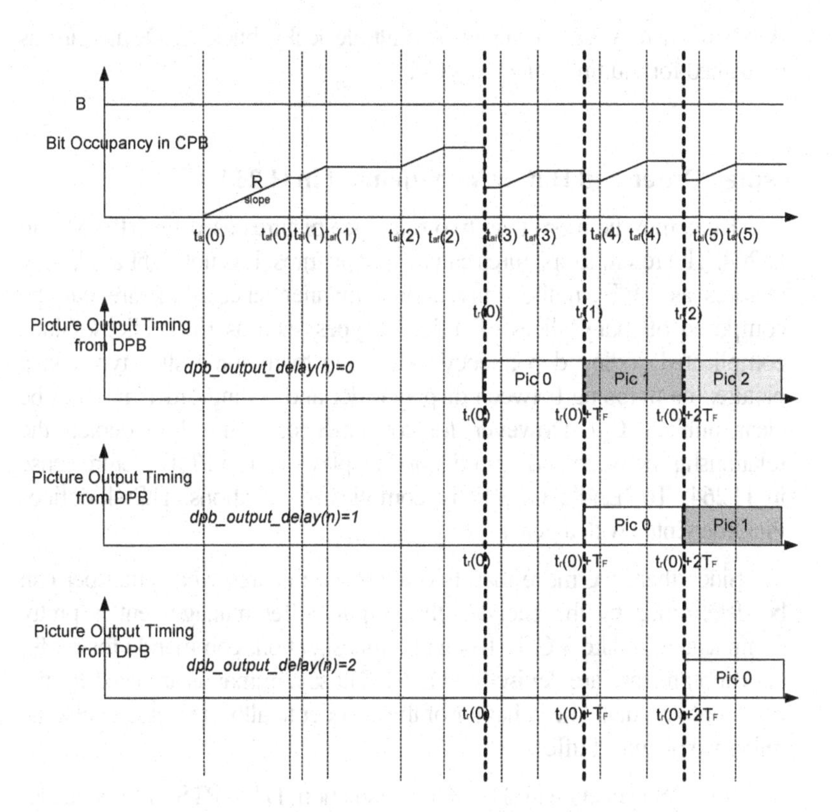


**Figure 3-12 Behavior Model of DPB in H.264 HRD**

The picture $n$ is output from the DPB with *dpb_output_delay(n)* clock ticks after its removal time $t_r(n)$ from the CPB. As shown in Figure 3-12, the display time can be controlled by the encoder with *dpb_output_delay* command. Note that a picture is not removed from the DPB at its output time when it is still marked as "used for short-term reference" or "used for long-term reference." The picture output order established by the values of this syntax element shall be the same order as established by the values of PicOrderCnt( ) that is explained in the next sections except that when two fields of a complementary reference field pair have the same value of PicOrderCnt( ), the two fields have different output times.

### Multiple HRD in H.264

A bitstream may be contained in multiple leaky buckets. Details are as explained for multiple HRD in VC-1.

### Display Order and Buffer Management in H.264

There may be more than two reference pictures used for ME/ MC in H.264. The reference pictures can be any pictures. It is not right to classify pictures as I/P/B in the conventional manner since a picture can be composed of many slices of different types such as I/P/B slices. Since complicated coding dependency exists due to multiple slice types in a picture, the mapping between display order and coding order may not be clear, unlike VC-1. However, for convenience, Figure 3-13 depicts the relationship between coding order and display order in I/P/B picture sense in H.264. To handle potentially complicated situations, H.264 defines PicOrderCnt( ) with a general way.

Since there are more than two reference pictures whose number can be determined by the encoder, the output buffer management is pretty complicated, unlike VC-1. To handle this situation, commands for buffer allocation/release are devised in H.264. Those commands are used for the encoder to regulate the behavior of the decoder to allocate/ release/ change reference pictures buffers.

For PES packets with H.264 encapsulation, DTS/ PTS in PES header can be directly used. Note that efforts for display re-ordering and buffer management cannot be significantly reduced even when the decoder interface is PES packets. Even after displaying some pictures, the encoder can decide to store them in DPB a pretty long time.

### Display Order and POC in H.264

In MPEG-2, coding order and display order have limited mapping capability between them for a GOP. The idea of Picture Order Count (POC) in H.264 is to generalize such a mapping between coding order and display order (more generally, reference order). Such natural renaming of coded (or decoded) pictures in POC helps management of reference pictures in the form of Lists. The POC values may represent temporal correlation best when the display order is chosen for them. Reference lists generally use POC (or PicNum, a special case of POC) to identify specific pictures for P/B slice list management in progressive or interlace pictures.

There are three cases designed for display order mapping and POC is nothing but a series of numbers in which display order takes place [JVT:H.264, richardson:white]. The first case is that POC is always indicated in each slice header. This makes the display order the most general and flexible. The second case is that POC is general, but periodic. This is more or less like that of MPEG-2. The third case is that the coding order and the display order happen to be the same.

The POC starts with 0 at an IDR picture, which implies that POC values are only meaningful until the next IDR picture. Also, the coding order "frame_num" starts with 0 at an IDR picture as well and it is decoded in each slice header. One difference is that frame_num increments by 1 from the previous reference picture. In contrast, POC typically increments by 2 for every complete picture. However, POC difference can be 1 for certain scenarios – in fact, the standard does not limit increments to be 1 or 2. In such cases, the important thing is its order, not absolute POC value. Note that POC increments by 2 for each top field or each bottom field with *TopFieldOrderCount* or *BottomFieldOrderCount*, respectively. From now on, a prefix *Prefix.xxx* is used to imply where the data is captured. For example, *Slice.xxx* is to indicate that the data xxx is captured from Slice header. As such, *SPS.xxx* means that the data xxx is captured from the selected SPS.

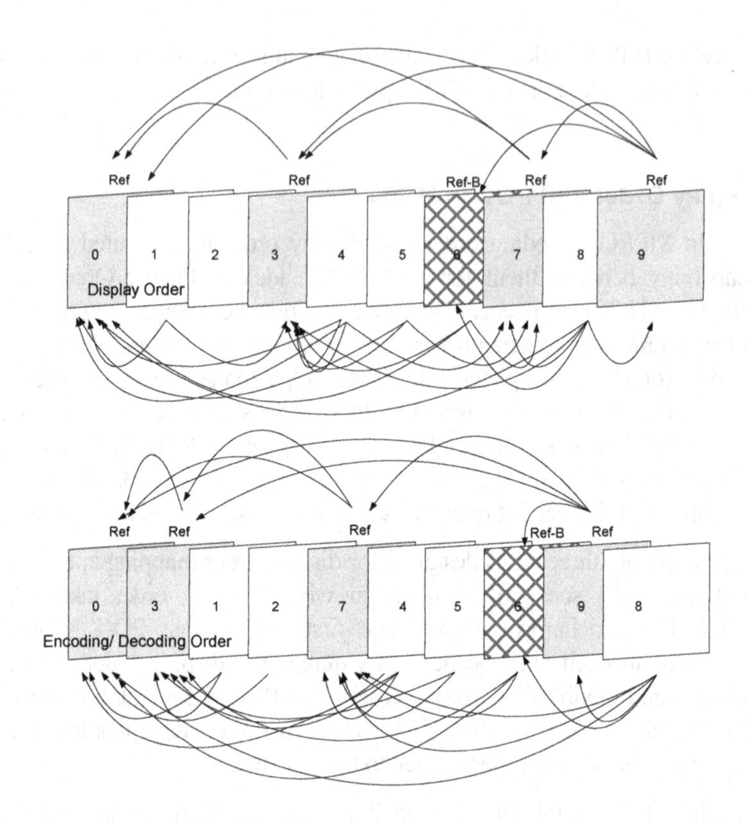


**Figure 3-13 Display Order and Coding Order Example in H.264**

*SPS.pic_order_cnt_type* "0" means that POC is explicitly specified in each slice header. The computation of POC is conceptually as follows:

$$TopFieldOrderCount = POCMsb + Slice.pic\_order\_cnt\_lsb \quad (3\text{-}11)$$

$$BottomFieldOrderCount = POCMsb + Slice.delta\_pic\_order\_cnt\_bottom \quad (3\text{-}12)$$

$$= POCMsb + Slice.pic\_order\_cnt\_lsb$$

(bottom_field_flag)

The *pic_order_cnt_lsb* is described in each slice header. However, *POCMsb* is an internal variable that increments by 1 when *pic_order_cnt_lsb* reaches its maximum value. The maximum value is calculated from log2_max_pic_order_cnt_lsb_minus4, which is specified in the SPS, to give $MaxPicOrderCntLsb = 2^{(log2\_max\_pic\_order\_cnt\_lsb\_minus4+4)}$.

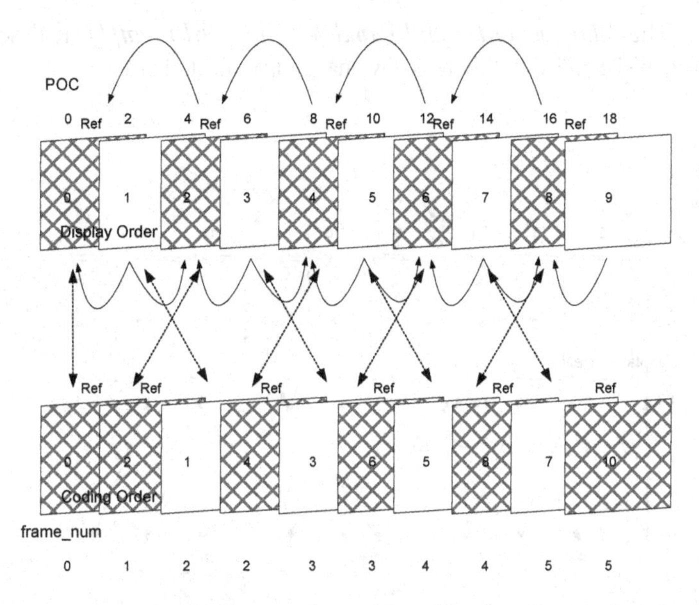


**Figure 3-14 H.264 Display Order and Coding Order Example (1B)**

*SPS.pic_order_cnt_type* "1" means that POC is explicitly specified in the SPS. The period defines the cycle of the mapping between display order and coding order. The number of reference pictures in a POC cycle is defined in the SPS. In addition, the offset to each reference picture in the cycle and the offset to each non-reference frame are also defined in the SPS.

The computation of POC is conceptually as follows:

$$TopFieldOrderCount = SPS.POC + Slice.delta\_pic\_order\_cnt[0] \quad (3\text{-}13)$$

$$BottomFieldOrderCount = SPS.POC + Slice.delta\_pic\_order\_cnt[1] \quad (3\text{-}14)$$

$$= SPS.POC + Slice.delta\_pic\_order\_cnt[0] + SPS.offset\_for\_top\_to\_bottom\_field$$

(bottom_field_flag)

The *delta_pic_order_cnt[0] and delta_pic_order_cnt[1]* are described in each slice header if there is any change in expected order.

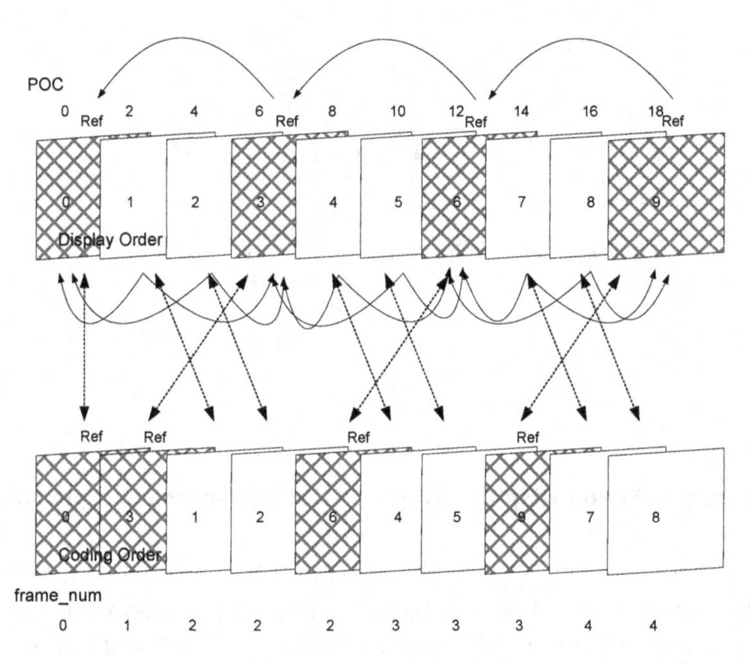


**Figure 3-15 H.264 Display Order and Coding Order Example (2B)**

*SPS.pic_order_cnt_type* "2" means that the coding order and display order are the same. The computation of POC is conceptually as follows:

If (the picture is used for a reference)

$$TopFieldOrderCount = 2 \times frame\_num$$

Else (3-15)

$$TopFieldOrderCount = 2 \times frame\_num - 1$$

Here, *BottomFieldOrderCount* in either a frame or its bottom field has the same value as *TopFieldOrderCount*. Note that "1" POC increment is allowed for every complete picture in this scenario. In other words, POC increments by 2 only for a reference picture, but the POC of a non-reference picture increments by 1. This implies that only one non-reference picture is allowed between reference pictures for this scheme.

Figure 3-14 and Figure 3-15 depict two examples where one B picture or two B pictures constitute a GOP cycle, respectively, and no B picture is used as a reference. Let's design POC related parameters in slice header based on three different types.

Let's consider 1B picture POC case, first. The example in Figure 3-14 can be described with two Tables as shown below based on display order and coding order.

**Table 3-1 H.264 Coding Parameter Construction-type 0 (1B)**

| Display order | 0 | 1 | 2 | 3 | 4 | 5 | 6 | 7 | 8 | 9 |
|---|---|---|---|---|---|---|---|---|---|---|
| POC expected | 0 | 2 | 4 | 6 | 8 | 10 | 12 | 14 | 16 | 18 |

| Coding order | 0 | 2 | 1 | 4 | 3 | 6 | 5 | 8 | 7 | 10 |
|---|---|---|---|---|---|---|---|---|---|---|
| Picture type | I | P | B | P | B | P | B | P | B | P |
| frame_num | 0 | 1 | 2 | 2 | 3 | 3 | 4 | 4 | 5 | 5 |

Note that the frame_num increments by one relative to the previous reference picture in Table 3-1 and POC increments by two due to potential non-reference picture handling in a combined manner. If the mapping designed in Figure 3-14 is valid, encoding parameters in each slice header can be represented in combinations shown in Table 3-1.

**Table 3-2 Encoding Parameters in Slice Header-type 0 (1B)**

| The order of AUs in coding order | 1st | 2nd | 3rd | 4th | 5th | 6th | 7th | 8th | 9th | 10th |
|---|---|---|---|---|---|---|---|---|---|---|
| frame_num in slice header | 0 | 1 | 2 | 2 | 3 | 3 | 4 | 4 | 5 | 5 |
| POC in slice header | 0 | 4 | 2 | 8 | 6 | 12 | 10 | 16 | 14 | 20 |

**Table 3-3 POC Derivation from SPS and delta_pic_order_cnt[0] Computation-type 1 (1B)**

| Actual POC (*) | 0 | 4 | 2 | 8 | 6 | 12 | 10 | 16 | 14 | 20 |
|---|---|---|---|---|---|---|---|---|---|---|
| Picture type | I | P | B | P | B | P | B | P | B | P |
| POC derivation with offset_for_ref _frame[0]=4 | 0 | 4 | | 8 | | 12 | | 16 | | 20 |
| POC derivation with offset_for_non _ref_pic=-2 | | | 2 | | 6 | | 10 | | 14 | |
| Overall POC derivation (**) | 0 | 4 | 2 | 8 | 6 | 12 | 10 | 16 | 14 | 20 |
| Computation for delta_pic_ order_cnt[0] | 0 | 0 | 0 | 0 | 0 | 0 | 0 | 0 | 0 | 0 |

If *SPS.pic_order_cnt_type*=0 is used for bitstream representation of Figure 3-14, direct values of frame_num and POC are included in slice headers as shown in Table 3-2.

If *SPS.pic_order_cnt_type*=1 is used for bitstream representation in the case of Figure 3-14, direct values of frame_num and *delta_pic_order_cnt[0]* are included in slice headers. Note that SPS should contain *SPS.num_ref_frames_in_pic_order_cnt_cycle*=1, *SPS.offset_for_ref_ frame [0]* =4, and *SPS.offset_for_non_ref_pic*=-2. With such SPS parameters, POC can be derived as shown in Table 3-3. The *delta_pic_order_cnt[0]* can be computed as a difference between the actual POC(*) and the derived POC(**) as shown in Table 3-3, and it is encoded within Slice headers.

If *SPS.pic_order_cnt_type*=2 is used for bitstream representation, direct values of frame_num are only included in Slice headers. However, this example doesn't apply to the case of Figure 3-14 since the coding order and the display order are not the same.

Let's consider the 2B picture POC case. The example in Figure 3-15 can be described with two Tables as shown below based on display order and coding order.

**Table 3-4 H.264 Coding Parameter Construction-type 0 (2B)**

| Display order | 0 | 1 | 2 | 3 | 4 | 5 | 6 | 7 | 8 | 9 |
|---|---|---|---|---|---|---|---|---|---|---|
| POC expected | 0 | 2 | 4 | 6 | 8 | 10 | 12 | 14 | 16 | 18 |

| Coding order | 0 | 3 | 1 | 2 | 6 | 4 | 5 | 9 | 7 | 8 |
|---|---|---|---|---|---|---|---|---|---|---|
| Picture type | I | P | B | B | P | B | B | P | B | B |
| frame_num | 0 | 1 | 2 | 2 | 2 | 3 | 3 | 3 | 4 | 4 |

If the mapping designed in Figure 3-15 is valid, encoding parameters in each slice header can be represented in combinations shown in Table 3-4.

**Table 3-5 Encoding Parameters in Slice Header-type 0 (2B)**

| The order of AUs in coding order | 1st | 2nd | 3rd | 4th | 5th | 6th | 7th | 8th | 9th | 10th |
|---|---|---|---|---|---|---|---|---|---|---|
| frame_num in slice header | 0 | 1 | 2 | 2 | 2 | 3 | 3 | 3 | 4 | 4 |
| POC in slice header | 0 | 6 | 2 | 4 | 12 | 8 | 10 | 18 | 14 | 16 |

If *SPS.pic_order_cnt_type*=0 is used for bitstream representation of Figure 3-15, direct values of frame_num and POC are included in slice headers as shown in Table 3-5.

If *SPS.pic_order_cnt_type*=1 is used for bitstream representation of Figure 3-15, direct values of frame_num and *delta_pic_order_cnt[0]* are included in slice headers. Note that SPS should contain *SPS.num_ref_frames_in_pic_order_cnt_cycle*=1, *SPS.offset_for_ref_frame [0]* =6, and *SPS.offset_for_non_ref_pic*=-4. With such SPS parameters, POC can be derived as shown in Table 3-6. The *delta_pic_order_cnt[0]* can be computed as a difference between the actual POC(*) and the derived POC(**) as shown in Table 3-6, and it is encoded in Slice headers.

If *SPS.pic_order_cnt_type*=2 is used for bitstream representation, direct values of frame_num are only included in Slice headers. However, this example doesn't apply to the case of Figure 3-15 since the coding order and the display order are not the same.

A case of *SPS.pic_order_cnt_type*=2 is depicted in Figure 3-16, where direct values of frame_num are only included in Slice headers. Note that only I and P slices are used for entire pictures, thus making the display order and the coding order the same. The dependency, however, is not ordinary since odd-numbered P pictures are not used as reference pictures in the example.

**Table 3-6 POC Derivation from SPS and delta_pic_order_cnt[0] Computation-type 1 (2B)**

| Actual POC (*) | 0 | 6 | 2 | 4 | 12 | 8 | 10 | 18 | 14 | 16 |
|---|---|---|---|---|---|---|---|---|---|---|
| Picture type | I | P | B | B | P | B | B | P | B | B |
| POC derivation with offset_for_ref _frame[0]=6 | 0 | 6 | | | 12 | | | 18 | | |
| POC derivation with offset_for_non _ref_pic=-4 | | | 2 | 2 | | 8 | 8 | | 14 | 14 |
| Overall POC derivation (**) | 0 | 6 | 2 | 2 | 12 | 8 | 8 | 18 | 14 | 14 |
| Computation for delta_pic_ order_ cnt[0] | 0 | 0 | 0 | 2 | 0 | 0 | 2 | 0 | 0 | 2 |

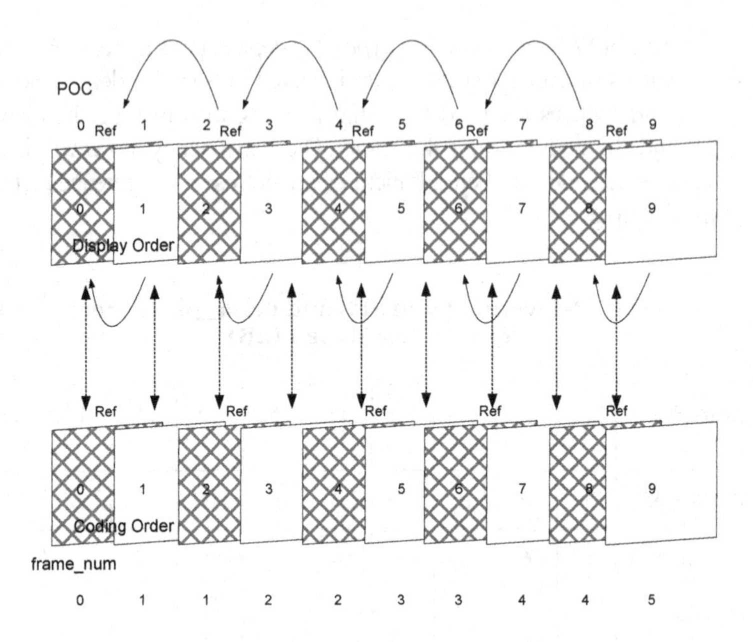


**Figure 3-16 H.264 Display Order and Coding Order Example (1P)**

The example of Figure 3-16 can be described with 1 Table shown below based on display order and coding order. If the mapping designed in Figure 3-16 is valid, the frame_num parameter in each slice header can be presented and corresponding POC values can be derived as follows:

**Table 3-7 H.264 POC Derivation-type 2 (1P)**

| Display or Coding order | 0 | 1 | 2 | 3 | 4 | 5 | 6 | 7 | 8 | 9 |
|---|---|---|---|---|---|---|---|---|---|---|
| Picture type | I* | P | P* | P | P* | P | P* | P | P* | P |
| frame_num | 0 | 1 | 1 | 2 | 2 | 3 | 3 | 4 | 4 | 5 |
| POC derived | 0 | 1 | 2 | 3 | 4 | 5 | 6 | 7 | 8 | 9 |

Note that picture(*) denotes a reference picture in Table 3-7. As seen in POC derivation, POC value of this case increments by 1. In other words, in the case of *SPS.pic_order_cnt_type*=0 or 1, the display order value shall be derived by dividing the POC value by 2. However, in the case of *SPS.pic_order_cnt_type*=2, the display order value shall be the same as the derived POC value.

## Reference Picture List Ordering

Pictures decoded previously are saved in the DPB in both the encoder and the decoder. The encoder and the decoder maintain a common list for reference pictures for the purpose of efficient referencing in ME/ MC in P or B slices. If absolute frame_num were used to reference pictures, the coding efficiency might be degraded. Therefore, instead of using absolute values for the frames, the most recently decoded frames are conceptually "re-titled" for compact representation, where the mapping between the indices of the list and actual frame numbers is handled in a pre-defined manner.

For P slice prediction, List0 is used to capture recently decoded frames with short-term reference pictures and long-term reference pictures. List0 for P slice contains PicNum in decreasing order, where PicNum is a variable "wrapped around" with MaxFrameNum. Here, MaxFrameNum is defined by $MaxFrameNum = 2^{(log2\_max\_frame\_mum\_minus4+4)}$. Note that PicNum is derived from FrameNum (a.k.a., frame_num). By default, an encoded picture is reconstructed by the encoder and marked as a short-term reference picture which is a recently-coded picture available for prediction. Long-term reference pictures are selected based on temporal correlation impact over a relatively long range in time and they are identified with a variable LongTermPicNum. Classification of short-term and long-term reference pictures is purely a task of the encoder. If the number of reference pictures is over the maximum, the oldest short-term picture (highest index) is removed from the buffer – hence called "sliding window" memory control. In contrast, adaptive memory control means sending commands from the encoder to any decoder to clearly identify/ manage the short-term reference pictures and the long-term reference pictures.

The Decoded Reference Picture Marking (DRPM) process is the process that handles reference picture management. The marking of a reference picture can be "unused for reference," "used for short-term reference," or "used for long-term reference," but only one among these three. When a reference picture is referred to as being marked as "used for reference," this collectively refers to the picture being marked as "used for short-term reference" or "used for long-term reference." A reference picture that is marked as "used for long-term reference" is referred to as a long-term reference picture. Note that the timing of the marking process is right after decoding of all slices for the current picture.

| List0.index | 0 | 1 | 2 | 3 | 4 |
|---|---|---|---|---|---|
| …. | | | | | |
| Encode 49 | 49 | 48 | 47 | 46 | 45 |
| Mark all references as "unused for reference" | - | - | - | - | - |
| Encode 50 | 50 | - | - | - | - |
| Encode 51 | 51 | 50 | - | - | - |
| Encode 52 | 52 | 51 | 50 | - | - |
| Encode 53 | 53 | 52 | 51 | 50 | - |
| Encode 54 | 54 | 53 | 52 | 51 | 50 |
| Encode 55 | 55 | 54 | 53 | 52 | 51 |
| Assign 54 to LongTermPicNum 0* | 55 | 53 | 52 | 51 | 0* |
| Encode 56 | 56 | 55 | 53 | 52 | 0* |
| Assign 56 to LongTermPicNum 1* | 55 | 53 | 52 | 1* | 0* |
| Encode 57 | 57 | 55 | 53 | 1* | 0* |
| …. | | | | | |

**Figure 3-17 Listo Ordering Example for P Slice**

The encoder manages reference picture numbers and usage of the DPB entirely on its own. That is why the size of the DPB is not fixed, but $DPB_{max}$ is defined unlike MPEG-2. The encoder issues commands for a short-term picture to be "used for a long-term picture" or for any reference picture to be marked as "unused for reference."

For example, let's say, maximum size of the DPB is five frames and the current frame_num is 50. Then, List0 can develop as follows, as time goes by, for the values of Display order as shown in Figure 3-17.

| List0.index\| List1.index | 0 | 1 | 2 | 3 | 4 |
|---|---|---|---|---|---|
| .... | | | | | |
| Encode 49 | 48 \| 49 | 47 \| 48 | 46 \| 47 | 49 \| 46 | - \| - |
| Mark all references as "unused for reference" | - \| - | - \| - | - \| - | - \| - | - \| - |
| Encode 50 | 50 \| - | - \| - | - \| - | - \| - | - \| - |
| Encode 56 | 50 \| 56 | 56 \| 50 | - \| - | - \| - | - \| - |
| Encode 51 | 50 \| 56 | 51 \| 51 | 52 \| 50 | - \| - | - \| - |
| Encode 52 | 51 \| 56 | 50 \| 52 | 52 \| 51 | 56 \| 50 | - \| - |
| Encode 53 | 52 \| 56 | 51 \| 53 | 53 \| 52 | 56 \| 51 | - \| - |
| Encode 54 | 53 \| 54 | 52 \| 53 | 51 \| 52 | 54 \| 51 | - \| - |
| Encode 55 | 55 \| - | 54 \| - | 53 \| - | 52 \| - | - \| - |
| Encode 62 | 55 \| 62 | 54 \| 55 | 53 \| 54 | 62 \| 53 | - \| - |
| Encode 57 | 55 \| 62 | 54 \| 57 | 57 \| 55 | 62 \| 54 | - \| - |
| .... | | | | | |

**Figure 3-18 List0 and List1 Ordering Example for B Slice**

For B slice prediction, List0 and List1 are used to capture recently decoded frames with short-term reference pictures and long-term reference pictures. List0 for B slice contains PicOrderCount in decreasing order for the first part, where PicOrderCount is POC earlier than current picture, and PicOrderCount in increasing order for the later part, where PicOrderCount is POC later than current picture. In contrast, List1 for B slice contains PicOrderCount in increasing order for the first part, where PicOrderCount is POC later than current picture, and PicOrderCount in decreasing order for the later part, where PicOrderCount is POC earlier than current picture. The main reason for the future-past composition for both List0 and List1 is that the same directional predictions (past and past; future and future) are possible in H.264. Long-term reference pictures can occur in List0 or List1.

For example, let's say, maximum size of DPB is four frames and the current frame_num is 50. Let's assume that there are five B frames between two references. Then, List0 and List1 can develop as follows, as time goes by, for the values of Display order (instead of POC to clearly compare with Figure 3-17) as shown in Figure 3-18.

#### Reference Picture List Re-ordering

The purpose of reference picture re-ordering is to change the default order of reference pictures in List0 for P slice or in List0/List1 for B slice temporarily to use a more compact representation for reference pictures. Let's say, there is a reference picture that is particularly useful for prediction of the current slice, but is not in position 0 in the default list. This process enables the encoder to place this reference picture at a low index in the list so that it costs fewer bits to signal prediction from this picture. Note that the re-ordering process occurs at the stage of slice decoding time, and is only in effect for the current slice. The syntax is shown in Figure 3-19.

The ref_pic_list_reordering_flag tells whether the list re-ordering is considered. If considered, the re-ordering process is repeatedly carried out until reordering_of_pic_nums_idc is 3. Otherwise, remapped_picture is computed from abs_diff_pic_num_minus_1 or long_term_pic_num. Note that a couple of command structures are in a single slice header repeatedly, where reordering_of_pic_num_idc and abs_diff_pic_num_minus_1 are repeatedly represented in ref_pic_list_reordering( ) syntax.

```
If (ref_pic_list_reordering_flag) {
   do{
      If(reordering_of_pic_num_idc==0)
            Remapped_picture = predicted_picture -
                      abs_diff_pic_num_minus_1+1);
      Elseif(reordering_of_pic_num_idc==1)
            Remapped_picture = predicted_picture +
                      (abs_diff_pic_num_minus_1+1);
      Elseif(reordering_of_pic_num_idc==2)
            Remapped_picture=long_term_pic_num;
      Else
            Exit;
   } while (1)
} else {
   Exit
}
```

**Figure 3-19 Syntax of Re-ordering Information**

When ref_pic_list_recording_flag_l0 is equal to 1, the following applies:

- Let refIdxL0 be an index into the reference picture List0. It is initially set equal to 0.
- The corresponding syntax elements reordering_of_pic_nums_idc are processed in the order they occur in the bitstream. For each of these syntax elements, the following applies:
  - If reordering_of_pic_nums_idc is equal to 0, 1 or 2, Figure 3-19 is invoked with refIdxL0 as input with output assigned to refIdxL0. Increment refIdxL0.
  - Otherwise, the reordering process for reference picture List0 is finished.

When ref_pic_list_recording_flag_l1 is equal to 1, the following applies:

- Let refIdxL1 be an index into the reference picture List1. It is initially set equal to 0.
- The corresponding syntax elements reordering_of_pic_nums_idc are processed in the order they occur in the

bitstream. For each of these syntax elements, the following applies:

- o If reordering_of_pic_nums_idc is equal to 0, 1 or 2, Figure 3-19 is invoked with refIdxL1 as input with output assigned to refIdxL1. Increment refIdxL1.
- o Otherwise, the reordering process for reference picture List1 is finished.

For example, let's say, the maximum size of the DPB is five frames and the current frame_num is 58. Also, as in figure 3-17, the reference list after encoding frame #57 is as follows:

57, 55, 53, 1*, 0* .

The following series of reference picture reordering commands are assumed to receive in a slice header as an example to proceed to 53, 55, 57, 1*, 0*:

- ref_pic_list_reordering_flag = 1 → Initiate re-ordering process
- Initial predicted_picture=58 and initial refIdxL0=0.

reordering_of_pic_nums_idc=0, abs_diff_pic_num_minus_1=4

remapped_picture=58-5=53

New list: 53, 57, 55, 1*, 0*

New predicted picture=53 and new refIdxL0=1

- reordering_of_pic_nums_idc=1, abs_diff_pic_num_minus_1=1

remapped_picture=53+2=55

New list: 53, 55, 57, 1*, 0*

New predicted picture=55 and new refIdxL0=2

- reordering_of_pic_nums_idc=3 → Exit reordering process

### Reference Picture Marking

The DRPM process is the only process that handles reference picture management. Sliding window decoded reference picture marking process and adaptive memory control decoded reference picture marking process are the only procedures to perform DRPM.

Sliding window reference marking mode provides a first-in first-out mechanism for short-term reference picture, while adaptive reference marking mode uses syntax elements to specify marking of reference pictures as "unused for reference," "used for short-term reference," or "used for long-term reference."

Sliding window reference marking process is on when adaptive_ref_pic_marking_ mode_flag=0. In this mode, a new picture is added to the short-term reference list at position 0 and the indices of the remaining short-term pictures increment by 1. If the number of all reference pictures exceeds the maximum number of references defined in the encoding process, the oldest short-term reference picture in the lists is eliminated from the DPB.

Adaptive reference marking process is on when adaptive_ref_pic_marking_mode_flag=1. In this mode, a new picture is added to the short-term reference list or an already existing short-term reference is transferred to the long-term reference list. Also, the reference pictures can be removed from either the short-term reference list or the long-term reference list. All possible operations are commanded from the encoder with memory management control operation commands listed in Table 3-8.

**Table 3-8 Memory Marking Process Commands**

| Memory_management_control_operation | Operation |
|---|---|
| 0 | End memory_managment_control_operation syntax element loop |
| 1 | Mark a short-term reference picture as "unused for reference" |
| 2 | Mark a long-term reference picture as "unused for reference" |

| | |
|---|---|
| 3 | Mark a short-term reference picture as "used for long-term reference" and assign a long-term frame index to it. |
| 4 | Specify the maximum long-term frame index and mark all long-term reference pictures having long-term frame indices greater than the maximum value as "unused for reference" |
| 5 | Mark all reference pictures as "unused for reference" and set the MaxLongTermFrameIdx variable to "no long-term frame indices" |
| 6 | Mark the current picture as "used for long-term reference" and assign a long-term frame index to it. |

Note that the syntax element adaptive_ref_pic_marking_mode_flag and the content of the decoded reference picture marking syntax structure shall be identical for all coded slices of a coded picture. The syntax flow is shown in Figure 3-20.

```
If (adaptive_ref_picture_marking_mode_flag) {
   do{
      If(memory_management_control_operation==1||3)
        picNumX=CurrPicNum -
                  (difference_of_pic_nums_minus1+1);
      If(memory_management_control_operation==1)
         De-allocate #(picNumX) from short-term
                  reference pictures
      Else
         Switch #(picNumX) short-term reference
         to a future coming  long-term
         reference #(picNumY);
      If(memory_management_control_operation==2)
         De-allocate (long_term_pic_num) from
         long term reference pictures;
      If(memory_management_control_operation==3||6)
         picNumY=long_term_frame_idx;
      If(memory_management_control_operation==3)
         Allocate #(picNumY) long-term reference
        with just released short-term
```

```
            reference picture (picNumX)
        Else
            Allocate #(picNumY) long-term reference
            with the current picture;
        If(memory_management_control_operation==4)
            MaxLongTermFrameIdx =
                            (max_long_term_frame_idx_plus1-1)
            De-allocate all long-term references
            whose LongTermFrameIdx are greater
            than MaxLongTermFrameIdx;
        If(memory_management_control_operation==5)
            De-allocate all references and
            reset MaxLongTermFrameIdx variable;
        If(memory_management_control_operation==0)
            Exit;
    } while(1)
} else {
    Exit
}
```

**Figure 3-20 Syntax Interpretation of Picture Marking Information**

The value difference_of_pic_nums_minus1 is used with memory_ management_control_ operation equal to 1 or 3 to mark a short-term reference picture as "unused for reference" or to assign a long-term frame index to a short-term reference picture, respectively. When the associated memory_management_control_operation is processed by the decoding process, the resulting picture number derived from difference_of_pic_nums_minus1 shall be a picture number assigned to one of the reference pictures marked as "used for reference" and not previously assigned to a long-term frame index.

Let picNumX be specified by:

picNumX=CurrPicNum-(difference_of_pic_nums_minus1+1).

For memory_management_control_operation equal to 1, the short-term reference frame or short-term complementary reference field pair specified by picNumX and both of its fields are marked as "unused for reference." For memory_management_control_ operation equal to 3, the variable picNumX shall refer to a frame or complementary reference field pair or non-paired reference field marked as "used for short-term reference" and not marked as "non-existing."

With memory_management_control_operation equal to 2, the value long_term_pic_num is used to mark a long-term reference picture as "unused for reference." When the associated memory_management_control_operation is processed by the decoding process, long_term_pic_num shall be equal to a long-term picture number assigned to one of the reference pictures that is currently marked as "used for long-term reference."

The value long_term_frame_idx is used with memory_management_control_operation equal to 3 or 6 to assign a long-term frame index to a picture. When the associated memory_ management_control_operation is processed by the decoding process, the value of long_term_frame_idx shall be in the range of 0 to MaxLongTermFrameIdx, inclusive.

With memory_management_control_operation equal to 4, the value max_long_term_frame_idx_plus1 minus 1 specifies the maximum value of long-term frame index allowed for long-term reference pictures (until receipt of another value of max_long_term_frame_idx_plus1). The value of max_long_term_frame_idx_plus1 shall be in the range of 0 to num_ref_frames, inclusive.

Let MaxLongTermFrameIdx be specified by:

MaxLongTermFrameIdx =(max_long_term_frame_idx_plus1-1).

All pictures for which LongTermFrameIdx is greater than max_long_term_ frame_idx_ plus1-1 and that are marked as "used for long-term reference" shall be marked as "unused for reference."

With memory_management_control_operation equal to 5, the variable MaxLongTermFrameIdx is reset with "no long-term frame indices" and all reference pictures are marked as "unused for reference."

With memory_management_control_operation equal to 6, the current picture is marked as "used for long-term reference" and assigned the variable LongTermFramdIdx equal to long_term_frame_idx.

Note that IDR pictures do not use memory_management_control_operation to mark themselves as a long-term reference. Instead, the long_term_reference_flag in dec_ref_pic_marketing( ) structure is used to declare it with the nal_unit_type==5. For example, the series of memory_management_control_operation commands are as follows:

| action \ syntax | nal_unit_type | Adaptive_ref_pic_making_mode_flag | Commands |
|---|---|---|---|
| …. | | | |
| Encode 49 | 1 | 1 | memory_management_control_operation =5 |
| Mark all references as “unused for reference” | | | |
| Encode 50 | 1 | 0 | - |
| Encode 51 | 1 | 0 | - |
| Encode 52 | 1 | 0 | - |
| Encode 53 | 1 | 0 | - |
| Encode 54 | 1 | 0 | - |
| Encode 55 | 1 | 1 | memory_management_control_operation =3,<br>difference_of_pic_nums_minus1=0,<br>long_term_frame_idx=0 |
| Assign 54 to LongTermPicNum 0* | | | |
| Encode 56 | 1 | 1 | memory_management_control_operation =6,<br>long_term_frame_idx=1 |
| Assign 56 to LongTermPicNum 1* | | | |
| Encode 57 | 1 | 0 | - |
| …. | | | |

**Figure 3-21 Memory Marking Commands for Figure 3-17**

## 3.4 Constant Delay CBR HRD Mirroring in Encoder Buffer

### Relationship between Actual Buffer and Virtual Buffer

There is an interesting relationship between the actual encoder output buffer at the encoder and the virtual decoder input buffer at the decoder. The encoder output buffer is actual since the video encoder can actually see and verify the occupancy of the contents in the buffer. In contrast, the decoder input buffer is virtual since the video encoder cannot actually see, but must guess the occupancy of the contents in the buffer based on pre-defined behaviors of the decoder.

As shown in Figure 3-22, the occupancy of the contents in the encoder output buffer is actually a mirrored image of the occupancy of the contents in the decoder output buffer. Also, the sum of the two occupancy values is always the buffer size of B at any point. The summed-up value of B is generally chosen by the encoder to make the communication delay smallest [ribas-corbera:generalized].

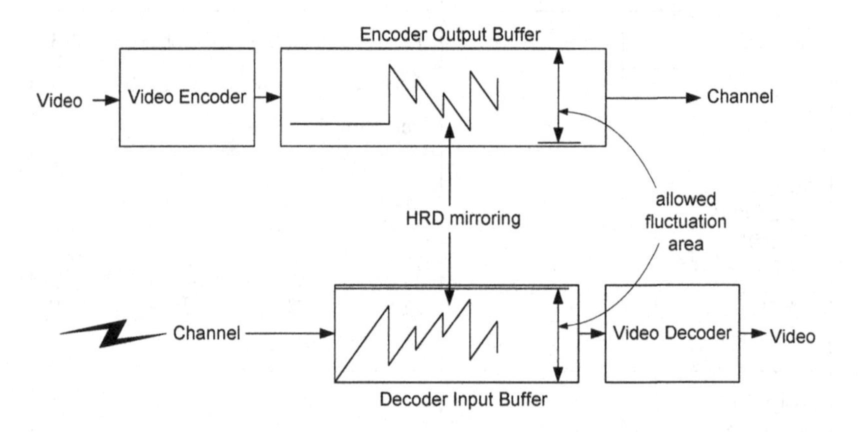


**Figure 3-22 HRD Buffer Mirroring in Encoder Output Buffer**

### Rate Control based on Encoder Actual Buffer

When B is chosen as the summed-up buffer size of the two buffers, an interesting characteristic can be observed as follows:

When the occupancy of the encoder output buffer reaches the "overflow" point, the occupancy of the decoder input buffer reaches the "underflow" point. When the occupancy of the encoder output buffer reaches at "underflow" point, the occupancy of the decoder input buffer reaches at "overflow" point.

Since one of key goals of rate control algorithm is to prevent the buffer from overflow or underflow, the rate control algorithm can be devised based on behavior of the encoder output buffer. However, the policy to assign Qp needs to be carefully revised if the rate control algorithm is based on the encoder buffer occupancy. The policy most likely is reversed – action initiation required with a high occupancy requires a low occupancy.

### Rate Control based on Encoder Virtual Buffer

When either the HRD buffer at the decoder or its mirror buffer at the encoder is used for rate control, the buffer occupancy information cannot be used for MB-level rate control. This is because the behavior of the HRD buffer is described only at each picture time. To get fine granularity of the rate control over each MB, an imaginary buffer of the encoder is sometimes defined for some rate control algorithms. This is why some literatures about rate control algorithms examine the occupancy of the encoder virtual buffer at each MB time. Such rate control algorithms include those of H.261 and MPEG-2. While examining the encoder virtual buffer, the HRD buffer can still be guaranteed not to overflow or underflow through a "guard zone" of the HRD buffer. In other words, actual occupancy violating the guard zone in the HRD buffer can send a signal to the rate control unit to change the policy of the rate control algorithm for a very short time to prevent it from such buffer struggles. Two major benefits of using the encoder virtual buffer for the rate control algorithm are to gain MB-level fine control and to obtain freedom of parameters value setting (such as buffer occupancy) at any time.

In this book, $V$ is introduced to indicate the occupancy of the HRD buffer, while $d$ is introduced to indicate the occupancy of the encoder virtual buffer to clearly distinguish the two buffer occupancies.

## 3.5 Rate Control Algorithms in Standard Test Models

In the development process of international standards, there were several test, simulation, and verification models where proposed algorithms were verified. In this section, we investigate the background ideas of several rate control schemes including some internal models of international standards [lee:ratecontrol]. There are two major probe points in the rate control schemes— buffer occupancy and the approximated rate distortion model.

The advantage of using the buffer occupancy measure is that overflow or underflow of the buffer can be clearly avoided. Generally buffer occupancy depends on the position of the picture in the GOP and the bitrate can also be directly controlled with respect to a target bitrate.

The advantage of using a rate distortion model to control bitrate relative to a target bitrate is that a specific target bitrate can be used for each picture type, and even for each individual picture in a GOP. In this approach, however, any excess or deficiency in bits must be carried over from one control unit to the next.

Many practical rate control algorithms consider these two measures and their advantages. Overall, five different rate control schemes are examined in this section.

### H.261

H.261 is a block-based, motion-compensated transform coding (DCT) design. The $p \times 64$ codec, H.261, uses only I-pictures and P-pictures. A FIFO buffer (a.k.a., the encoder virtual buffer) is assumed between encoder and channel, and several times per picture is monitored for fullness. The quantization parameter, $Q_p$, (proportional to quantization step size, or MQUANT as it is referred to in the standard) is increased or decreased according to whether the buffer is relatively full or empty. It prescribes the quantization of DCT coefficients using identical uniform quantizers with dead zones for all AC coefficients – 8-bit uniform quantization with no perceptual frequency weighting is used. The AC coefficient quantizer step size is determined as twice the value of a parameter $Q_p$, which can be indicated at up to the MB level.

Reference Model 8 (RM8) is a "reference implementation" of an H.261 encoder that was developed and used internally by the H.261 working group, with the purpose of providing a common environment in which experiments could be conducted. In RM8, a proper $Q_p$ corresponds to where the current buffer occupancy is in the encoder virtual buffer. In other words, the buffer is uniformly divided into 31 fragments, and each of them is named for fragment 1,..., fragment 31 along with the increase of buffer size as shown in Figure 3-23 [eleftheriadis:auto1].

If current buffer occupancy falls into a certain fragment, then the name of the fragment is assigned to the value of $Q_p$. For example, a larger $Q_p$ value is given with buffer occupancy increasing to fullness, while a smaller $Q_p$ value is assigned as buffer occupancy decreases to emptiness. Note that the relation between buffer occupancy and $Q_p$ is linear and that the $Q_p$ value is only dependent on current virtual buffer occupancy as shown in Equation (3-16). The very first picture, which is an I-picture, is coded with a constant $Q_p$ of 16 with the virtual buffer being initialized at 50% occupancy.

For the remaining pictures, $Q_p$ is adapted at the start of each line of MBs within a group of blocks (GOB), where a rectangular array of $11 \times 3$ MBs defines a GOB. In other words, $Q_p$ is adapted three times within each GOB. The buffer occupancy is examined after the transmission of each MB and, if overflow occurs, the next MB is skipped. Note that this does not result in a temporary buffer overflow since the MB that caused the overflow is in the virtual encoder buffer. However, this trespassing of the "guard zone" in the HRD buffer can signal the rate control unit to adjust the rate control scheme to prevent it from actual overflow or underflow.

$Q_p$ is updated "linearly" with the buffer occupancy according to the relation

$$Q_{pi} = \min\{31, \left\lfloor \frac{d_i}{B/32} \right\rfloor + 1\} \tag{3-16}$$

where $Q_{pi}$ is the value of $Q_p$ selected for MB I, $d_i$ is the virtual buffer occupancy at the encoder just prior to coding MB i, and $B$ is the size of the virtual buffer size. A buffer size of $6400 \times q$ bits is used, given a bit rate of $q \times 64$ kbps of the video signal only.

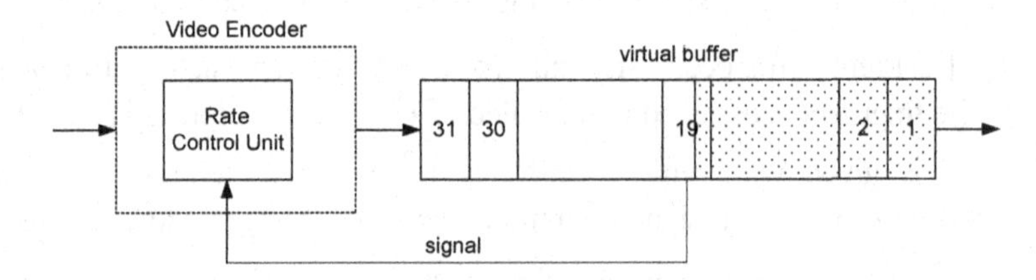


**Figure 3-23 H.261 Rate Control**

### H.263 (MPEG-4 Part 2 Baseline)

H.263 is a block-based, motion-compensated transform coding (DCT) design. TMN5 is a test model to validate its algorithms, and uses motion compensation with accuracy of half-pixel (like that of MPEG-1, whereas H.261 supports only integer-pixel accuracy) as well as optional modes such as PB-frames (a substitute for B-pictures of MPEG-1), unrestricted motion vector, advanced $8 \times 8$ block prediction, and syntax based arithmetic coding. Incidentally, these modes are options that are negotiated between a decoder and an encoder. The ITU-T has continued work on further embellishing H.263 by adding yet many more features and optional modes for the upcoming H.263+.

The idea of TMN5 rate control is to use an approximation of a general rate distortion function. A general rate distortion function is a convex function as is known in Information Theory. The first step of the derivation of a rate control formula is to approximate the rate distortion function by an inverse proportional curve as shown in Figure 3-24. If the rate distortion property is stationary, X is expected to be pretty consistent through blocks and frames. If X is given as a constant, we can see that

$$B_{previous} \times Q_{previous} = B_{current} \times Q_{current} = X \,. \tag{3-17}$$

Equation (3-17) can be represented by:

$$Q_{current} = Q_{previous}(1 + \frac{B_{previous} - B_{current}}{B_{current}}) \,. \tag{3-18}$$

If we define $\Delta B$ as $B_{previous} - B_{current}$, then Equation (3-18) can be written as:.

$$Q_{current} = Q_{previous}(1 + \frac{\Delta B}{B_{current}}) \,. \tag{3-19}$$

In H.263, the rate distortion characteristic is quite stationary since generally there is only one picture type (i.e., P) except the first frame (i.e., I). So, this can be used as a quantizer parameter prediction from one frame to another frame. However, performing solely frame-based rate control is not a good idea since a limited buffer can be easily overflowed or underflowed due to incoming sequences' randomness. Therefore, we break the Equation into two parts once more to handle MB level rate control as follows:

$$Q_{current} = Q_{previous}(1 + \frac{\Delta B}{2B_{current}} + \frac{\Delta B}{2B_{current}}) \,. \tag{3-20}$$

A good rate control algorithm usually adapts itself for a global and local utilization of a resource simultaneously. To consider a global and local utilization of a resource, we intentionally assign the second term to adjustment of difference between target and actual bits for frames (i.e., inter-frames bit budget adaptation), and assign the third term to adjustment of the difference between target and actual bits for MBs (i.e., intra-frame bit budget adaptation). A potential candidate for the rate control algorithm which assigns resources globally and locally can be of the following form:

$$Q_{current,MB} = Q_{previous,frame}(1 + \frac{\Delta B_{frame}}{2B_{current,frame}} + \frac{\Delta B_{MB}}{2B_{current,MB}}). \quad (3\text{-}21)$$

The last step is to match the constant X by comparing the inverse proportional graph with the actual data. This should be obtained by experiments. To take the experimental factor into consideration, the above equation is modified with K as follows:

$$Q_{current,MB} = Q_{previous,frame}(1 + \frac{\Delta B_{frame}}{2B_{current,frame}} + K\frac{\Delta B_{MB}}{2B_{current,MB}}). \quad (3\text{-}22\text{-}1)$$

There may be two different approaches for the experimental factor K. For each frame, first of all, K can be calculated based on previous frames' data to localize the formula for the consideration of randomness in incoming sequences at every frame. Secondly, we fix the K factor through frames as a constant, but we control the frame rate based on buffer occupancy. The second approach was taken in TMN5 rate control. When K is taken as 12/5 and "variable frame rate" is allowed for target rate trace, the TMN5 rate control algorithm is obtained.

To be more specific, TMN5 rate control can be re-written in detail as follows:

$$Q_{pnew} = \overline{Q}_{pi-1}(1 + \frac{\Delta B_{frame}}{2\overline{B}} + \frac{12\Delta B_{MB}}{2R}) \quad (3\text{-}22\text{-}2)$$

with $\Delta B_{frame} = B_{i-1} - \overline{B}$ and $\Delta B_{MB} = (B_a) - \frac{a}{A}\overline{B}$

where:

$\overline{Q}_{pi-1}$ is the mean quantization parameter for the previous picture, $\overline{B}$ is the target number of bits for picture, $R$ is the target bit rate (at about 10 fps),

$B_{i-1}$ is the number of bits spent for the previous (i-1) picture, $a$ is the index of the current MB, $A$ is the number of MBs in a picture, $B_a$ is the number of bits spent for the picture prior to coding the $a$ th MB.

The first picture is coded with $Q_p = 16$ and after the first picture, the initial buffer occupancy is set to $\frac{R}{f_{target}} + 3 \times \frac{R}{F}$, where $f_{target} = 10 - \frac{\overline{Q}_{pi-1}}{4}$ (therefore, $\overline{B} = \frac{R}{f_{target}}$) and $F$ is the frame rate of the source material.

The occupancy of the encoder virtual buffer is updated after each complete picture in the following way:

```
buffer_content=buffer_content+B_finalMB ;
while (buffer_content > 3×R/F) {
        buffer_content = buffer_content -R/F;
        frame_incr++:
}
```

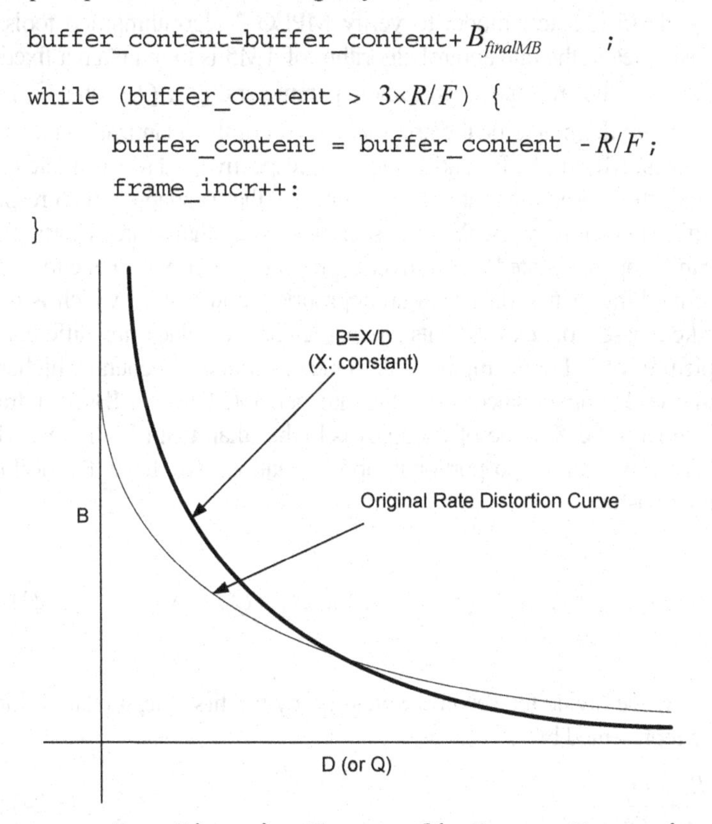


**Figure 3-24 Rate Distortion Curve and its Inverse Proportional Approximation**

In other words, when the buffer occupancy exceeds an average of three frames, subsequent frames are skipped until the occupancy drops below a pre-determined threshold.

## MPEG-2

MPEG-2 offers little benefit over MPEG-1 for programming material that was initially recorded on film. In fact, nearly all movies and television programs with high production values (budgets) are shot at 24 celluloid frames per second. What MPEG-2 does offer is a more efficient means to code interlaced video signals, such as those that originate from electronic cameras.

TM5 is a test model to verify MPEG-2 algorithms and tools. The basic idea in the rate control algorithm of TM5 is to maintain a fixed ratio between the average quantization parameters, $Q_I$, $Q_P$, and $Q_B$, where these are defined to be the arithmetic mean of the quantization parameters over all MBs in I-, P-, and B-pictures, respectively. TM5 is a rate control algorithm based on a rate distortion model that is adaptive with respect to different picture types. We have seen previously that a rate distortion curve can be approximated by an inverse proportional curve. This curve can be refined into actual data with an appropriate constant X, which is usually taken based on experiments. In particular, X values are different with picture types. For example, the X value of I frames is generally higher than that of P frames since more bits are generated from MBs in I frames. Similarly the X value of P frames is higher than that of B frames. If we take the inverse proportional approximation, we have the following relationship:

$$B_I \times Q_I = X_I,\ B_P \times Q_P = X_P, \text{ and } B_B \times Q_B = X_B. \tag{3-23}$$

If we divide the second expression by the first one, a relationship can be represented by:

$$\frac{B_P \times Q_P}{B_I \times Q_I} = \frac{X_P}{X_I}. \tag{3-24}$$

If we define $K_P$ as $Q_P / Q_I$, the above expression is re-written by:

$$\frac{B_P}{B_I} = \frac{X_P}{X_I} \times \frac{1}{K_P}. \tag{3-25}$$

Note that $K_P$ is defined by the ratio of "frame-averaged" quantization parameters since $B_P$, $B_I$, $X_P$ and $X_I$ are frame-based expressions. The above ratio is expected to be less than 1, considering I frames generate more bits. A similar relation holds between B and I frames as follows:

$$\frac{B_B}{B_I} = \frac{X_B}{X_I} \times \frac{1}{K_B} \tag{3-26}$$

where the definition of $K_B$ as $Q_B / Q_I$. With the same reason, the above ratio is expected to be less than 1.

Since we already defined $K_P$ and $K_B$, we can define the ratio of average quantization parameters between P and B frames with $K_P$ and $K_B$ as follows:

$$\frac{B_P \times Q_P}{B_B \times Q_B} = \frac{X_P}{X_B}. \tag{3-27}$$

Then, it can be re-written by:

$$\frac{B_P}{B_B} \times \frac{Q_P}{Q_I} \times \frac{Q_I}{Q_B} = \frac{B_P}{B_B} \times K_P \times \frac{1}{K_B} = \frac{X_P}{X_B}. \tag{3-28}$$

Therefore, we conclude that

$$\frac{B_P}{B_B} = \frac{X_P}{X_B} \times \frac{K_B}{K_P}. \tag{3-29}$$

To understand that $K_P$ and $K_B$ are maintained as constants is very important. That is, the purpose of rate control in TM5 is to maintain a certain quality ratio through different picture types (i.e., I, P, and B frames). The quality of movies is acceptable to human perception based on experiments when the constants are taken as $K_P = 1$ and $K_B = 1.4$. To maintain the quality ratios, we intentionally fluctuate each frame's bit budget. Note that the value X, the constant value which refines its approximation (i.e., an inverse proportional curve) to the actual data, is considered as a constant in the current picture, but is calculated from the closest picture of the same type.

TM5 consists of three steps: target picture bit budget allocation, MB quantization parameter assignment, and activity modulation at MB quantization parameters. To allocate the target bit budget for a picture is easy if the same coding technique is applied to each picture and input pictures are stationary. For example, if the target bit rate is 1.5 Mbps and input is frame-based video, then each frame bit budget is supposed to be 50Kbps (i.e., 1.5 Mbps/30 since the input is a video of 30 frames per second). This is, however, not the case in MPEG-video since each picture type such as I, P, and B uses different coding techniques, thus generating different amount of output bits. Therefore we should expect a different target bit budget for a different picture type. Generally we expect a larger picture bit budget for I pictures, while we allocate a smallest bit budget for B pictures. We define R as a left over bits expected in a GOP through the current picture, so R is given with $R := R - B_{I,P,B}$ after encoding a picture, as shown in Figure 3-25. Since R is a left over bits, the target picture bit budget, T, should be the following form:

$$T = \frac{R}{N_{left}} \tag{3-30}$$

where $N_{left}$ is the number of frames left over through current picture in a GOP. Note that the $R$ represents both target bitrate (i.e., target bits in 1 second) and the left over bits due to the representation of recursion $R := R - B_{I,P,B}$.

For I frames, the target bit budget for a picture is expected to be higher than other two types. Thus, $N_{left}$ is expected to be lower than others proportionally. Considering an I frame is the first in a GOP, $N_{left} = 1 + \alpha_P N_I + \beta_I N_B$. (3-31)

The values of $N_P$ and $N_B$ should be interpreted to be lower than the actual number (i.e. $\alpha_I$ and $\beta_I$ factors) from the I frames' point of view, because P and B frames generate less bits as such. Note that the factors, $\alpha_I$ and $\beta_I$, can be taken as the aforementioned bit budget ratios, $B_P / B_I$ and $B_B / B_I$. Therefore, we can set the picture target bit rate by:

$$T_I = \max\{\frac{R}{1 + \frac{B_P}{B_I} N_P + \frac{B_B}{B_I} N_B}, \frac{\overline{B}}{8}\} \tag{3-32}$$

for I frames,

$$T_P = \max\{\frac{R}{N_P + \frac{B_B}{B_P} N_B}, \frac{\overline{B}}{8}\} \tag{3-33}$$

for P frames, and

$$T_B = \max\{\frac{R}{\frac{B_P}{B_B} N_P + N_B}, \frac{\overline{B}}{8}\} \tag{3-34}$$

for B frames, where $\overline{B}$ is the average picture bit budget (i.e., target bitrate/ picture per second).

Note that to avoid buffer underflow, one eighth of the average picture bit budget is selected when the target bit budget allocation is too small.

The next step is to assign a quantization parameter to the current MB. In TM5 the quantization parameter, $Q_j$, is taken to be proportional to virtual buffer fullness, $d_j$,

$$Q_j = Ad_j \tag{3-35}$$

where A is constant and j is the current MB number. Note that virtual buffer introduced in this formula has nothing to do with the HRD buffer, but is a conveniently taken imaginary buffer at the encoder to determine $Q_j$ value linearly.

In the virtual buffer, a target MB bit budget is constantly going out by the value of T/MB_count, where MB_count is the total number of MBs in a picture. The purpose of MB level rate control is to spend the target picture bit budget entirely through MBs in the current picture. Therefore, the virtual buffer occupancy is updated after encoding each MB as follows:

$$d_j^I = d_0^I + b_{j-1} - [\frac{T_I \times (j-1)}{MB\_count}] \quad (3\text{-}36)$$

for I frames,

$$d_j^P = d_0^P + b_{j-1} - [\frac{T_P \times (j-1)}{MB\_count}] \quad (3\text{-}37)$$

for P frames, and

$$d_j^B = d_0^B + b_{j-1} - [\frac{T_B \times (j-1)}{MB\_count}] \quad (3\text{-}38)$$

for B frames, where $d_0^I$, $d_0^B$ and $d_0^B$ are initial virtual buffer fullness at the start of each picture and $b_j$ is the number of bits generated by encoding all MBs in the picture up to and including j. In experiments of the TM5 development process, the maximum size of the virtual buffer, $r_{virmax}$, was taken as "two times" the average picture bit budget (i.e., $2 \times \overline{B}$). Therefore, the relation in Equation (3-35) can be re-written as follows:

$$Q_j = (\frac{31}{virtual\_buffer\_\max})d_j = (\frac{31}{r_{virmax}})d_j. \quad (3\text{-}39)$$

For example, 31 is given to a quantization parameter when the current buffer occupancy, $d_j$, is equal to the maximum size of the virtual buffer, $r_{virmax}$. In that case, it is reasonable to give the highest value for a quantization parameter since the virtual buffer is about to overflow. Note that $Q_j$ is updated "linearly" with the buffer occupancy according to Equation (3-39) when $d_j$ is lower than $r_{virmax}$.

The last step is to modulate $Q_j$ with a measure of the local spatial activity based on the minimum variance of the four luma blocks in the MB in TM5. As a rule of thumb, a given stimulus will be harder to see in a very active block (i.e., one with very complex and strong spatially-varying intensities). This characteristic is so called the "spatial masking effect." For example, in a portion of a picture where strong spatially-varying intensities or rapid intensity changes occur, the eye has difficulty discerning small variations in intensity. Conversely, in very smooth areas, the same small changes may be quite visible.

This aspect is taken into account in TM5.

There are a lot of ways to define an activity measure. Activity measure in TM5 is obtained by taking a minimum value out of several activity candidates as follows:

$$ACT_j = 1 + \min\{vblk1, vblk2, ...., vblk8\} \quad (3\text{-}40)$$

where the first half (i.e., *vblk1, ..., vblk4*) of the total eight candidates are variances of luminance frame-organized sub-blocks and the last half (i.e., *vblk5, ..., vblk8*) are luminance field-organized sub-blocks.

The spatial activity, ACT, is normalized by an average value, AVG_ACT, obtained for the previous picture and mapped to the range 0.5 to 2.0 by a nonlinear function:

$$N\_ACT_j = \frac{2 \times ACT_j + AVG\_ACT}{ACT_j + 2 \times AVG\_ACT}. \quad (3\text{-}41)$$

Note that $N\_ACT_j$ is the weighting factor for predetermined quantization parameters. If we take $ACT_j = k \times AVG\_ACT$ and put it in Equation (3-41), it can be represented by:

$$N\_ACT_j = \frac{2 \times k + 1}{k + 2}. \quad (3\text{-}42)$$

Note that k=1 (i.e., the current MB's activity is the same to the average activity) makes the weighting factor 1. Further, $k \to 0$ makes the value 0.5, while $k \to \infty$ makes it 2.0. That is, if the current MB's activity is lower than the average activity, the quality of the MB is enhanced on purpose. On the other hand, if the current MB's activity is higher than the average activity, the quality of the MB is degraded to take advantage of the spatial masking effect.

If we define a new quantization parameter $Q_{pj}$, which is actually used for rate control, it can be found by the following relationship in TM5 for activity measure consideration:

$$Q_{pj} = N\_ACT_j \times Q_j . \quad (3\text{-}43)$$

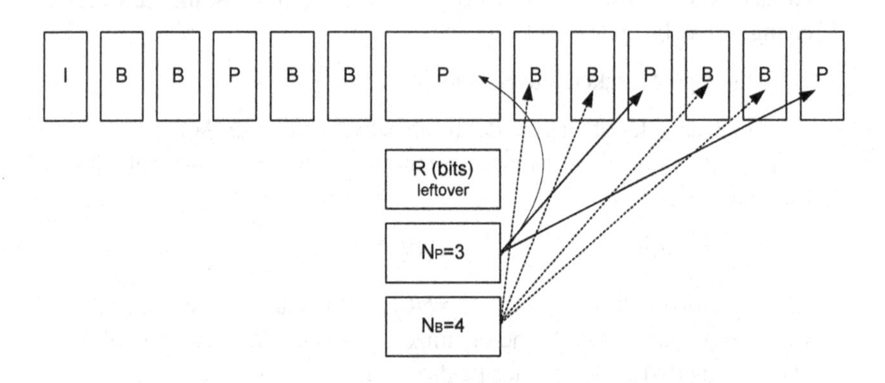


**Figure 3-25 Implications of Parameters in MPEG-2 Rate Control**

### MPEG-4 Part 2

Recently a new rate control scheme is proposed in VM5 for MPEG-4 video [chiang:newrate]. The idea of VM5 is to use a quadratic approximation for the rate distortion function. The rate control scheme proposes a way to allocate the target bit budget for each frame based on that approximation. The distortion measure is assumed to be the average quantization scale of a frame. The rate distortion function is modeled as a second-order inverse proportional function of the distortion measure, and this scheme has a closed form solution for the target bit allocation which includes MPEG-2 TM5 rate control scheme as a special case.

As we see in Figure 3-26, the rate distortion function is approximated by a second order inverse proportional function as follows:

$$B = \frac{a}{Q} + \frac{b}{Q^2} . \quad (3\text{-}44)$$

In this formulation, $a$ and $b$ are constants in a picture, and the distortion measure is represented as the average quantization scale of a frame with a specific picture prediction type.

The encoder collects the bit rate and average quantization parameter for each type of picture at the end of encoding each frame. Then, the model parameters, $a_I$, $a_P$, $a_B$, $b_I$, $b_P$ and $b_B$ can be obtained through Equation (3-45) and Equation (3-46). Since statistics of previous frames are known to the encoder, linear regression analysis and the formula below can be used to find optimized parameters $a$ and $b$ in Equation (3-44).

$$b_k = \frac{n\sum_{i=1}^{n} B_i - (\sum_{i=1}^{n} Q^{-1}{}_i)(\sum_{i=1}^{n} Q_i B_i)}{n\sum_{i=1}^{n} Q^{-2}{}_i - (\sum_{i=1}^{n} Q^{-1}{}_i)^2} \quad (3\text{-}45)$$

$$a_k = \frac{\sum_{i=1}^{n} Q_i B_i - b_k Q_i^{-1}}{n} \quad (3\text{-}46)$$

where n is the number of frames in the past, $Q_i$ and $B_i$ actual encoding average quantization parameter and bit count in the past with k=I, P, B, respectively. The scheme is not limited by the method of finding the parameters. They can be used to solve for target bit rates $T_I$, $T_P$, and $T_B$ based on the following six equations:

$$K_P \times Q_I = Q_P \quad (3\text{-}47)$$

$$K_B \times Q_P = K_P \times Q_B \quad (3\text{-}48)$$

$$R = T_I + N_P \times T_P + N_B \times T_B \quad (3\text{-}49)$$

$$T_I = a_I \times Q_I^{-1} + b_I \times Q_I^{-2} \quad (3\text{-}50)$$

$$T_P = a_P \times Q_P^{-1} + b_P \times Q_P^{-2} \quad (3\text{-}51)$$

$$T_B = a_B \times Q_B^{-1} + b_B \times Q_B^{-2} \quad (3\text{-}52)$$

where $K_P$, $K_B$ are constant ratios for I, P, and B pictures, $T_I$, $T_P$, and $T_B$ target bit rates for I, P, and B pictures, $Q_I$, $Q_P$, and $Q_B$ average quantization parameters for I, P, and B pictures, $N_I$, $N_P$, and $N_B$ frames to be encoded for I, P, and B pictures, and $R$ remaining number of bits in the current GOP.

The first two equations make the assumption that we want to make ratios $K_P$, $K_B$ constant between the average quantization parameters of individual picture types as was in the case of TM5. The constant $K_P$ and $K_B$ are taken as 1, and 1.4, respectively, as were the exact same values in TM5. The variable $R$ is updated for every GOP according to the assigned bit rate. The last three equations state that a second order rate distortion function is assumed with the distortion measure as the average quantization parameters of the encoded picture. In the formulation, the distortion measure $Q_i$ and the actual picture bit count $B_i$ in the past are found after the encoding of each picture. Using this information, the model parameters for the last three equations can be derived. Based on the model found in parameter estimation, we can allocate the target bit budget before encoding. The closed form solutions of the above six equations for the unknown $T_I$, $T_P$, and $T_B$ are as follows:

Equation (3-51) ~ Equation (3-52) are plugged into Equation (3-49), thus making it:

$$R = T_I + N_P \times (a_P \times Q_P^{-1} + b_P \times Q_P^{-2}) + N_B \times (a_B \times Q_B^{-1} + b_B \times Q_B^{-2}). \quad (3\text{-}53)$$

Furthermore, we can plug Equation (3-47) and Equation (3-48) into the above Equation (3-53) to find:

$$R = T_I + N_P \times (a_P \times (K_P \times Q_I)^{-1} + b_P \times (K_P \times Q_I)^{-2}) + N_B \times (a_B \times (K_B \times Q_I)^{-1} + b_B \times (K_B \times Q_I)^{-2}). \quad (3\text{-}54)$$

If we put Equation (3-50) into the above Equation (3-54), we obtain the following Equation:

$$0 = (b_I + N_P \times b_P \times \frac{1}{K_P^2} + N_B \times b_B \times \frac{1}{K_B^2})Q_I^{-2} + (a_I + N_P \times a_P \times \frac{1}{K_P} + N_B \times a_B \times \frac{1}{K_B})Q_I^{-1} - R. \quad (3\text{-}55)$$

The rest of the problem is to solve this quadratic equation for $Q_I^{-1}$. The formulas are different for I, P, and B frames. To find $T_I$ (Target bit rate for I frame), we define the following terms, which were used in the previous derivation.

$$\alpha = b_I + N_P \times b_P \times \frac{1}{K_P^2} + N_B \times b_B \times \frac{1}{K_B^2} \quad (3\text{-}56)$$

$$\beta = a_I + N_P \times a_P \times \frac{1}{K_P} + N_B \times a_B \times \frac{1}{K_B} \quad (3\text{-}57)$$

$$\gamma = -R \quad (3\text{-}58)$$

$$\delta = \beta^2 - 4\alpha\gamma. \quad (3\text{-}59)$$

If $\delta < 0$, there is no real value solution for $Q_I^{-1}$ for the quadratic equation. In this case, we take the first order inverse proportional approximation as follows:

$$T_I = a_I \times Q_I^{-1} \quad (3\text{-}60)$$

where $Q_I^{-1} = -\gamma / \beta$. (3-61)

If $\delta$ is not negative, there exists a real value solution for the quadratic equation. Therefore, $T_I$ can be found by:

$$T_I = a_I \times Q_I^{-1} + a_P \times Q_I^{-2} \quad (3\text{-}62)$$

where $Q_I^{-1} = (\sqrt{\delta} - \beta) / 2\alpha$. (3-63)

The same derivation is applied to $T_P$ and $T_B$ cases. To find $T_P$ (Target bit rate for P frame),

$$\alpha = N_P \times b_P + N_B \times b_B \times \frac{K_P^2}{K_B^2} \quad (3\text{-}64)$$

$$\beta = N_P \times a_P + N_B \times a_B \times \frac{K_P}{K_B} \quad (3\text{-}65)$$

$$\gamma = -R \quad (3\text{-}66)$$

$$\delta = \beta^2 - 4\alpha\gamma . \quad (3\text{-}67)$$

If $\delta < 0$, then

$$T_P = a_I \times Q_P^{-1} \quad (3\text{-}68)$$

where $Q_P^{-1} = -\gamma / \beta$. (3-69)

Otherwise,

$$T_P = a_I \times Q_P^{-1} + a_P \times Q_P^{-2} \quad (3\text{-}70)$$

where $Q_P^{-1} = (\sqrt{\delta} - \beta) / 2\alpha$. (3-71)

To find $T_B$ (Target bit rate for B frame),

$$\alpha = N_P \times b_P \times \frac{K_B^2}{K_P^2} + N_B \times b_B \quad (3\text{-}72)$$

$$\beta = N_P \times a_P \times \frac{K_B}{K_P} + N_B \times a_B \quad (3\text{-}73)$$

$$\gamma = -R \quad (3\text{-}74)$$

$$\delta = \beta^2 - 4\alpha\gamma . \quad (3\text{-}75)$$

If $\delta < 0$, then

$$T_B = a_I \times Q_B^{-1} \quad (3\text{-}76)$$

where $Q_B^{-1} = -\gamma / \beta$. (3-77)

Otherwise,

$$T_B = a_I \times Q_B^{-1} + a_P \times Q_B^{-2} \quad (3\text{-}78)$$

where $Q_B^{-1} = (\sqrt{\delta} - \beta)/2\alpha$. (3-79)

The above solution can be reduced to the TM5 target bit allocation when the formula for $\delta < 0$ is always used.

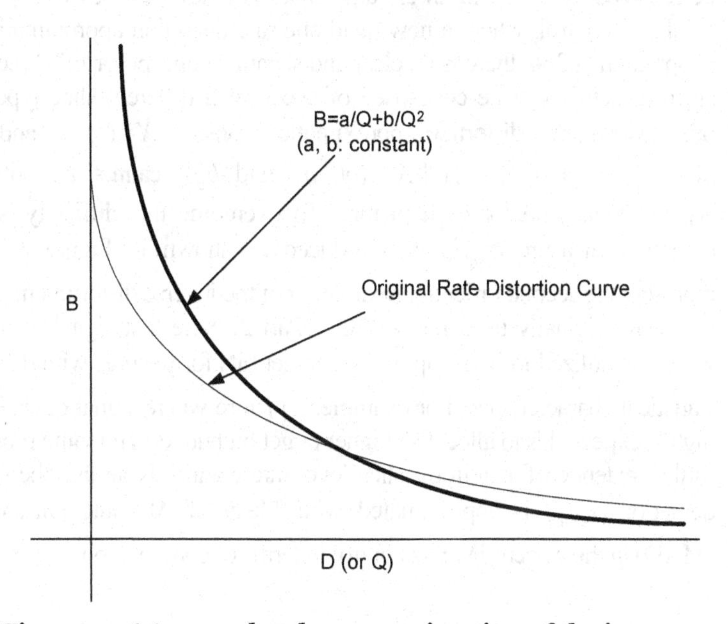


**Figure 3-26 A second order approximation of the inverse distortion measure**

## VC-1

No public document is available at this moment for VC-1 encoder details.

**MPEG-4 Part 10 (H.264)**

This section presents the rate control algorithm of the JM reference software implementation for H.264 codec [lim:rate]. The tested rate control in JM is a special case of possible H.264 rate control algorithms based on an assumption that a GOP is formed. Note that a GOP is not a necessary concept in H.264. The rate control algorithm consists of three tightly consecutive components – GOP level rate control, picture level rate control and "uniform slice" level rate control. The uniform slice unit can be replaced by a MB in an extreme case. The key idea lies with picture level rate control, where a new quadratic rate distortion approximation is adopted. In H.264, there is no clear and separate concept for I/P/B pictures since a picture can be composed of slices with different slice types. In other words, rate distortion approximation constant $X_I$ (or, $a_I$ and $b_I$), $X_P$ (or, $a_P$ and $b_P$) and $X_B$ (or, $a_B$ and $b_B$) cannot be obtained separately to tune for each picture. To overcome this difficulty, signal deviation measure $\sigma_{signal}$ is introduced as shown in Figure 3-27 to represent the second order approximation of the inverse distortion measure that was originally used for MPEG-4 Part 2. Note that signal deviation $\sigma_{signal}$ is utilized to boost up/ reduce target bits for pictures with different statistical characteristics. For example, a picture where signal deviation is high is expected is to allocate a higher target bit budget. The same model is further extended for uniform slice level rate control. Note that the signal deviation $\sigma_{signal}$ is approximated with Mean of Absolute Differences (MAD) in the current JM rate control in terms of computation.

GOP level rate control allocates the total bits for the rest of pictures in this GOP with the initial quantization parameter of the IDR picture and that of the first stored picture. An important point of the GOP level rate control is to limit the variation of quantization parameters in a small range from GOP to GOP so that the quality difference is not very noticeable.

Let's say that the j$^{th}$ picture in the i$^{th}$ GOP is being encoded. $R$, remaining number of bits in the current GOP, are given as follows:

$$R := \begin{cases} \frac{R_i(j)}{f_{FR}} \times N_i - V_i(j) & j=1 \\ R + \frac{R_i(j) - R_i(j-1)}{f_{FR}} \times (N_i - j + 1) - b_i(j-1) & j = 2,3,...,N_i. \end{cases}$$

(3-80)

For the first picture in a GOP (i.e. j = 1), the total bits are calculated from Equation (3-80), where $f_{FR}$ is the pre-defined coding frame rate. $N_i$ is the total number of pictures in the i[th] GOP. $R_i(j)$ and $V_i(j)$ are the instant target bit rate and the occupancy of the HRD buffer, respectively.

For other pictures, the total bits are calculated from Equation (3-80), where $b_i(j-1)$ is the actual generated bits in the (j-1)[th] picture. $R_i(j)$ may vary at different frames and GOPs for VBR case, while $R_i(j)$ is always equal to $R_i(j-1)$ for CBR case. The formula for CBR case can be simplified as:

$$R := R - b_i(j-1) \; . \tag{3-81}$$

The occupancy of the HRD buffer $V_i(j)$ is updated after coding each picture as :

$$V_i(1) = \begin{cases} 0 & i = 1 \\ V_{i-1}(N_{i-1}) & otherwise \end{cases}$$

$$V_i(j) = V_i(j-1) + b_i(j-1) - \frac{R_i(j-1)}{f_{FR}} \qquad j = 2,3,...,N_i \,.$$

(3-82)

An initial quantization parameter $Q_{pi}(1)$ is set for the IDR picture and the first stored picture of the i[th] GOP.

For the first GOP, $Q_{p1}(1)$ is pre-defined based on the target compress factor as follows:

$$Q_{p1}(1) = \begin{cases} 40 & bpp \le l1 \\ 30 & l1 < bpp \le l2 \\ 20 & l2 < bpp \le l3 \\ 10 & bpp > l3 \end{cases} \tag{3-83}$$

where $bpp = \frac{R_1(1)}{f_{FR} \times N_{pixel}}$ . (3-84)

$N_{pixel}$ is the number of pixel in a picture. Note that the test model recommends that $l1$=0.15, $l2$=0.45, $l3$=0.9 be used for QCIF/CIF, while $l1$=0.6, $l2$=1.4, $l3$=2.4 be used for picture size larger than CIF.

For the other GOPs, the variance of $Q_{pi}(1)$ from GOP to GOP is limited in [$Q_{pi-1}(1)$ -2, $Q_{pi-1}(1)$ +2] as follows:

$$Q_{pi}(1) = \max\left\{Q_{pi-1}(1)-2, \min\left\{\begin{array}{l} Q_{pi-1}(1)+2, \\ \dfrac{SumPQ_p(i-1)}{N_S(i-1)} - \min\left\{2, \dfrac{N_{i-1}}{15}\right\} \end{array}\right\}\right\}. \tag{3-85}$$

$N_S(i\text{-}1)$ is the total number of stored pictures in the (i-1)$^{th}$ GOP, and $SumPQ_p(i\text{-}1)$ is the sum of average picture quantization parameters for all stored pictures in the (i-1)$^{th}$ GOP. When $Q_{pi-1}(1)$ is still big, it's further adjusted by:

$$Q_{pi}(1) = Q_{pi}(1) - 1 \quad if \quad Q_{pi}(1) > Q_{pi-1}(N_{i-1} - L) - 2. \tag{3-86}$$

$Q_{pi-1}(N_{i-1} - L)$ is the quantization parameter of the last stored picture in the previous GOP, and L is the number of successive non-stored pictures between two stored pictures.

The picture level rate control consists of two stages—pre-encoding and post-encoding.

The objective of the pre-encoding stage is to compute a quantization parameter for each picture, while the post-encoding stage recalculates for uniform slices or MB units to refine quantization values for an allocated target bit rate.

Let's consider the pre-encoding stage, first. Different methods are proposed for stored and non-stored pictures.

For non-stored pictures, the quantization parameters are computed with interpolation as follows:

Suppose that the j$^{th}$ and $(j+L+1)^{th}$ pictures are stored pictures and the quantization parameters for these stored pictures are $Q_{pi}(j)$ and

$Q_{pi}(j+L+1)$, respectively. The quantization parameter of the i[th] non-stored pictures is proposed in the following two cases:

If there is only one non-stored picture between two stored pictures (i.e., L=1), the quantization parameter is given by:

$$Q_{pi}(j+1) = \begin{cases} \frac{Q_{pi}(j)+Q_{pi}(j+2)+2}{2} & \text{if } Q_{pi}(j) \neq Q_{pi}(j+2) \\ Q_{pi}(j)+2 & \text{Otherwise.} \end{cases} \quad \text{(3-87)}$$

If there are more than one non-stored pictures between two stored pictures (i.e., L>1), the quantization parameters are given by:

$$Q_{pi}(j+k) = Q_{pi}(j) + \alpha + \max\{\min\{\frac{(Q_{pi}(j+L+1) - Q_{pi}(j)}{L-1}, 2\times(k-1)\}, -2\times(k-1)\} \quad \text{(3-88)}$$

where $k = 1, \ldots, L$, and $\alpha$ is proposed as:

$$\alpha = \begin{cases} -3 & Q_{pi}(j+L+1) - Q_{pi}(j) \leq -2\times L - 3 \\ -2 & Q_{pi}(j+L+1) - Q_{pi}(j) = -2\times L - 2 \\ -1 & Q_{pi}(j+L+1) - Q_{pi}(j) = -2\times L - 1 \\ 0 & Q_{pi}(j+L+1) - Q_{pi}(j) = -2\times L \\ 1 & Q_{pi}(j+L+1) - Q_{pi}(j) = -2\times L + 1 \\ 2 & \text{Otherwise.} \end{cases} \quad \text{(3-89)}$$

The quantization parameter $Q_{pi}(j+k)$ is further bounded by 0 and 51.

For stored pictures, the quantization parameter is computed through the following two steps – target bits computation and quantization computation.

As the first step, the target bit budget for each stored picture is determined in terms of expected picture bits and expected buffer level from current buffer occupancy as follows:

The target buffer level $S_i(j)$ is pre-defined for each stored picture according to the coded bits of the first IDR picture/ the first stored picture and the average picture complexity. After coding the first stored picture in the i[th] GOP, the initial value of target buffer level is set to:

$$S_i(2) = V_i(2). \tag{3-90}$$

The target buffer level for the subsequent stored picture is determined by:

$$S_i(j+1) = S_i(j) - \frac{S_i(2)}{N_S(i)-1} + \frac{\overline{W}_{P,i}(j)\times(L+1)\times R_i(j)}{f_{FR}\times(\overline{W}_{P,i}(j)+\overline{W}_{B,i}(j)\times L)} - \frac{R_i(j)}{f_{FR}} \tag{3-91}$$

where $\overline{W}_{P,i}(j)$ and $\overline{W}_{B,i}(j)$ are the average complexity weight of stored pictures and the average complexity weight of non-stored pictures, respectively. They are given by:

$$\begin{aligned}
\overline{W}_{P,i}(j) &= \frac{W_{P,i}(j)}{8} + \frac{7\times\overline{W}_{P,i}(j-1)}{8} \\
\overline{W}_{B,i}(j) &= \frac{W_{B,i}(j)}{8} + \frac{7\times\overline{W}_{B,i}(j-1)}{8} \\
W_{P,i}(j) &= b_i(j)\times Q_{pP,i}(j) \\
W_{B,i}(j) &= \frac{b_i(j)\times Q_{pB,i}(j)}{1.3636}.
\end{aligned} \tag{3-92}$$

When there is no non-stored picture between two stored pictures, Equation (3-91) is simplified as

$$S_i(j+1) = S_i(j) - \frac{S_i(2)}{N_S(i)-1}. \tag{3-93}$$

The target bits $T_i(j)$ for the current stored picture can be determined based on the target buffer level $S_i(j)$ derived. The target bits $\widetilde{T}_i(j)$ allocated for the j[th] stored picture in the i[th] GOP are determined based on the target buffer level Equation (3-94), the frame rate $f_{FR}$, the target bit rate $R_i(j)$, and the HRD buffer occupancy $V_i(j)$ as follows:

$$\widetilde{T}_i(j) = \frac{R_i(j)}{f_{FR}} + \gamma\times(S_i(j)-V_i(j)) \tag{3-94}$$

where γ is a constant and its typical value is 0.5 if there are no non-stored pictures. Otherwise, γ is 0.25.

The target bits $\hat{T}_i(j)$ considered with the number of remaining bits are given as follows:

$$\hat{T}_i(j) = \frac{W_{P,i}(j-1) \times R}{W_{P,i}(j-1) \times N_{S,r} + W_{B,i}(j-1) \times N_{N,r}} \tag{3-95}$$

where $N_{S,r}$ and $N_{N,r}$ are the number of the remaining stored pictures and the number of the remaining non-stored pictures, respectively.

The target bits in effect are proposed as a weighted combination of $\tilde{T}_i(j)$ and $\hat{T}_i(j)$ as follows:

$$T_i(j) = \beta \times \hat{T}_i(j) + (1-\beta) \times \tilde{T}_i(j) \tag{3-96}$$

where $\beta$ is a constant and its typical value is 0.5 for the no non-stored picture case. Otherwise, $\beta$ is 0.9.

To conform with the HRD requirement, the target bits are bounded by

$$\begin{aligned} T_i(j) &= \max\{Z_i(j), \mathrm{T_i}(j)\} \\ T_i(j) &= \min\{U_i(j), \mathrm{T_i}(j)\} \end{aligned} \tag{3-97}$$

where $Z_i(j)$ and $U_i(j)$ are given by:

$$Z_i(j) = \begin{cases} R_{left} + \dfrac{R_i(j)}{f_{FR}} & j=1 \\ Z_i(j-1) + \dfrac{R_i(j)}{f_{FR}} - b_i(j) & otherwise \end{cases} \tag{3-98}$$

$$U_i(j) = \begin{cases} (R_{left} + t_{r,1}(1)) \times \varpi & j=1 \\ U_i(j-1) + (\dfrac{R_i(j)}{f_{FR}} - b_i(j)) \times \varpi & otherwise. \end{cases} \tag{3-99}$$

$R_{left}$ is the leftover bits from the last GOP to the current GOP at the first picture in the current GOP, and $t_{r,1}(1)$ is the removal time of the first picture from the coded picture buffer. $\varpi$ is a constant with typical value of 0.9.

As the second step, the quantization parameter is determined and Rate Distortion Optimization (RDO) is performed as follows:

The MAD of the current stored picture, $\tilde{\sigma}_i(j)$, is predicted by a linear model Equation (3-100) using the actual MAD of the previous stored picture, $\sigma_i(j-1-L)$.

$$\tilde{\sigma}_i(j) = a_1 \times \sigma_i(j-1-L) + a_2 \tag{3-100}$$

where $a_1$ and $a_2$ are two coefficients. The initial value of $a_1$ and $a_2$ are set to 1 and 0, respectively. They are updated by a linear regression method similar to that of MPEG-4 Part 2 after coding each picture.

The quantization step corresponding to the target bits is then computed by using the following quadratic as shown in Figure 3-27:

$$T_i(j) = c_1 \times \frac{\tilde{\sigma}_i(j)}{Q_{step,i}(j)} + c_2 \times \frac{\tilde{\sigma}_i(j)}{Q^2_{step,i}(j)} - m_{h,i}(j) \tag{3-101}$$

where $m_{h,i}(j)$ is the total number of header bits and motion vector bits, $c_1$ and $c_2$ are two coefficients.

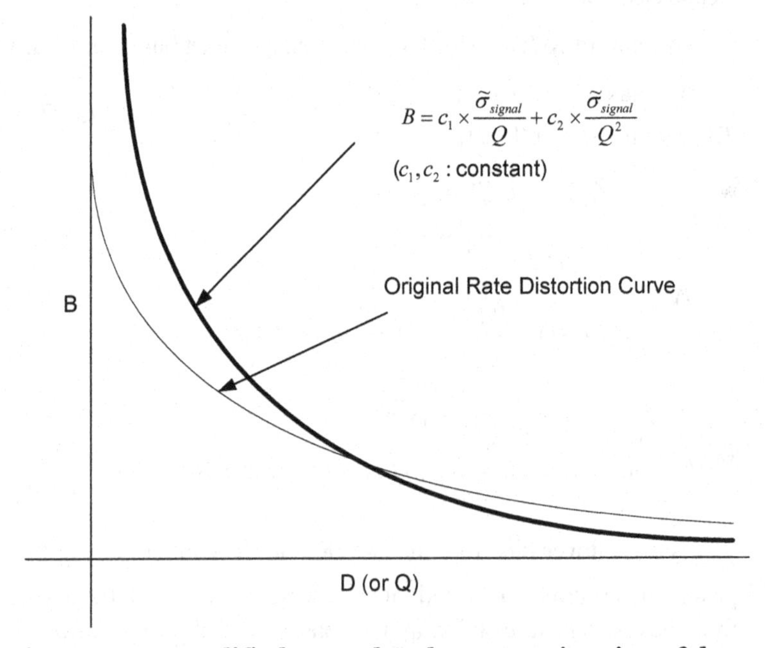


**Figure 3-27 A Modified Second Order Approximation of the Inverse Distortion Measure**

The corresponding quantization parameter $Q_{pi}(j)$ is computed by using the relationship between the quantization step and the quantization parameter of H.264. To maintain the smoothness of visual quality among successive frames, the quantization parameter $Q_{pi}(j)$ is limited in $[Q_{pi}(j-L-1)\text{-}2,\ Q_{pi}(j-L-1)\text{+}2]$ as follows:

$$Q_{pi}(j) = \min\{Q_{pi}(j-L-1)+2, \max\{Q_{pi}(j-L-1)-2, Q_{pi}(j)\}\}. \tag{3-102}$$

The final quantization parameter is further bounded by 0 and 51. The quantization parameter is then used to perform RDO for each MB in the current frame.

After pre-encoding a picture, the parameters *a1* and *a2* of linear prediction model, as well as $c_1$ and $c_2$ of quadratic rate distortion model are updated for post-encoding stage. A linear regression method similar to MPEG-4 Part 2 is used to update these parameters. The actual bits generated are added to the buffer with consideration of the updated buffer occupancy to be not too high.

Let's consider the post-encoding stage from now on. Uniform slice rate control handles bits under the picture level. Suppose that a picture is composed of $N_{mbpic}$ MBs. A uniform slice is defined to be a group of continuous MBs, and consists of $N_{mbunit}$ MBs, where $N_{mbunit}$ is a fraction of $N_{mbpic}$. If $N_{mbunit}$ equals to $N_{mbpic}$, it would be a picture level rate control, and if $N_{mbunit}$ equals to 1, it falls back to a MB level rate control.

The total number of uniform slices in a frame, $N_{unit}$, is given by

$$N_{unit} = \frac{N_{mbpic}}{N_{mbunit}}. \tag{3-103}$$

The idea of uniform slice level rate control is the same as that of picture level rate control except that the concept of $Q_{pi}(j)$ and $Q_{pi}(j+L+1)$ is replaced by the average values of quantization parameters of all uniform slices in the corresponding picture. The uniform slice level rate control selects the values of quantization parameters of all uniform slices in a frame, so that the sum of generated bits is close to the frame target $T_i(j)$.

To do so, the prediction of the MADs, $\tilde{\sigma}_{l,i}(j)$, of the remaining uniform slices in the current stored picture is performed by the model Equation (3-100) using the actual MADs of the co-located uniform slices in the previous stored picture. Then, computation of the number of texture bits $\hat{b}_l$ for the $l^{th}$ uniform slice is performed. The idea is to distribute the leftover bits of a frame in the remaining uniform slices based on a square distortion measure due to general characteristics assumed as $B \propto 1/D^2$. To compute the target bits for the $l^{th}$ uniform slice, let $T_r$ denote the number of remaining bits for the current frame. Initially, $T_r$ is set to $T_i(j)$. The target bits for the $l^{th}$ uniform slice are approximated by:

$$\tilde{b}_l = T_r \times \frac{\tilde{\sigma}_{l,i}^2(l)}{\sum_{k=l}^{N_{unit}} \tilde{\sigma}_{k,i}^2(j)}. \tag{3-104}$$

Since header portion is to be subtracted from $\tilde{b}_l$, the average number of header bits generated by all coded uniform slices is computed as follows:

$$\begin{aligned} \tilde{m}_{hdr,l} &= \tilde{m}_{hdr,l-1} \times (1 - \frac{1}{l}) + \frac{\hat{m}_{hdr,l}}{l} \\ m_{hdr,l} &= \tilde{m}_{hdr,l} \times \frac{l}{N_{unit}} + \overline{m}_{hdr,1} \times (1 - \frac{l}{N_{unit}}), \qquad 1 \le l \le N_{unit} \end{aligned} \tag{3-105}$$

where $\hat{m}_{hdr,l}$ is the actual number of header bits generated by the $l^{th}$ uniform slice in the current stored picture. $\overline{m}_{hdr,1}$ is the estimation from all uniform slices in the previous stored picture. The number of texture bits in effect for the $l^{th}$ uniform slice is given by:

$$\hat{b}_l = \tilde{b}_l - m_{hdr,l}. \tag{3-106}$$

Finally, the quantization step for the $l^{th}$ uniform slice of $j^{th}$ picture in $i^{th}$ GOP is computed by using the new quadratic rate distortion model Equation (3-101) and it is converted to the corresponding quantization parameter $Q_{pl,i}(j)$ by using the method provided by JM of H.264. For the first uniform slice in the current frame,

$$Q_{pl.i}(j) = \overline{Q}_{pi}(j - L - 1) \quad (3\text{-}107)$$

where $\overline{Q}_{pi}(j - L - 1)$ is the average value of quantization parameters for all uniform slices in the previous stored picture. When the number of remaining bits is less than 0, the quantization parameter should be greater than that of previous uniform slice such that the sum of generated bits is close to the target bits, i.e.

$$Q_{pl,i}(j) = Q_{pl-1,i}(j) + \Delta_{Bu} \quad (3\text{-}108)$$

where $\Delta_{Bu}$ is the varying range of the quantization parameter among the uniform slices, the initial value of $\Delta_{Bu}$ is 1 if $N_{unit}$ is greater than 8, and 2 otherwise. It is updated after coding each uniform slice as follows:

$$\Delta_{Bu} = \begin{cases} 1; & \text{if } Q_{pl-1,i}(j) > 25 \\ 2; & \text{otherwise.} \end{cases} \quad (3\text{-}109)$$

To maintain the smoothness of perceptual quality, the quantization parameter is further bounded by:

$$Q_{pl,i}(j) = \max\{0, \overline{Q}_{pi}(j - L - 1) - \Delta_{Fr}, \min\{51, \overline{Q}_{pi}(j - L - 1) + \Delta_{Fr}, Q_{pl,i}(j)\}\} \quad (3\text{-}110)$$

where $\Delta_{Fr}$ is the varying range of quantization parameter among frames, and is defined by:

$$\Delta_{Fr} = \begin{cases} 2; & \text{if } N_{unit} > 18 \\ 4; & \text{if } 18 \geq N_{unit} > 9 \\ 6; & \text{otherwise.} \end{cases} \quad (3\text{-}111)$$

Otherwise, we shall first compute a quantization step by using the quadratic model Equation (3-101) and convert it into the corresponding quantization parameter $Q_{pl,i}(j)$. To provide smoothness of the quality, it is bounded by:

$$Q_{pl,i}(j) = \max\{Q_{pl-1,i}(j) - \Delta_{Bu}, \min\{Q_{pl.i}(j), Q_{pl-1,i}(j) + \Delta_{Bu}\}\}. \quad (3\text{-}112)$$

Once parameter $Q_{pl,i}(j)$ is obtained, RDO for all MBs in the current picture is performed [wiegand:rate, ortega:optimal, ramchandran:optimal]. To obtain a good trade-off between average PSNR and bit fluctuation, $N_{mbunit}$ is recommended to be the number of MBs in a row for field coding, adaptive field/frame coding, or MB-AFF coding, and $N_{unit}$ is recommended to be 9 for other cases.

The main means of JM RDO for MB level rate control is the *Lagrangian multiplier* method. Consider K source samples that are collected in the *K-tuple* $S=(S_1, S_2, ...., S_K)$. In practice, the elements $S_1, S_2$...etc could be thought of as the 1st coding block, the 2nd coding block,.. etc. Generally, a source sample $S_K$ can be a scalar or vector. Each source sample $S_K$ can be quantized using several possible coding options that are indicated by an index out of the set $O_k=(O_{k1}, O_{k2}, ...., O_{kN_k})$. In practice, the elements $O_{k1}, O_{k2}$ ....etc can be thought of as "intra 16x16 prediction mode," "intra 4x4 prediction mode," "inter 8x4 prediction mode,"...etc. Let $I_k \in O_k$ be the selected index to code $S_K$. Then, the coding options assigned to the elements in S are given by the components in the *K-tuple* $I=(I_1, I_2, ...., I_K)$. In practice, the elements $I_1, I_2$ ....etc are the coding mode index mapped for blocks from the first coding block (i.e., block 1) to the last coding block (i.e., block K).

The problem of finding the combination of coding option set *I* that minimizes the distortion for the given sequence of source samples subject to a given rate constraint $R_C$ can be formulated as:

$$\begin{aligned} &\min_{I} \quad D(S,I) \\ &subject \quad to \quad R(S,I) \le R_C . \end{aligned} \tag{3-113}$$

Here, $D(S,I)$ and $R(S,I)$ represent the total distortion and rate, respectively, resulting from quantization of *S* with particular combination of coding options *I*. Note that quantization parameter $Q_{pl,i}(j)$ is excluded from this formulation because the proposed RDO performs only after a specific quantization parameter is selected in higher level rate control portions.

The constrained optimization problem defined in Equation (3-113) can be re-formed to an equivalent unconstrained optimization problem as follows:

$$I^* = \arg\min J(S, I \mid \lambda)$$
$$with \quad J(S, I \mid \lambda) = D(S, I) + \lambda \times R(S, I). \qquad (3\text{-}114)$$

Here, the multiplier $\lambda \geq 0$ is called "*Lagrange multiplier.*"

When distortion and rate measures can be assumed "additive" and "independent" (i.e., only dependent on the choice of parameter corresponding to each sample), then a much simpler *Lagrangian* cost function can be used as follows:

$$J(S_k, I \mid \lambda) = J(S_k, I_k \mid \lambda). \qquad (3\text{-}115)$$

Then, an important reduction in the additive measure takes place in the optimization problem as follows:

$$\min_I \sum_{k=1}^{K} J(S_k, I \mid \lambda) = \sum_{k=1}^{K} \min_{I_k} J(S_k, I_k \mid \lambda). \qquad (3\text{-}116)$$

This implies that independent optimized selection of coding option for each block is equivalent to global optimized selection of coding option for the entire blocks set.

The formulation customized in JM with MB mode $I_k$ is as follows:

$$J_{MODE}(S_k, I_k \mid Q_p, \lambda_k) = D_{REC}(S_k, I_k \mid Q_p) + \lambda_{MODE} \times R_{REC}(S_k, I_k \mid Q_p). \qquad (3\text{-}117)$$

Here, $\lambda_{MODE}$ and $Q_p$ are given as prior knowledge and the elements of coding mode options are provided mainly with INTRA-4x4, INTRA-16x16, SKIP, INTER-16x16, INTER-16x8, INTER-8x16 and INTER-8x8. JM additionally provides the following set of sub-MB types for each 8x8 sub-MB of a P slice MB that is coded in INTER-8x8 mode: INTER-8x8, INTER-8x4, INTER-4x8 and INTER-4x4.

The actual distortion measure for Intra mode is the Sum of Squared Differences (SSD) as follows:

$$SSD = \sum_{(x,y)\in A} | s[x,y,t] - s'[x,y,t] |^2 . \qquad (3\text{-}118)$$

The actual distortion measure for Inter mode is the Displaced Frame Differences (DFD) as follows:

$$DFD = \sum_{(x,y)\in A} | s[x,y,t] - s'[x-m_x, y-m_y, t-m_t] |^p . \quad (3\text{-}119)$$

The value $p$ can be chosen as a coding parameter.

The choice of $\lambda_{MODE}$ is suggested based on experiments as follows:

$$\lambda_{MODE} = 0.85 \times 2^{(Q_p - 12)/3}$$
$$\lambda_{MOTION} = \begin{cases} \sqrt{\lambda_{MODE}} & p = 1 \\ \lambda_{MODE} & p = 2. \end{cases} \quad (3\text{-}120)$$

Note that $R_{REC}(S_k, SKIP \mid Q_p)$ and $D_{REC}(S_k, SKIP \mid Q_p)$ do not depend on the current quantization parameter.

## 3.6 Bandwidth Panic Mode in VC-1

Since VC-1 originated from WMV-9, it has inherited features for bandwidth adaptation under non-QoS scenarios. Microsoft had to develop a kind of rate control techniques to handle urgent bandwidth panic situations because the Internet in nature cannot guarantee allocated bandwidth during a session time for video and audio communications. Two tools are designed for urgent bandwidth hazard – Range Reduction and Multi-resolution. With these two tools, extreme low bitrate of VC-1 bitstreams can be generated almost immediately.

### Range Reduction (or PreProc)

The key idea of Range Reduction (or sometimes called PreProc through previous several versions of WMV-9) is to modify the probability distribution of input Symbols. When the dynamic range of input intensity is smaller, a better compression performance is generally achieved. If urgent bandwidth shrink happens, the mapping of original video to modified video is performed before encoding as shown in Figure 3-28. This process is performed in the encoder and the RANGEREDFRM flag in Picture header is set equal to 1 in Simple Profile and Main Profile. The inverse operation to scale up the decoded current picture shall be performed at the decoder, while keeping the current reconstructed picture intact. The standard defines the inverse operation as follows:

$$Y[n] = Clip((Y[n]-128)\times 2+128), \quad (3\text{-}121)$$

$$C_b[n] = Clip((C_b[n]-128)\times 2+128), \quad (3\text{-}122)$$

$$C_r[n] = Clip((C_r[n]-128)\times 2+128). \quad (3\text{-}123)$$

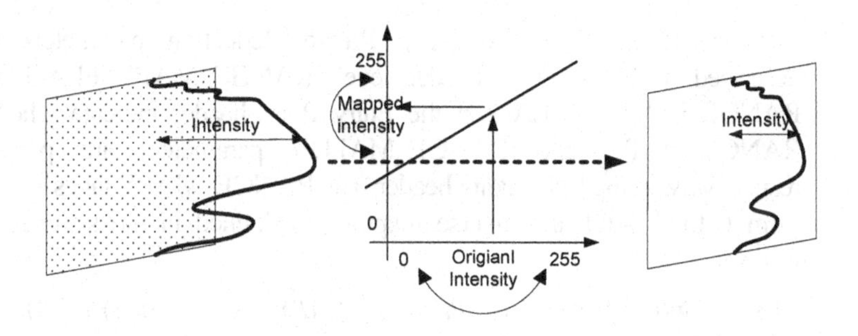


**Figure 3-28 Range Reduction**

In addition, the previously reconstructed reference pictures shall be scaled through Equation (3-121) ~ Equation (3-126) prior to using it for Motion Compensation (MC). Since the RANGEREDFRM flag is described in Picture header of SP or MP, the Range Reduction parameter must be considered picture-by-picture. The scaling process shall be applied to the reconstructed picture as the first stage of decoding – the standard defines following two operations to handle these cases:

- The current picture's RANGEREDFRM is signaled and the previous picture's RANGEREDFRM is not signaled. In this case, the previously reconstructed picture shall be scaled down as follows:

$$Y[n] = ((Y[n] - 128) >> 1) + 128, \tag{3-124}$$

$$C_b[n] = ((C_b[n] - 128) >> 1) + 128, \tag{3-125}$$

$$C_r[n] = ((C_r[n] - 128) >> 1) + 128. \tag{3-126}$$

- The current picture's RANGEREDFRM is not signaled and the previous picture's RANGEREDFRM is signaled. In this case, the previously reconstructed picture shall be scaled up through Equation (3-121) ~ Equation (3-123).

Note that entire process can be lossy since more than two different values can be mapped to a single value in the encoder due to the range reduction function as shown in Figure 3-28.

For Advanced Profile (AP), Range Reduction parameters are described in Entry Point header level. RANGE_MAPY_FLAG and RANGE_MAPUV_FLAG in the Entry Point header indicate whether RANGE_MAPY and RANGE_MAPUV parameters are present, respectively, in the Entry Point header. The RANGE_MAPY takes a value from 0 to 7. And, the inverse mapping is defined in the standard as follows:

$$Y[n] = Clip((((Y[n] - 128) \times (RANGE\_MAPY + 9) + 4) >> 3) + 128). \tag{3-127}$$

The RANGE_MAPUV takes a value from 0 to 7. And, the inverse mapping is defined in the standard as follows:

$$C_b[n] = Clip((((C_b[n] - 128) \times (RANGE\_MAPUV + 9) + 4) >> 3) + 128), \tag{3-128}$$

$$C_r[n] = Clip((((C_r[n] - 128) \times (RANGE\_MAPUV + 9) + 4) >> 3) + 128). \tag{3-129}$$

Note that AP generalizes the range mapping functions as shown in Equation (3-127) ~ Equation (3-129). When RANGE_MAPY=2 and RANGE_MAPUV=2, Equation (3-127) ~ Equation (3-129) correspond conceptually to Equation (3-121) ~ Equation (3-123), respectively.

### Multi-resolution

The key idea of Multi-resolution is to down-size the input video before compression. If urgent bandwidth shrink happens, the mapping of original video to modified video is performed before encoding as shown in Figure 3-29. This process is performed in the encoder and the MULTIRES flag in Picture header is set equal to 1 in SP and MP. There are four options to change the size – no down scaling, horizontal down scaling only, vertical down scaling only, and both horizontal and vertical down scaling. The decision about a proper mode is made in the encoder based on urgent bandwidth needs. When input video is scaled down at the beginning, the size of generated bitstreams can be pretty small. The quality is expected to be low, though. The inverse operation to scale up the decoded current picture shall be performed at the decoder.

RESPIC is a 2-bit syntax element that shall be present in progressive I and P pictures, in SP and MP, if MULTIRES=1. This syntax element specifies the scaling factor of current picture relative to the full resolution picture as shown in Table 3-9. The RESPIC syntax element of a P picture header shall carry the same value as the RSPIC syntax element of the closest preceding I picture. In other words, the resolution of an I picture determines the resolution of all subsequent P pictures until the next I picture.

**Table 3-9 Progressive RESPIC Code Table**

| RESPIC FLC | Horizontal Scale | Vertical Scale |
|---|---|---|
| 00 | Full | Full |
| 01 | Half | Full |
| 10 | Full | Half |
| 11 | Half | Half |

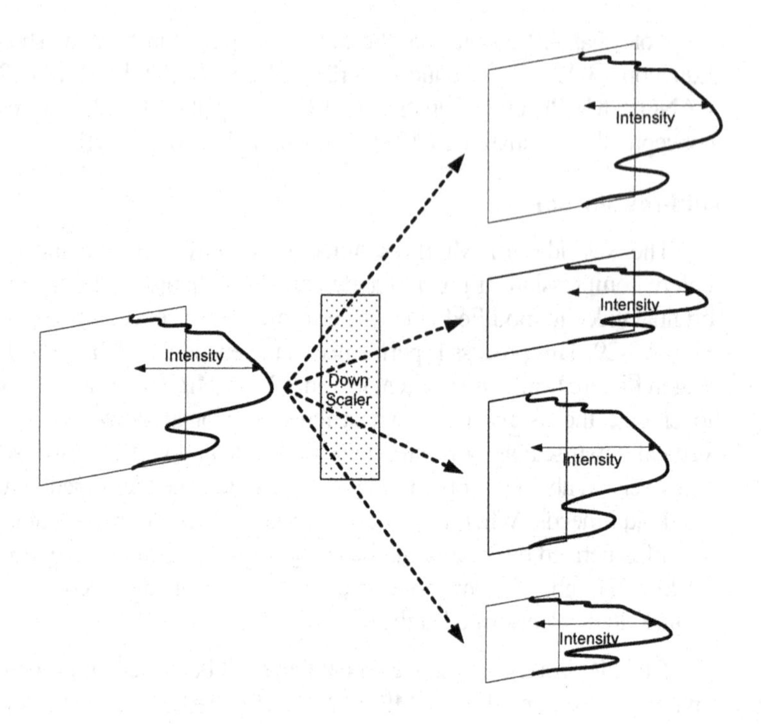


**Figure 3-29 Multi-resolution**

# 4. Transform and Quantization

## 4.1 Transform Coding

### Signal Decomposition and Contrast Sensitivity

Modern video compression technology takes advantage of characteristics of human visual systems [netravali:DIP, haskell:MPEG2, rao:techniques, mitchell:MPEG]. The key to this approach is to decompose input signals as shown in Figure 4-1, Figure 4-2 and Figure 4-3. Figure 4-1 depicts a way to decompose a signal into the first and second halves of the time domain. Figure 4-2 illustrates a signal decomposed into its DC and differential. Figure 4-3 shows the decomposition of a signal into several pre-defined sinusoidal basis signals, where the amplitudes of each basis constitute a linear combination to form the original signal. Without signal decomposition, compression would depend heavily on quantization, where extraction of characteristics in human visual systems cannot be systematically organized. Therefore, the stage of decomposing a signal is a must-have block in modern compression technology. Then, the real question now becomes which decomposition is most desirable.

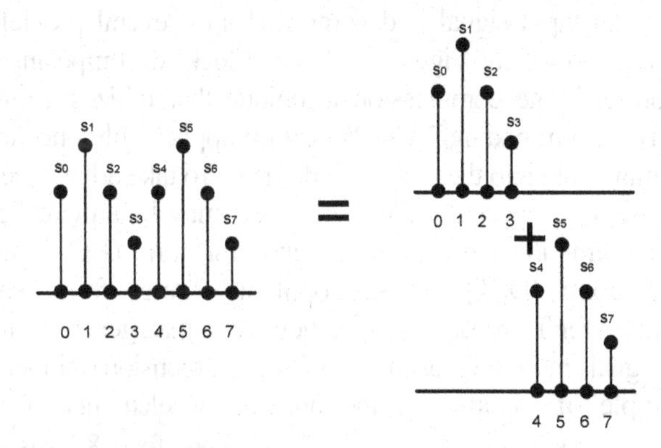


**Figure 4-1 An Example of Signal Vector Decomposition (1)**

Recent research about human visual systems shows that human observers are much more sensitive to lower frequencies than higher ones, as is often indicated by experiments in "Contrast Sensitivity." In this sense, Figure 4-2 type decomposition and Figure 4-3 type decomposition are good candidates. Note that "DC" or "average" means low frequency signal, while "differential" means high frequency signal in Figure 4-2. Decomposition schemes such as pyramid or wavelet fall into this category, where the low frequency component is repeatedly decomposed further [vetterli:wavelet, martucci:zerotree].

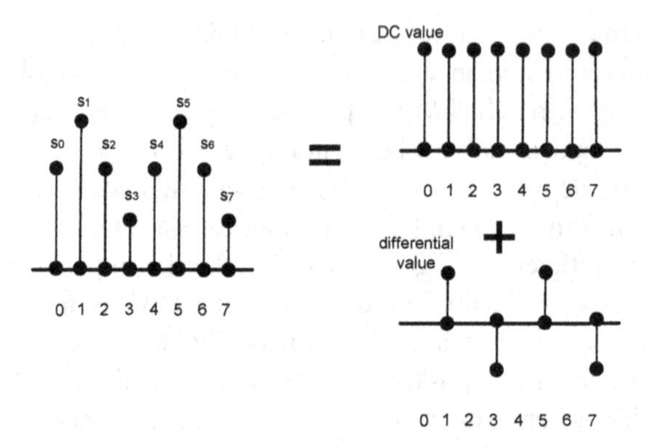


**Figure 4-2 An Example of Signal Vector Decomposition (2)**

When an input signal is decomposed into several pre-defined basis signals as shown in Figure 4-3, the block decomposing is called "Transformer." The compression technique that utilizes Transformer is called "Transform coding." The Transform applies only once to bring the pixel domain data into the transform domain. To take advantage of human visual systems, a spectrally analyzing basis may be an excellent choice. Such transforms include Discrete Cosine Transform (DCT) and Discrete Sine Transform (DST), whose popularity comes from real number manipulation for Transform computation. Note that, generally, the number of basis signals required is as the same as the dimension of input signals. In the example of Figure 4-3, the number of elements of the signal $(s_0 \quad s_1 \quad s_2 \quad s_3 \quad s_4 \quad s_5 \quad s_6 \quad s_7)$ is 8. Therefore, 8 basis signals are required for the decomposition of any given signal to be complete.

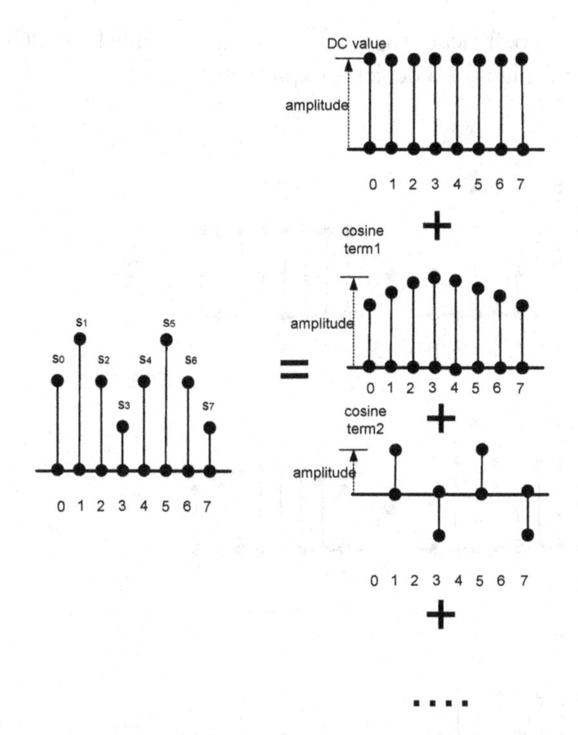


**Figure 4-3 An Example of Signal Vector Decomposition (3)**

### Basis and Extraction of Frequency Components

The frequency components can be extracted through an inner product operation between the input signal and the basis signal as shown in Figure 4-4. Generally, the outcome of the inner product operation can be thought of as the correlation coefficient that tells how close the patterns of two data series look. If a basis looks similar to the input signal, the result of the inner product would be a relatively large number. If not, such a value would be relatively small. On top of that, if the cross inner product between any two different basis signals becomes "0," such a set of basis signals can be thought of as being independently extracting the essence of each frequency component from the input signal. In other words, the process can be interpreted to extract each frequency component with each "corresponding" basis from the input signal. When a basis set gives the perfectly matched "frequency component" in the Transform domain as the correlation coefficient, such a Transform is called an "Orthonormal Transform." Figure 4-4 depicts this property.

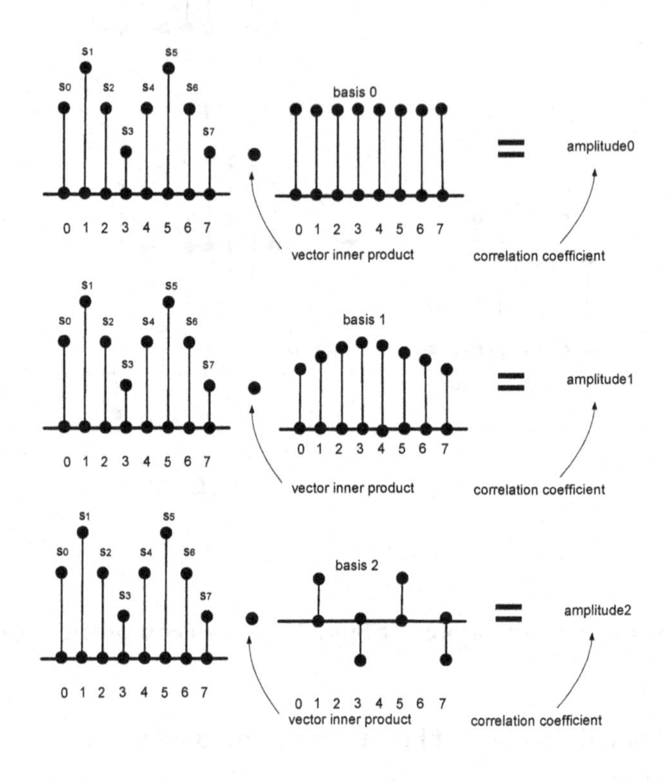


**Figure 4-4 Extraction of Frequency Components based on Inner Product**

An Orthonormal Transform is defined to have the inverse Transform as its transposed Transform. In other words:

$$T^{-1} = T^{t}. \tag{4-1}$$

Such Transforms have interesting characteristics. First, orthonormal Transforms preserve the norm of the original signal even in the Transform domain. Note that the norm measure is the square root of the energy measure. Second, the linear combination of each weighted basis with corresponding Transform coefficient results in perfect recovery of the original signal as shown in Figure 4-4. Third, when a certain number of coefficients are discarded from the Transform domain, the recovered (i.e., inverse Transformed) data with such loss gives the minimum Mean Square Error (MSE) compared with the original data among all Transform candidates.

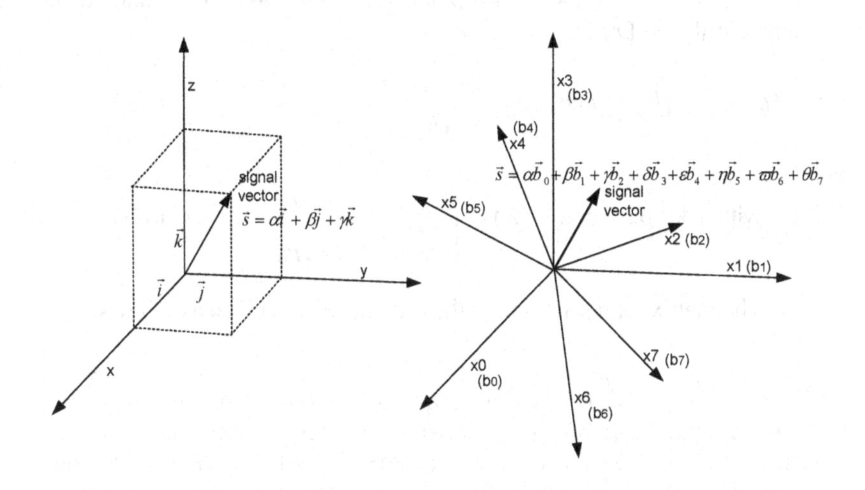


**Figure 4-5 Dimension and the Number of Basis Vectors**

The 3-dimensional signal vectors require three independent basis signals. Figure 4-5 illustrates any vector represented in the space with $\vec{i}$ =(1 0 0), $\vec{j}$ =(0 1 0) and $\vec{k}$ = (0 0 1). Unit vectors can be towards any direction as long as the three of them are linearly independent. The following three vectors can also span 3-dimensional space – $\vec{i}' = \frac{1}{\sqrt{6}}$(2 1 1), $\vec{j}' = \frac{1}{\sqrt{6}}$(1 2 1) and $\vec{k}' = \frac{1}{\sqrt{6}}$ (1 1 2). The same argument holds true with 8-dimensional signal vectors as shown in Figure 4-1, Figure 4-2, Figure 4-3 and Figure 4-4. Any signal vector $(s_0 \ s_1 \ s_2 \ s_3 \ s_4 \ s_5 \ s_6 \ s_7)$ can be represented in the 8-dimensional space with eight linearly independent basis vectors

$\vec{b}_0$, $\vec{b}_1$, $\vec{b}_2$, $\vec{b}_3$, $\vec{b}_4$, $\vec{b}_5$, $\vec{b}_6$, $\vec{b}_7$. In most international video or image compression standards , DCT is selected as such a basis set due to its superior energy compaction property into low signal bands in the case of typical visual signals.

## Discrete Cosine Transform

The definition of DCT for 8-points is proposed as follows [ahmed:dct, ahmed:orth, rao:DCT]:

$$F(k) = \frac{1}{2} C[k] \sum_{n=0}^{7} f(n) \cos \frac{(2n+1)k\pi}{16} \tag{4-2}$$

$$\text{with n, k =0,..,7 and } C(k) = \begin{cases} \frac{1}{\sqrt{2}} & k = 0 \\ 1 & otherwise. \end{cases} \tag{4-3}$$

The matrix representation of the 8-point DCT is given as follows:

$$\begin{bmatrix} \frac{1}{\sqrt{2}} & \frac{1}{\sqrt{2}} & \frac{1}{\sqrt{2}} & \frac{1}{\sqrt{2}} & \frac{1}{\sqrt{2}} & \frac{1}{\sqrt{2}} & \frac{1}{\sqrt{2}} & \frac{1}{\sqrt{2}} \\ c(\pi/16) & c(3\pi/16) & c(5\pi/16) & c(7\pi/16) & c(9\pi/16) & c(11\pi/16) & c(13\pi/16) & c(15\pi/16) \\ c(2\pi/16) & c(6\pi/16) & c(10\pi/16) & c(14\pi/16) & c(18\pi/16) & c(22\pi/16) & c(26\pi/16) & c(30\pi/16) \\ c(3\pi/16) & c(9\pi/16) & c(15\pi/16) & c(21\pi/16) & c(27\pi/16) & c(33\pi/16) & c(39\pi/16) & c(45\pi/16) \\ c(4\pi/16) & c(12\pi/16) & c(20\pi/16) & c(28\pi/16) & c(36\pi/16) & c(44\pi/16) & c(52\pi/16) & c(60\pi/16) \\ c(5\pi/16) & c(15\pi/16) & c(25\pi/16) & c(35\pi/16) & c(45\pi/16) & c(55\pi/16) & c(65\pi/16) & c(75\pi/16) \\ c(6\pi/16) & c(18\pi/16) & c(30\pi/16) & c(42\pi/16) & c(54\pi/16) & c(66\pi/16) & c(78\pi/16) & c(90\pi/16) \\ c(7\pi/16) & c(21\pi/16) & c(35\pi/16) & c(49\pi/16) & c(63\pi/16) & c(77\pi/16) & c(91\pi/16) & c(105\pi/16) \end{bmatrix}$$

Here the notation $c(x)$ means $\cos(x)$. The pictorial description of such basis signals is as shown in Figure 4-6. Note that the basis set is arranged in such a way that the lowest frequency component maps to the index 0 Transform coefficient, while the highest frequency component maps to the index 7 Transform coefficient – this is actually the order of human visual systems' sensitivity.

There are two main benefits to adopting DCT. First, DCT is asymptotically equivalent to Karhunen-Loeve Transform (KLT) for a stationary first-order Markov process source with its correlation coefficient approaching one (i.e., a good real world assumption for typical video signals). Second, DCT is still one of best Transforms after Motion Estimation (i.e., residual data distribution of Laplacian). This implies that

DCT is one of the best basis signals in Motion Compensated Transform coding schemes.

The 8-point IDCT is defined analytically in conjunction with the definition of the aforementioned DCT as:

$$f(n) = \frac{1}{2}\sum_{k=0}^{7} C[k]F(k)\cos\frac{(2n+1)k\pi}{16}. \tag{4-4}$$

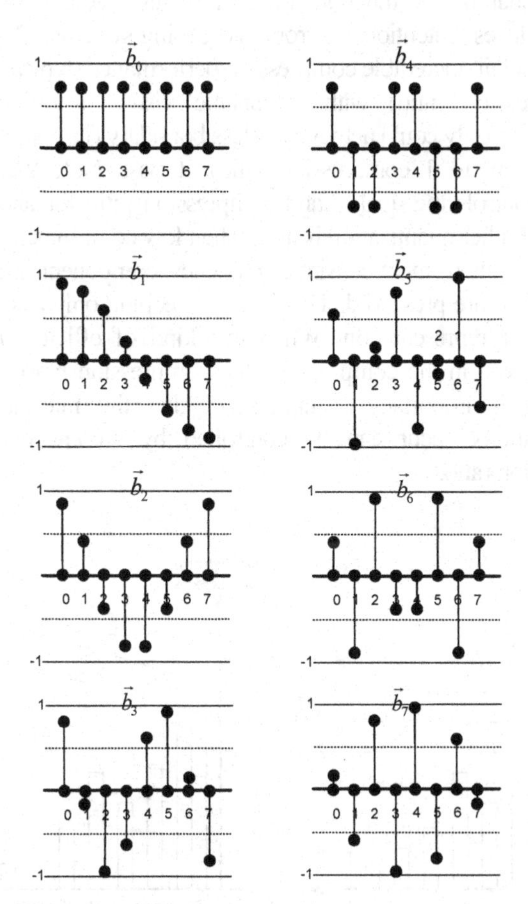

**Figure 4-6 8-point DCT Basis Vectors**

## Quantization and Visual Weighting

In conjunction with signal decomposition methods, the visual weighting unit is another key part of modern video coding systems. In other words, quantization needs to apply proportionally to different frequency components since the signal is already decomposed in the order of sensitivity to human visual systems. The quantization process provides a means for visual weighting.

The quantization function is the only legitimate part of the algorithm that introduces "intentional" errors into coding systems. The reason to do this is to obtain incredible compression performance (typically 100~200 to 1 compression ratio) with reasonable visual distortion. Otherwise, compression only could achieve lossless boundary (i.e., entropy – which is typically few to 1 compression ratio) at best. Note that quantization directly controls the quality and compression ratio for applications. For example, higher quantization is used when low communication bandwidth is allowed. Then, most sensitive frequency components (low frequency components) are preserved. This leads to keeping only a few coefficients for efficient representation. When this kind of efficient representation happens more in the compression, the compression ratio goes up much higher. It is important to understand that the rate at which such representations occur can be controlled by an encoder to improve compression ratio.

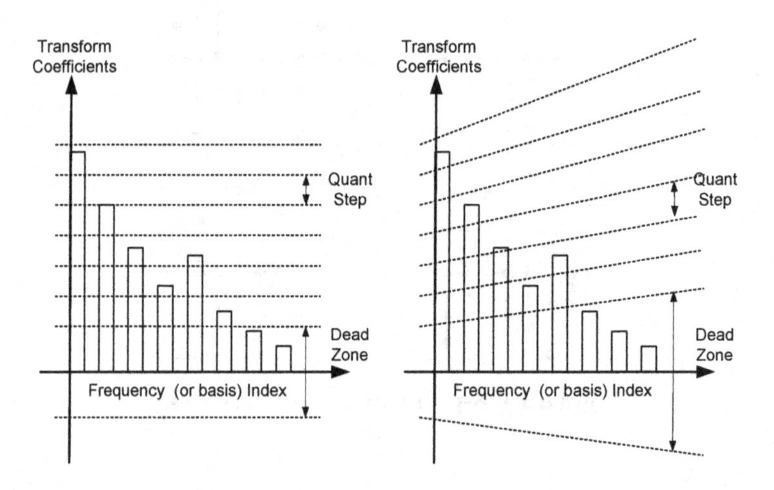


**Figure 4-7 Visual Weighting on Different Frequency Bands**

The quantized values of Transform coefficients are obtained with nearest integer( $F(k)/Q_{step}$ ). To this end, there are two factors considered as shown in Figure 4-7 – (Quantization Stepsize, DeadZone). Any Transform coefficient that is between two lines in Figure 4-7 would correspond to the same integer value. With visual weighting factors that are differently defined with each frequency component, the visually weighted quantized values of Transform coefficients can be obtained with nearest integer( $(F(k)\times c)/(Q_{step}\times w(k))$ ), where $w(k)$ increases w.r.t. $k$ and is pre-defined as shown in Figure 4-7. Here, constant $c$ is conveniently used to be able to represent weighting factors $w(k)$ only in the integer domain. The effect of such processing is to introduce more quantization errors into higher frequency components, but also allows components to be represented by far fewer bits. Visualization of such processing on 2-dimemsional data is shown in Figure 4-8.

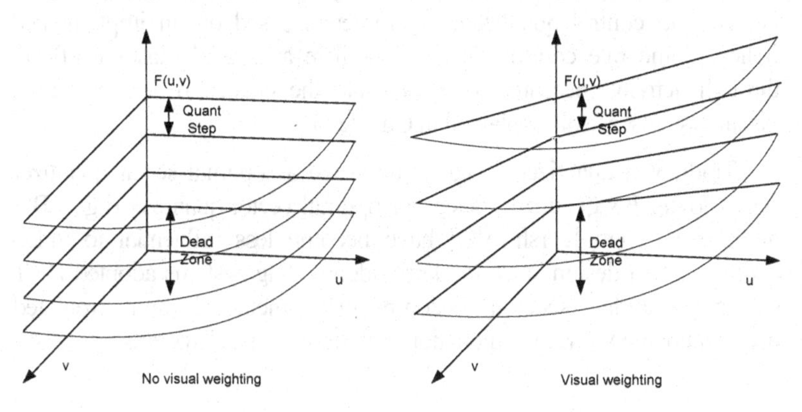


**Figure 4-8 Visual Weighting on 2D Transform Domain Data**

Note that the quantization step size is not fixed for all MBs. Rather, the quantization step size changes potentially at every MB. To be more specific, there are 31 different quantization methods in MPEG-2 for example, each of which is a uniform quantizer. Any specific quantization method is selected based on Quantization (Q) value at every MB. This idea came from "adaptive quantization" from DPCM. Since statistics of an image are not stationary, selecting quantization step size adaptively based on local statistics often produces better quality. Therefore, the Q value is selected at every MB independently. A Q=1 produces the smallest quantization errors, while Q=31 produces the largest quantization errors on average. The encoder and decoder pair adaptively selects one of 31 uniform quantization methods at each MB based on a rate control policy. In fact, a key issue of how to achieve a given compression ratio is connected with the problem of how to allocate a given bit budget into the spatio-temporal domain of the input video. A major means to resolve these issues is to control quantization parameters based on an implemented policy. Some rate control algorithms utilize a "spatial masking effect" through activity measures to incorporate the characteristics of human visual systems, as was explained in Chapter 3.

Traditional quantizer design issues such as optimal scalar quantizer (i.g., Lloyd-Max quantizer) design or optimal vector quantizer (i.g., LBG quantizer) design [gersho:VQ] have become less influential to video coding system design since modern video coding systems adopted DCT (or any similar spectral decomposition methods) and weighted quantization methods in conjunction with human visual systems.

## DCT and IDCT in MPEG-2

The DCT is separable. In other words, the 2-dimensional DCT can be obtained by computing 1-dimensional DCT in x and y directions separately.

The 8x8 DCT is defined as:

$$F(u,v) = \frac{1}{4} C[u]C[v] \sum_{y=0}^{7} \sum_{x=0}^{7} f(x,y) \cos\frac{(2x+1)u\pi}{16} \cos\frac{(2y+1)v\pi}{16} \quad (4\text{-}5)$$

with u, v, x, y =0,..,7 and $C(u), C(v) = \begin{cases} \frac{1}{\sqrt{2}} & u, v = 0 \\ 1 & otherwise. \end{cases}$

(4-6)

The input to the forward transform and output from the inverse transform is represented with 9 bits, where the dynamic range of the pixels is [-256:255].

The 8x8 IDCT is defined in conjunction with the definition of aforementioned DCT as:

$$f(x,y) = \frac{1}{4}\sum_{u=0}^{7}\sum_{v=0}^{7} C[u]C[v]F(u,v)\cos\frac{(2x+1)u\pi}{16}\cos\frac{(2y+1)v\pi}{16}.$$

(4-7)

## Fast Implementation of DCT and IDCT

The computation involved in DCT is huge since 30 frames a second, for example, must be handled. Therefore, it has been an important subject to reduce the DCT computation, while the same DCT function is performed. Interestingly, the original DCT can be implemented with a couple of successive stages of sparse matrices, thus leading to a much smaller number of multiplications and additions. Such a method is called "Fast DCT." Let's define *x(n)* and *X(k)* as input and output of a 1-dimensional DCT block. New variables y and z (and d) are defined as intermediate variables for convenience as shown in Figure 4-9. Then,

$$\begin{bmatrix} X(0) \\ X(2) \\ X(4) \\ X(6) \end{bmatrix} = \begin{bmatrix} a & a & a & a \\ c & f & -f & -c \\ a & -a & -a & a \\ f & -c & c & -f \end{bmatrix} \begin{bmatrix} x(0)+x(7) \\ x(1)+x(6) \\ x(2)+x(5) \\ x(3)+x(4) \end{bmatrix} \qquad (4\text{-}8)$$

and

$$\begin{bmatrix} X(1) \\ X(3) \\ X(5) \\ X(7) \end{bmatrix} = \begin{bmatrix} b & d & e & g \\ d & -g & -b & -e \\ e & -b & g & d \\ g & -e & d & -b \end{bmatrix} \begin{bmatrix} x(0)-x(7) \\ x(1)-x(6) \\ x(2)-x(5) \\ x(3)-x(4) \end{bmatrix} \qquad (4\text{-}9)$$

where $a = \frac{1}{2}\cos(\frac{\pi}{4})$, $b = \frac{1}{2}\cos(\frac{\pi}{16})$, $c = \frac{1}{2}\cos(\frac{\pi}{8})$, $d = \frac{1}{2}\cos(\frac{3\pi}{16})$, $e = \frac{1}{2}\cos(\frac{5\pi}{16})$, $f = \frac{1}{2}\cos(\frac{3\pi}{8})$, $g = \frac{1}{2}\cos(\frac{7\pi}{16})$.

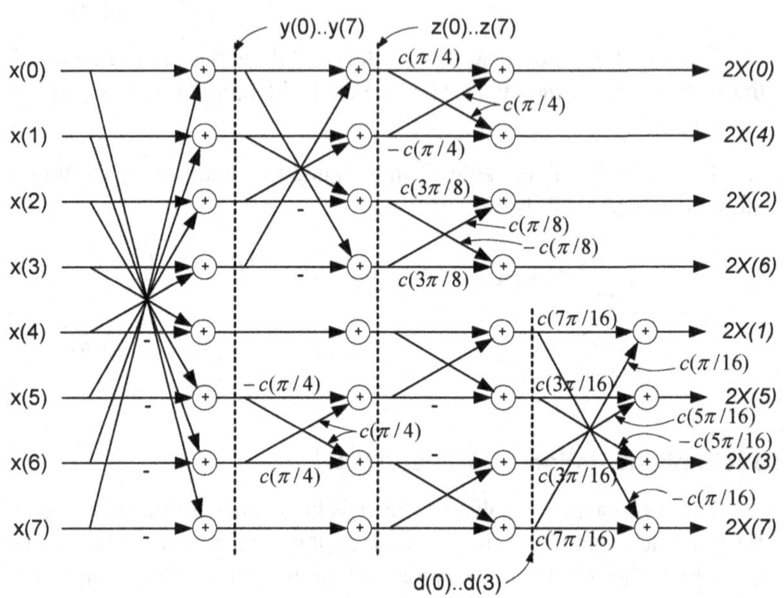


**Figure 4-9 Fast Implementation of 8-point Forward DCT based on Chen's Algorithm**

The four elements of the first matrix equation can lead to:

$$X(0) + X(4) = 2ay(0) + 2ay(3), \quad (4\text{-}10)$$

$$X(0) - X(4) = 2ay(1) + 2ay(2), \quad (4\text{-}11)$$

$$X(2) + X(6) = (f + c)(y(0) - y(3)) + (f - c)(y(1) - y(2)), \quad (4\text{-}12)$$

$$X(2) - X(6) = (c - f)(y(0) - y(3)) + (f + c)(y(1) - y(2)), \quad (4\text{-}13)$$

where $y(0) \equiv x(0) + x(7)$, $y(1) \equiv x(1) + x(6)$, $y(2) \equiv x(2) + x(5)$ and $y(3) \equiv x(3) + x(4)$. (4-14)

In matrix convention, these can be represented with:

$$\begin{bmatrix} X(0)+X(4) \\ X(0)-X(4) \end{bmatrix} = \begin{bmatrix} 2a & 0 \\ 0 & 2a \end{bmatrix} \begin{bmatrix} y(0)+y(3) \\ y(1)+y(2) \end{bmatrix} \tag{4-15}$$

and $$\begin{bmatrix} X(2)+X(6) \\ X(2)-X(6) \end{bmatrix} = \begin{bmatrix} c+f & f-c \\ c-f & c+f \end{bmatrix} \begin{bmatrix} y(0)-y(3) \\ y(1)-y(2) \end{bmatrix}. \tag{4-16}$$

The first two lines of the linear equations are added to get $X(0)$, while the same two lines of linear equations are subtracted to get $X(4)$. Then,

$2X(0) = 2a \cdot z(0) + 2a \cdot z(1)$ and $2X(4) = 2a \cdot z(0) - 2a \cdot z(1)$ , where $z(0) \equiv y(0) + y(3)$ and $z(1) \equiv y(1) + y(2)$. (4-17)

Similar modification for the later two lines of linear equations are performed to obtain:

$2X(6) = 2c \cdot z(3) + 2c \cdot z(2)$ and $2X(6) = 2f \cdot z(3) - 2f \cdot z(2)$ , where $z(3) \equiv y(0) - y(3)$ and $z(2) \equiv y(1) - y(2)$. (4-18)

Similar derivation for the four elements of the second matrix equation can be performed with $y(4) \equiv x(3) - x(4)$ , $y(5) \equiv x(2) - x(5)$ , $y(6) \equiv x(1) - x(6)$ , $y(7) \equiv x(0) - x(7)$ , $z(4) \equiv x(4)$ , $z(5) \equiv -2a \cdot y(5) + 2a \cdot y(6)$ , $z(6) \equiv 2a \cdot y(5) + 2a \cdot y(6)$ , $z(7) \equiv y(7)$ , $d(0) \equiv z(4) + z(5)$ , $d(1) \equiv z(4) - z(5)$ , $d(2) \equiv z(6) + z(7)$ and $d(3) \equiv z(6) - z(7)$. (4-19)

These all relationships can be drawn in a flow graph as shown in Figure 4-9.

Note that the Transform coefficients are generated with a multiplication factor of two through Chen's Fast DCT algorithm as shown in Figure 4-9. This is because certain early fast algorithms utilized the definition of the forward DCT with the scale factor $\frac{1}{2}$ dropped, while such a scale factor is compensated for the Inverse DCT. Fast DCT algorithms such as Chen or Lee fall into this category [chen:fastdct, lee:fastdct].

A fast IDCT can be easily obtained though its transposed graph as shown in Figure 4-10. This is true since the original transform is linear and orthonormal. The linear flow graph, where Mason's Gain formula holds, can lead to the transposed graph by reversing arrow direction with a signal branch becoming signal addition and signal addition becoming a signal branch in the transposed graph, respectively. Since DCT is an orthonormal transform where its inverse transform can be easily obtained with a transposed Transform of original one, the transposed graph directly implies the IDCT fast algorithm.

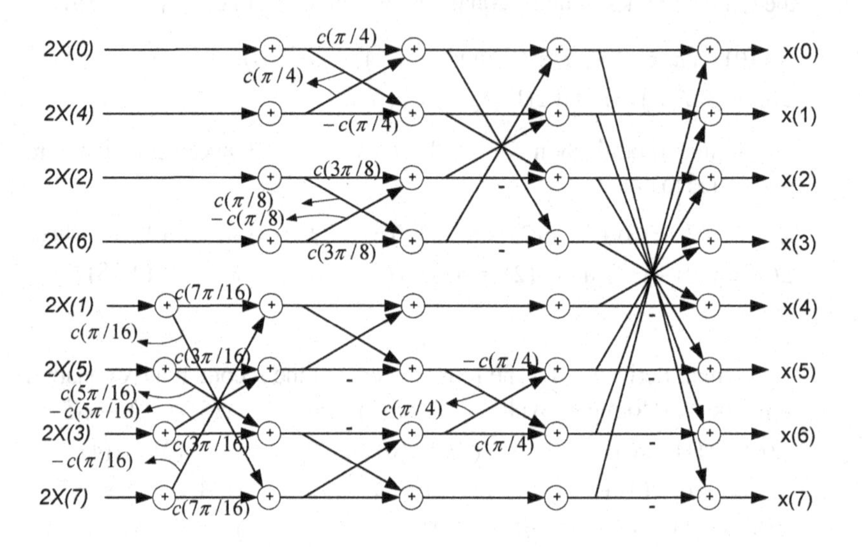


**Figure 4-10 Fast Implementation of 8-point Inverse DCT based on Chen's Algorithm**

Since the RTL implementation of Chen's algorithm is highly regular, it is a preferred architecture in hardware implementation. Even though the implementation requires more multiplications than other flavors of the algorithm, the regularity helps design complexity and provides better resource sharing.

### Encoder and Decoder Drift

The prediction mechanism generally requires reference data that is used to generate a predictor. To make the same reference available at both the encoder side and the decoder side, the encoder must handle/ generate reference data very carefully – hence, the encoder contains the decoder in the loop as depicted in Figure 4-11. Otherwise, both ends use drifted (i.e., different) reference data for the prediction. This kind of potential problem, where the output and reference data of the encoder are different from the output and reference data of the decoder, is called "drift."

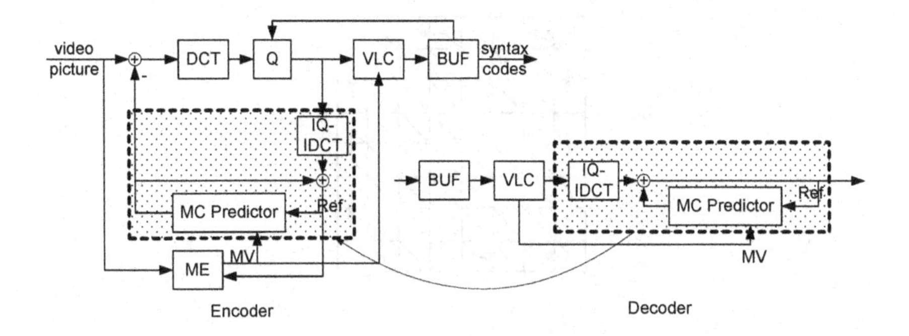


**Figure 4-11 Drift in MPEG-2**

However, there are potential problems in DCT, MC, etc. regarding drift issues in conventional video coding systems. DCT and IDCT are defined in the standards as "closed" forms that involve infinite resolution of cosine basis computations. In real world implementation, the basis must have finite resolutions by rounding or truncation. If the IDCT block in the encoder loop is different from that in the decoder (block diagrams shown in Figure 4-11), reference data generated from the encoder and the decoder might be drifting. This aspect is much amplified when the fast IDCT algorithm in the decoder is different from that of the encoder. This is quite possible since there are so many fast IDCT implementations in the market. The interpolation operation in MC implementation could be a potential source for drift as well, if not manipulated cautiously. Some standards utilize digital FIR filters, whose filter coefficients require floating operation, to interpolate artificial pixels in a higher resolution. High resolution data should be rounded or truncated in the integer pixel domain. Such an operation in the decoder should be exactly matched with that in the encoder. Otherwise, the encoder and the decoder will drift from each other.

## Zig-zag Scan and Inverse Zig-zag Scan

Zig-zag scan maps 2-D transform coefficients to a series of pairs (zero-run, level) or triplets (zero-run, level, last), while inverse zig-zag scan interprets pairs or triplets into 2-D transform coefficients. Figure 4-12 illustrates the MPEG-2 scanning pattern for progressive video input.

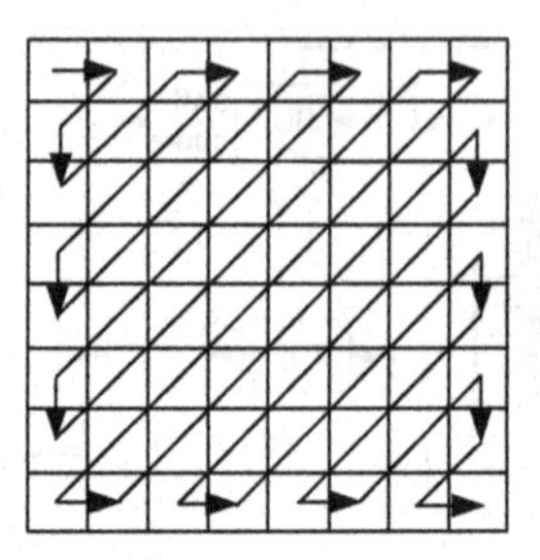

**Figure 4-12 Zig-zag Scan in MPEG-2 Progressive Video**

## Quantization and Inverse Quantization Process

The video standards define only the Inverse Quantization (IQ) Process at the decoder. Therefore, the Quantization (Q) process in the encoder should be inferred from the IQ definition for the decoder. While there is not much freedom for an encoder to utilize different Quantization Arithmetic, the possibility is still open to the encoder algorithm. For example, special truncation methods such as truncation or rounding at 0.4 can be adopted in the encoder for certain statistical input video, while truncation or rounding at 0.5 is typically the most popular one.

The IQ process employs three consecutive stages, generally, as shown in Figure 4-13. Some of the steps can be omitted in certain standards. For example, VC-1 standard doesn't utilize visual weighting and mismatch control.

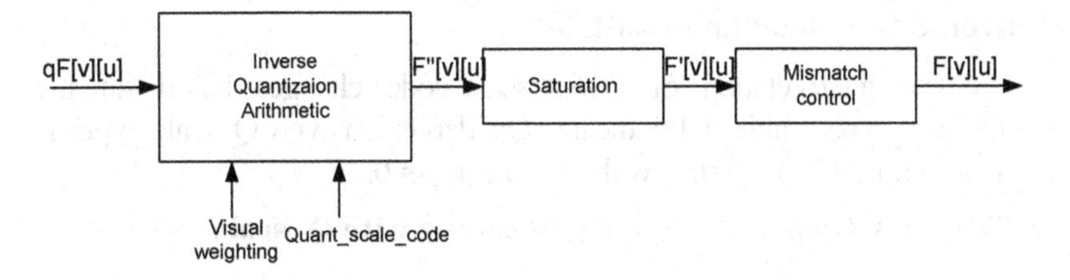


**Figure 4-13 Inverse Quantization Process in MPEG-2**

The coefficients resulting from the IQ Arithmetic are saturated to lie in the range [-2048:+2047]. Thus:

$$F' = \begin{cases} 2047 & F'' > 2047 \\ F'' & -2048 \le F'' \le 2047 \\ -2048 & F'' < -2048 \end{cases} \tag{4-20}$$

For mismatch control, all of the reconstructed, saturated coefficients, $F'$ in the block will be summed. This value will be tested to determine whether it is odd or even. If the sum is even, then a correction will be made to F[7][7]. Thus:

when $sum = \sum_{v=0}^{v<8} \sum_{u=0}^{u<8} F'[v][u]$, (4-21)

$F[v][u] = F'[v][u]$ for all u, v except u=v=7 (4-22)

where

$$F[7][7] = \begin{cases} F'[7][7] & \text{if } sum \text{ is odd} \\ \left.\begin{cases} F'[7][7]-1 & \text{if } F'[7][7] \text{ is odd} \\ F'[7][7]+1 & \text{if } F'[7][7] \text{ is even} \end{cases}\right\} & \text{if } sum \text{ is even} \end{cases}$$

(4-23)

### Inverse Quantization in MPEG-2

The interpretation of Quant_scale_code changes based on the Q_scale_type. Table 4-1 indicates Qp derivation with Q_scale_type=1, while Table 4-2 shows this with Q_scale_type=0.

**Table 4-1 Quant_scale_code Decoding with Q_scale_type=1**

| Quant_scale_code | Qstep | Quant_scale_code | Qstep | Quant_scale_code | Qstep |
|---|---|---|---|---|---|
| 0 | Reserved | 11 | 22 | 22 | 44 |
| 1 | 2 | 12 | 24 | 23 | 46 |
| 2 | 4 | 13 | 26 | 24 | 48 |
| 3 | 6 | 14 | 28 | 25 | 50 |
| 4 | 8 | 15 | 30 | 26 | 52 |
| 5 | 10 | 16 | 32 | 27 | 54 |
| 6 | 12 | 17 | 34 | 28 | 56 |
| 7 | 14 | 18 | 36 | 29 | 58 |
| 8 | 16 | 19 | 38 | 30 | 60 |
| 9 | 18 | 20 | 40 | 31 | 62 |
| 10 | 20 | 21 | 42 | | |

**Table 4-2 Quant_scale_code Decoding with Q_scale_type=0**

| Quant_scale _code | Qstep | Quant_scale _code | Qstep | Quant_scale _code | Qstep |
|---|---|---|---|---|---|
| 0 | Reserved | 11 | 14 | 22 | 48 |
| 1 | 1 | 12 | 16 | 23 | 52 |
| 2 | 2 | 13 | 18 | 24 | 56 |
| 3 | 3 | 14 | 20 | 25 | 64 |
| 4 | 4 | 15 | 22 | 26 | 72 |
| 5 | 5 | 16 | 24 | 27 | 80 |
| 6 | 6 | 17 | 28 | 28 | 88 |
| 7 | 7 | 18 | 32 | 29 | 96 |
| 8 | 8 | 19 | 36 | 30 | 104 |
| 9 | 10 | 20 | 40 | 31 | 112 |
| 10 | 12 | 21 | 44 | | |

Intra-DC Differentials are quantized with a uniform mid-step quantizer with several stepsizes. The Intra-DC is reconstructed with $differential\ DC = qDC \times stepsize$, where $qDC$ is a quantized DC differential value and intra_dc_precision can be 0, 1, 2 and 3 corresponding to stepsizes for 8, 4, 2 and 1, respectively.

The Intra-AC is reconstructed as follows:

$$F'' = (2 \times qF \times Q_{step} \times IntraQmatrx) / 32. \tag{4-24}$$

The non-Intra is reconstructed as follows:

$$F'' = ((2 \times qF + sign(qF)) \times Q_{step} \times NonIntraQmatrx) / 32. \tag{4-25}$$

The inverse quantization method utilizes two weighting matrices – one for Intra, the other for non-Intra.

The default weighting matrix for an Intra and non-intra blocks are shown in Figure 4-14 and Figure 4-15, respectively.

| | | | | | | | |
|---|---|---|---|---|---|---|---|
| 8 | 16 | 19 | 22 | 26 | 27 | 29 | 34 |
| 16 | 16 | 22 | 24 | 27 | 29 | 34 | 37 |
| 19 | 22 | 26 | 27 | 29 | 34 | 34 | 38 |
| 22 | 22 | 26 | 27 | 29 | 34 | 37 | 40 |
| 22 | 26 | 27 | 29 | 32 | 35 | 40 | 48 |
| 26 | 27 | 29 | 32 | 35 | 40 | 48 | 58 |
| 26 | 27 | 29 | 34 | 38 | 46 | 56 | 69 |
| 27 | 29 | 35 | 38 | 46 | 56 | 69 | 83 |

**Figure 4-14 MPEG-2 Intra Visual Weighting Matrix**

| | | | | | | | |
|---|---|---|---|---|---|---|---|
| 16 | 16 | 16 | 16 | 16 | 16 | 16 | 16 |
| 16 | 16 | 16 | 16 | 16 | 16 | 16 | 16 |
| 16 | 16 | 16 | 16 | 16 | 16 | 16 | 16 |
| 16 | 16 | 16 | 16 | 16 | 16 | 16 | 16 |
| 16 | 16 | 16 | 16 | 16 | 16 | 16 | 16 |
| 16 | 16 | 16 | 16 | 16 | 16 | 16 | 16 |
| 16 | 16 | 16 | 16 | 16 | 16 | 16 | 16 |
| 16 | 16 | 16 | 16 | 16 | 16 | 16 | 16 |

**Figure 4-15 MPEG-2 Non-Intra Visual Weighting Matrix**

## 4.2 VC-1 Transform and Quantization

### VC-1 Transform

Unlike the popular 8x8 DCT used in previous standards, the Transforms in VC-1 can be computed exactly in integer arithmetic, thus avoiding inverse Transform mismatch problems. A possible approach is to round the scaled entries of the DCT matrix to nearest integers:

$$H = round\{\alpha H_{DCT}\} \tag{4-26}$$

where $H_{DCT}$ is the DCT matrix.

However, VC-1 targets the capability of a fast algorithm implementation for the "inverse" Transform. A pattern similar to that of Chen's algorithm is preferred in order to implement a fast algorithm. In addition, integer approximation with the scaled entries of the IDCT matrix to nearest integers as follows:

$$H_{inv} = round\{\alpha H_{IDCT}\} \tag{4-27}$$

where $H_{IDCT}$ is the IDCT matrix.

Such an inverse transform was chosen for 1D 8-point transform with:

$$\hat{H}_{8inv} = \begin{bmatrix} 12 & 16 & 12 & 6 \\ 12 & 6 & -12 & -16 \\ 12 & -6 & -12 & 16 \\ 12 & -16 & 12 & -6 \end{bmatrix} \text{ and } \tilde{H}_{8inv} = \begin{bmatrix} 16 & 15 & 9 & 4 \\ 15 & -4 & -16 & -9 \\ 9 & -16 & 4 & 15 \\ 4 & -9 & 15 & -16 \end{bmatrix} \text{ in}$$

$$\begin{bmatrix} x(0) \\ x(1) \\ x(2) \\ x(3) \end{bmatrix} = \hat{H}_{8inv}/8 \begin{bmatrix} X(0) \\ X(2) \\ X(4) \\ X(6) \end{bmatrix} + \tilde{H}_{8inv}/8 \begin{bmatrix} X(1) \\ X(3) \\ X(5) \\ X(7) \end{bmatrix} + \begin{bmatrix} 4 \\ 4 \\ 4 \\ 4 \end{bmatrix} /8 \tag{4-28}$$

$$\text{and } \begin{bmatrix} x(7) \\ x(6) \\ x(5) \\ x(4) \end{bmatrix} = \hat{H}_{8inv}/8 \begin{bmatrix} X(0) \\ X(2) \\ X(4) \\ X(6) \end{bmatrix} - \tilde{H}_{8inv}/8 \begin{bmatrix} X(1) \\ X(3) \\ X(5) \\ X(7) \end{bmatrix} + \begin{bmatrix} 4 \\ 4 \\ 4 \\ 4 \end{bmatrix} /8 \cdot \tag{4-29}$$

Note that the 3rd term is a rounding operation.

Also, an inverse transform was chosen for 1D 4-point transform with:

$$\hat{H}_{4inv} = \begin{bmatrix} 17 & 17 \\ 17 & -17 \end{bmatrix} \text{ and } \tilde{H}_{4inv} = \begin{bmatrix} 22 & 10 \\ 10 & -22 \end{bmatrix} \text{ in} \tag{4-30}$$

$$\begin{bmatrix} x(0) \\ x(1) \end{bmatrix} = \hat{H}_{4inv}/8 \begin{bmatrix} X(0) \\ X(2) \end{bmatrix} + \tilde{H}_{4inv}/8 \begin{bmatrix} X(1) \\ X(3) \end{bmatrix} + \begin{bmatrix} 4 \\ 4 \end{bmatrix} /8 \tag{4-31}$$

$$\text{and } \begin{bmatrix} x(3) \\ x(2) \end{bmatrix} = \hat{H}_{4inv}/8 \begin{bmatrix} X(0) \\ X(2) \end{bmatrix} - \tilde{H}_{4inv}/8 \begin{bmatrix} X(1) \\ X(3) \end{bmatrix} + \begin{bmatrix} 4 \\ 4 \end{bmatrix} /8. \tag{4-32}$$

Note that the 3rd term is a rounding operation.

This leads to a formal "inverse" 2D transform definition that is not identical to IDCT, but has a fast implementation at the decoder as follows:

$$T_8 = \begin{bmatrix} 12 & 12 & 12 & 12 & 12 & 12 & 12 & 12 \\ 16 & 15 & 9 & 4 & -4 & -9 & -15 & -16 \\ 16 & 6 & -6 & -16 & -16 & -6 & 6 & 16 \\ 15 & -4 & -16 & -9 & 9 & 16 & 4 & -15 \\ 12 & -12 & -12 & 12 & 12 & -12 & -12 & 12 \\ 9 & -16 & 4 & 15 & -15 & -4 & 16 & -9 \\ 6 & -16 & 16 & -6 & -6 & 16 & -16 & 6 \\ 4 & -9 & 15 & -16 & 16 & -15 & 9 & -4 \end{bmatrix} \tag{4-33}$$

$$\text{and } T_4 = \begin{bmatrix} 17 & 17 & 17 & 17 \\ 22 & 10 & -10 & -22 \\ 17 & -17 & -17 & 17 \\ 10 & -22 & 22 & -10 \end{bmatrix}. \tag{4-34}$$

The first step to get the inverse Transform is to compute $E_{M\times N} = (D_{M\times N} \cdot T_M + 4) >> 3$, where M and N can be either 4 or 8. The second step is to compute $R_{M\times N} = (T_N^t \cdot E_{M\times N} + C_N \cdot 1_M + 64) >> 7$. Note that $D_{M\times N}$ is input transform domain data that shall not exceed the signed 12-bit range (i.e., $-2048 \leq$ entries of $D_{M\times N} < 2047$ ), while $E_{M\times N}$ is an intermediate matrix whose entries are in the 13-bit range (i.e.,

$-4096 \leq$ entries of $E_{M\times N} < 4095$). $R_{M\times N}$ is output pixel domain data that shall not exceed the signed 10-bit range (i.e., $-512 \leq$ entries of $R_{M\times N} < 511$). The definition of $C_N$ varies with N value as follows: $C_8^t = (0\ 0\ 0\ 0\ 1\ 1\ 1\ 1)$ and $C_4^t = (0\ 0\ 0\ 0)$. The definition of $1_M$ is an M length row vector of "1"s.

The forward Transform is not considered for fast implementation. It can be implemented in scaled integer arithmetic or using a floating point (or other) representation. Unlike the inverse transform, the matrix-multiplication representation of it is purely an analytical representation as follows: $\hat{D} = (T_4 D T_4^t) \circ N_{4\times 4}$, $\hat{D} = (T_8 D T_4^t) \circ N_{8\times 4}$, $\hat{D} = (T_4 D T_8^t) \circ N_{4\times 8}$ and $\hat{D} = (T_8 D T_8^t) \circ N_{8\times 8}$, where the operator $\circ$ is a component-wise multiplication and $\hat{D}$ represents the transform coefficient matrix. The entries of normalization matrix $N_{M\times N}$ are given by $N_{ij} = n_j \cdot n_i^t$, where the column vectors $n$ are given by $n_4^t = (\frac{8}{289}\ \frac{8}{292}\ \frac{8}{289}\ \frac{8}{292})$ and $n_8^t = (\frac{8}{288}\ \frac{8}{289}\ \frac{8}{292}\ \frac{8}{289}\ \frac{8}{288}\ \frac{8}{289}\ \frac{8}{292}\ \frac{8}{289})$. Different normalization factors for each basis signal imply that VC-1 defines Transforms that are only orthogonal, but not orthonormal. Therefore, conventional arguments about orthonormal Transforms do not hold.

### Inverse Quantization in VC-1

There are two inverse quantization methods. The first one is used at the Uniform quantizer, while the second one is used at the Non-uniform quantizer.

First, the Uniform Quantization method is composed of two sub-cases – one for Intra-DCs, the other for all other-ACs.

The Intra-DC is reconstructed as $DC = qDC \times DCStepSize$, (4-35-1)

where $qDC$ is a quantized DC differential value. Here, DCStepSize is as follows:

**Table 4-3 Derivation of DCStepSize for Uniform or Non-Uniform Quantization**

| Block type | MQUANT $\leq 2$ | $3 \leq$ MQUANT $\leq 4$ | $5 \leq$ MQUANT |
|---|---|---|---|
| Luma | $2\times$ MQUANT | 8 | MQUANT/2+6 |
| Chroma | $2\times$ MQUANT | 8 | MQUANT/2+6 |

All other-ACs are reconstructed as follows:

$$F'' = qF \times (2 \times MQUANT + \Delta(syntax)). \quad (4\text{-}36)$$

Here, $\Delta(syntax) =$ HALFQP when PQUANT based decoding is performed, whereas $\Delta(syntax) =$ 0 when VOPDQUANT based decoding is performed. No visual weighting matrices are used.

Second, the Non-Uniform Quantization method is composed of two sub-cases – one for Intra-DC, the other for all other-ACs.

The Intra-DC is reconstructed as $DC = qDC \times DCStepSize$, (4-35-2)

where $qDC$ is a quantized DC differential value. Here, DCStepSize is the same as in Table 4-3.

All other-ACs are reconstructed as follows:

$$F'' = qF \times (2 \times MQUANT + \Delta(syntax)) + sign(qF) \times MQUANT. \quad (4\text{-}37)$$

No visual weighting matrices are used.

The coefficients resulting from the IQ Arithmetic are rounded to the nearest integer and saturated to lie in the range [-2048:2047]. Thus:

$$F = \begin{cases} 2047 & F'' > 2047 \\ F'' & -2048 \le F'' \le 2047 \\ -2048 & F'' < -2048. \end{cases} \quad (4\text{-}38)$$

No Mismatch Control is used for VC-1.

### Inverse Zig-zag Scan in VC-1

The inverse zig-zag scan interprets (zero-run, level, last) triplets into 2-D transform coefficients. Figure 4-16 and Figure 4-17 illustrate VC-1 scanning patterns for progressive video input. In VC-1, there are three different zig-zag patterns used for Intra blocks depending on ACPRED on/off and DC prediction direction as shown in Figure 4-16 – Normal/ Horizontal/ Vertical scan types. However, there is only one scan pattern for each type of Inter blocks – one each for 8x8 Inter, 8x4 Inter, 4x8 Inter, and 4x4 Inter blocks as shown in Figure 4-17.

There are more scan patterns defined in the VC-1 standard, depending on Profile/ Levels. Note that Intra DC is specially treated like in most video coding standards.

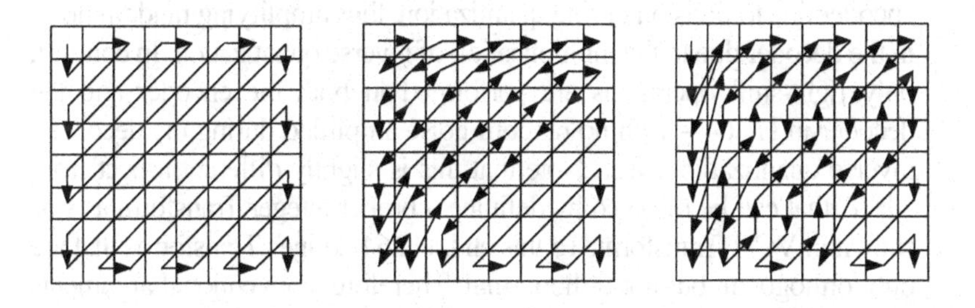

**Figure 4-16 Intra Zig-zag Scan in VC-1 Progressive Video**

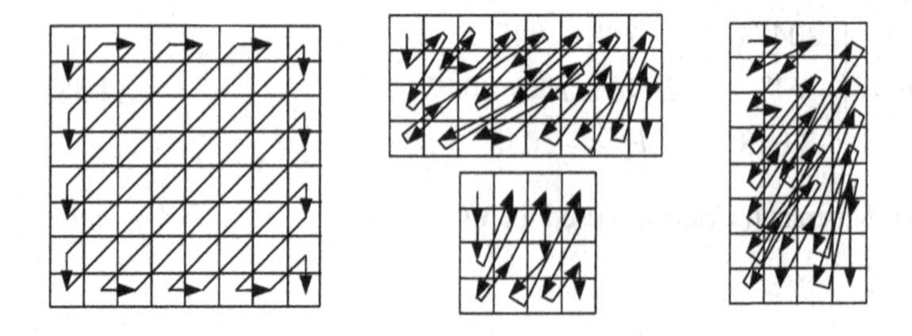

**Figure 4-17 Inter Zig-zag Scan in VC-1 Progressive Video**

## 4.3 H.264 Transform and Quantization

### Transform and Quantization in H.264

A key approach to Transform and Quantization in H.264 is to intentionally magnify transform coefficients in the middle (just before division part of quantization) to involve only right-shift operations in both the encoder and the decoder. This can be thought of as a part of the quantization process. Typically quantization loses some resolution in the encoder due to division during quantization, thus amplifying random noise in the decoder due to the multiplication of inverse quantization. In contrast, only right-shift operations are performed in both the encoder and the decoder in H.264 – with no random noise amplified during the decoder's inverse quantization stage. Note that this is slightly different aspect from "drift" that can be resolved by defining a proper integer Transform as was shown in VC-1 Transform. To this end, H.264 defines Transforms that are only orthogonal, but not orthonormal. Therefore, conventional arguments about orthonormal Transforms do not hold in this section.

### 4x4 Transform of H.264

Unlike the popular 8x8 DCT used in previous standards, the 4x4 transforms in H.264 can be computed exactly in integer arithmetic, thus avoiding inverse transform mismatch problems. The approach is to round the scaled entries of the DCT matrix to nearest integers:

$$H = round\{\alpha H_{DCT}\} \tag{4-39}$$

where $H_{DCT}$ is the DCT matrix. The selection of $\alpha = 2.5$ was proposed in the standard, which leads to a set of coefficients {a=1, b=2, c=1} with consideration of shift-domain approximation, as follows:

$$H_4 = \begin{bmatrix} 1 & 1 & 1 & 1 \\ 2 & 1 & -1 & -2 \\ 1 & -1 & -1 & 1 \\ 1 & -2 & 2 & -1 \end{bmatrix}. \quad (4\text{-}40)$$

This implies that storage of X(k) needs only 6 more bits than x(n) (i.e., $\log_2(6^2) = 5.17$). Note that the rows of $H_4$ are orthogonal but do not have the same norm. However, that can be easily compensated for in the quantization process.

In the decoder, we could use just the transpose of $H_4$ as long as we take care of scaling the reconstructed transform coefficients appropriately, to compensate for the different row norms. The inverse transform matrix is then defined by:

$$H_{4inv} = \begin{bmatrix} 1 & 1 & 1 & 1/2 \\ 1 & 1/2 & -1 & -1 \\ 1 & -1/2 & -1 & 1 \\ 1 & -1 & 1 & -1/2 \end{bmatrix} \quad (4\text{-}41)$$

where $H_{4inv}$ is a scaled inverse of $H_4$. In other words,

$$H_{4inv} diag\{\tfrac{1}{4}, \tfrac{1}{5}, \tfrac{1}{4}, \tfrac{1}{5}\} H_4 = I. \quad (4\text{-}42)$$

Figure 4-18 depicts forward and Inverse transforms of H.264.

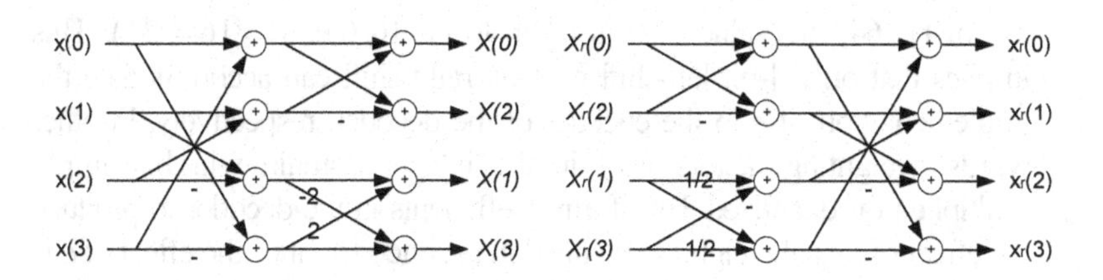


**Figure 4-18 H.264 4-point Forward and Inverse Transforms**

### Visual Weighting of H.264

Visual weighting is included in the High Profile (and above) of H.264 standard dated with March, 2005.

Without visual weighting, the quantized values of Transform coefficients can be obtained in an encoder as follows:

$$qF(k) = \text{Nearest integer}\,(F(k)/Q_{step}), \tag{4-43}$$

where $qF(k)$ is the quantized value of a Transform coefficient $F(k)$ and $k$ is the frequency index.

The inverse operation to obtain a Transform coefficient in a decoder is performed as follows:

$$F(k) = qF(k) \times Q_{step}. \tag{4-44}$$

With visual weighting factors that are differently defined with each frequency component, the visually weighted quantized values of Transform coefficients can be obtained in an encoder as follows:

$$qF(k) = \text{Nearest integer}\,((F(k) \times c)/(Q_{step} \times w(k))), \tag{4-45}$$

where $w(k)$ is weighting factors. Here, constant $c$ is conveniently used to be able to represent weighting factors $w(k)$ only in the integer domain. However, overall effect of a combination of $c$ and $w(k)$ takes on a rational number.

The inverse operation to obtain a Transform coefficient in a decoder is performed as follows:

$$F(k) = qF(k) \times (Q_{step} \times w(k))/c. \tag{4-46}$$

Note that the visual weighting effect can be nullified when $w(k) = c$.

In H.264, the value of $c$ is taken to be 16 (i.e., $c = 16 = 2^4$). This implies that only 4-bit left-shift or 4-bit right-shift can accommodate the implementation of $c$ in the encoder or the decoder, respectively. In other words, weighting factors $w(k)$ in the integer domain will be simply multiplied on quantized Transform coefficients in the decoder to produce the effect of visual weighting planned in the encoder since the effect of $c$ is already accommodated in the shift operation. Such weighting factors, which are referred to as "Scaling List" in the standard, are described in either Sequence level or Picture level.

The seq_scaling_matrix_present_flag equal to 1 specifies that the flags seq_scaling_list_present_flag[i] for i=0..7 are present. The seq_scaling_matrix_present_flag equal to 0 specifies that these flags are not present and the Sequence level scaling list specified by Flat_4x4_16 shall be inferred for i=0..5 and the Sequence level scaling list specified by Flat_8x8_16 shall be inferred for i=6..7. When seq_scaling_matrix_present flag is not present, it shall be inferred to be equal to 0. Here, Flat_4x4_16 and Flat_8x8_16 are defined as follows:

Flat_4x4_16[idx]=16, with idx=0..15, (4-47)

Flat_8x8_16[idx]=16, with idx=0..63. (4-48)

The seq_scaling_list_present_flag[i] in the SPS equal to 1 specifies that the syntax structure for scaling list i is present in the SPS. The flag equal to 0 specifies that the syntax structure for scaling list i is not present in the SPS and the scaling list fall-back rule set A specified in Table 4-4 shall be used to infer the Sequence level scaling list for index i.

**Table 4-4 Assignment of Mnemonic Names to Scaling List Indices and Specification of Fall-back Rule (IS: Table 7-2)**

| Value of scaling list index | Mnemonic Name | Block size | MB pred type | Comp. | Scaling list fall back rule set A | Scaling list fall back rule set B | Default scaling list |
|---|---|---|---|---|---|---|---|
| 0 | SI_4x4_Intra_Y | 4x4 | Intra | Y | default | Seq level | Default_4x4_Intra |
| 1 | SI_4x4_Intra_Cb | 4x4 | Intra | Cb | Scaling list for i = 0 | Scaling list for i = 0 | Default_4x4_Intra |
| 2 | SI_4x4_Intra_Cr | 4x4 | Intra | Cr | Scaling list for i = 1 | Scaling list for i = 1 | Default_4x4_Intra |
| 3 | SI_4x4_Inter_Y | 4x4 | Inter | Y | default | Seq level | Default_4x4_Inter |
| 4 | SI_4x4_Inter_Cb | 4x4 | Inter | Cb | Scaling list for i = 3 | Scaling list for i = 3 | Default_4x4_Inter |
| 5 | SI_4x4 | 4x4 | Inter | Cr | Scaling list | Scaling list | Default_ |

| | _Inter_Cr | | | | for i = 4 | for i = 4 | 4x4_Inter |
|---|---|---|---|---|---|---|---|
| 6 | SI_8x8 _Intra_Y | 8x8 | Intra | Y | default | Seq level | Default_ 8x8_Intra |
| 7 | SI_8x8 _Inter_Y | 8x8 | Inter | Y | default | Seq level | Default_ 8x8_Inter |

Figure 4-19 and Figure 4-20 specify the default scaling matrices Default_4x4_Intra and Default_4x4_Inter, while Figure 4-21 and Figure 4-22 define the default scaling matrices Default_8x8_Intra and Default_8x8_Inter. These may vary in Field or Frame MBs since the construction of visual scaling matrices is scanning pattern dependent.

| | | | |
|---|---|---|---|
| 6 | 13 | 20 | 28 |
| 13 | 20 | 28 | 32 |
| 20 | 28 | 32 | 37 |
| 28 | 32 | 37 | 42 |

**Figure 4-19 H.264 4x4 Intra Visual Scaling Matrix**

| | | | |
|---|---|---|---|
| 10 | 14 | 20 | 24 |
| 14 | 20 | 24 | 27 |
| 20 | 24 | 27 | 30 |
| 24 | 27 | 30 | 34 |

**Figure 4-20 H.264 4x4 Inter Visual Scaling Matrix**

| | | | | | | | |
|---|---|---|---|---|---|---|---|
| 6 | 10 | 13 | 16 | 18 | 23 | 25 | 27 |
| 10 | 11 | 16 | 18 | 23 | 25 | 27 | 29 |
| 13 | 16 | 18 | 23 | 25 | 27 | 29 | 31 |

| 16 | 18 | 23 | 25 | 27 | 29 | 31 | 33 |
|---|---|---|---|---|---|---|---|
| 18 | 23 | 25 | 27 | 29 | 31 | 33 | 36 |
| 23 | 25 | 27 | 29 | 31 | 33 | 36 | 38 |
| 25 | 27 | 29 | 31 | 33 | 36 | 38 | 40 |
| 27 | 29 | 31 | 33 | 36 | 38 | 40 | 42 |

**Figure 4-21 H.264 8x8 Intra Visual Scaling Matrix**

| 9 | 13 | 15 | 17 | 19 | 21 | 22 | 24 |
|---|---|---|---|---|---|---|---|
| 13 | 13 | 17 | 19 | 21 | 22 | 24 | 25 |
| 15 | 17 | 19 | 21 | 22 | 24 | 25 | 27 |
| 17 | 19 | 21 | 22 | 24 | 25 | 27 | 28 |
| 19 | 21 | 22 | 24 | 25 | 27 | 28 | 30 |
| 21 | 22 | 24 | 25 | 27 | 28 | 30 | 32 |
| 22 | 24 | 25 | 27 | 28 | 30 | 32 | 33 |
| 24 | 25 | 27 | 28 | 30 | 32 | 33 | 35 |

**Figure 4-22 H.264 8x8 Non-Intra Visual Scaling Matrix**

The pic_scaling_matrix_present_flag equal to 1 specifies that parameters are present to modify the scaling lists specified in the SPS. The pic_scaling_matrix_present_flag equal to 0 specifies that the scaling lists used for the picture shall be inferred to be equal to those specified by the SPS. When pic_scaling_matrix_present_flag is not present, it shall be inferred to be equal to 0.

The pic_scaling_list_present_flag[i] in the PPS equal to 1 specifies that the scaling list syntax structure is present to specify the scaling list for index i. The flag equal to 0 specifies that the syntax structure for scaling list i is not present in the PPS and that the scaling list fall-back rule set A specified in Table 4-4 shall be used to infer the Picture level scaling list for index i under seq_scaling_matrix_present_flag equal to 0. The scaling list fall-back rule set B specified in Table 4-4 shall be used to infer the Picture level scaling list for index i under seq_scaling_matrix_present_flag equal to 1.

The quantization and the inverse quantization processes are separately defined as explained in the next sections, where division and multiplication of weighting factors in the encoder and the decoder, respectively, are not shown explicitly. Instead, $\widetilde{X}_q(i,j)$ is used in the next sections to represent Transform coefficients multiplied by scaling matrices.

### Quantization of 4x4 Transform

For a given step size, the encoder can perform quantization by:

$$X_q(i,j) = sign\{X(i,j)\}[(|X(i,j)|A(Q_M,i,j) + f \cdot 2^{(15+Q_E)}) \gg (15+Q_E)] \tag{4-49}$$

where i and j are the row and column indices and $Q_M \equiv Q \bmod 6$ with $Q_E \equiv Q/6$. This periodicity enables us to define a large range of quantization parameters without increasing the memory requirements. For example, Q=1 means that $Q_M = 1$ and $Q_E = 0$, while Q=7 implies that $Q_M = 1$ and $Q_E = 1$. The value of $A(Q_M,i,j)$ stays as the same, but one right-shift is triggered with Q of 7. In other words, quantization only requires simple fixed Table for $A(Q_M,i,j)$, while its periodicity has something to do merely with right-shift operation. The parameter $f$ controls the quantization width near the origin (a.k.a., dead zone). This parameter is chosen by the encoder and is typically in the range of 0 to ½. The decoder can perform inverse quantization by:

$$X_r(i,j) = \widetilde{X}_q(i,j)B(Q_M,i,j) \ll Q_E. \tag{4-50}$$

The quantization and reconstruction factors $A(Q_M, i, j)$ and $B(Q_M, i, j)$ depend on the transform coefficient position {i, j} inside the block. This is necessary to compensate for the different scaling factor in the 2-D version of the non-orthonormal property. Their values are given by:

$A(Q_M, i, j) = M(Q_M, r)$ and $B(Q_M, i, j) = S(Q_M, r)$ , where r=0 for (i,j) = {(0,0), (0,2), (2,0), (2,2)}, r=1 for (i,j) = {(1,1), (1,3), (3,1), (3,3)} and r=2 otherwise, with:

$$M = \begin{bmatrix} 13107 & 5243 & 8066 \\ 11916 & 4660 & 7490 \\ 10082 & 4194 & 6554 \\ 9362 & 3647 & 5825 \\ 8192 & 3355 & 5243 \\ 7282 & 2893 & 4559 \end{bmatrix} \text{ and } S = \begin{bmatrix} 10 & 16 & 13 \\ 11 & 18 & 14 \\ 13 & 20 & 16 \\ 14 & 23 & 18 \\ 16 & 25 & 20 \\ 18 & 29 & 23 \end{bmatrix}. \quad (4\text{-}51)$$

These matrices were designed to maximize dynamic range of amplified transform coefficients and to satisfy a relationship of:

$$M(Q_M, r)S(Q_M, r)v(r) \cong 2^{21} \quad (4\text{-}52)$$

where $v(r) = \{4^2, 5^2, 4 \times 5\}$. (4-53)

For example, the first element of M and S gives

$$M(0,0) \times S(0,0) \times 4^2 = 2097120. \quad (4\text{-}54)$$

This number is close enough to $2^{21} = 2097152$. The difference of the two numbers is small enough to cause no effect after right-shift operations both at the encoder and the decoder. All entries of matrices M and S were designed to have the same property. This implies that the total number of right-shift operations taking places at the encoder and the decoder must be 21. For this specific example, the design distributes 15 right-shift operations to the encoder and 6 shift-right operations to the decoder, thus making it 21 shift-right operations overall as shown in Equation (4-49) and (4-55). Note that shift operation involved with $Q_E$ has nothing to do with actual quantization, but something to do with quantization table periodicity.

The final scaling after reconstruction becomes:

$$x_r = (H_{inv} X_r + 2^5 e) >> 6 \quad (4\text{-}55)$$

where $e = [1111]^T$.

From the implementation point of view, total 21 right-shifts can be re-distributed in "three" serial modules without any problems – quantization, inverse quantization and final reconstruction, for example, with 11, 4 and 6, respectively. Then, the equations can be re-written as follows:

$$X_q(i,j) = sign\{X(i,j)\}[(|X(i,j)| A(Q_M, i, j) + f \cdot 2^{(11+Q_E)}) >> (11 + Q_E)] \quad (4\text{-}56)$$

$$X_r(i,j) = \widetilde{X}_q(i,j) B(Q_M, i, j) << (Q_E - 4), \text{ if } Q_E \geq 4, \quad (4\text{-}57)$$

$$X_r(i,j) = \widetilde{X}_q(i,j) B(Q_M, i, j) + 2^{(3-Q_E)} >> (4 - Q_E) \text{ if } Q_E < 4. \quad (4\text{-}58)$$

Since $Q_E \equiv Q/6$, the conditions, $Q_E \geq 4$ and $Q_E < 4$, mean that $Q \geq 24$ and $Q < 24$, respectively. Note that Equation (4-55) remains the same with this re-organization. There are many possible re-organizations that do not impact on final reconstruction Equation (4-55).

It is important to understand that actual quantization implementation in the encoder can take many different forms without disrupting structures of the inverse quantization and the final reconstruction scaling. For example, the quantization Equation (4-49) can still work with the inverse quantization Equation (4-57), Equation (4-58) and the reconstruction Equation (4-55) when certain signal scale processing is carefully considered. Since four more right-shifts are performed in Equation (4-49) compared with Equation (4-56), multiplication of input data by $2^4$ at any point in the signal path through the encoder can justify the use of Equation (4-49) with Equation (4-57), Equation (4-58) and Equation (4-55).

**Table 4-5 Q and its Implication with Qstep**

| Q | Qstep | Q | Qstep | Q | Qstep |
|---|---|---|---|---|---|
| 0 | 0.625 | 9 | 1.75 | 30 | 20 |
| 1 | 0.6875 | 10 | 2 | ... | ... |
| 2 | 0.8125 | 11 | 2.25 | 36 | 40 |
| 3 | 0.875 | 12 | 2.5 | ... | ... |
| 4 | 1 | ... | ... | 42 | 80 |
| 5 | 1.125 | 18 | 5 | ... | ... |
| 6 | 1.25 | ... | ... | 48 | 160 |
| 7 | 1.375 | 24 | 10 | ... | ... |
| 8 | 1.625 | ... | ... | 51 | 224 |

Note that in the H.264 standard only the reconstruction formulas are specified since the standard specifies only the decoder, not the encoder. Practical implication of Q is shown in Table 4-5.

## 8x8 Transform of H.264

The 4-point DCT decomposes input signals into 4 equidistant spectral bands, while the 8-point DCT decomposes input signals into 8 equidistant spectral bands. Since the quantization introduces errors in selected spectral bands, 8-point DCT has twice quantization immunity against quantization errors at a specific band compared with 4-point DCT. This is an important aspect that helps high quality reconstruction at high target bitrates. Let's say that the highest frequency component in theory gets damaged due to quantization. Then, equidistant 1/8 band at the highest spectral position would get distorted through 8-point DCT, while equidistant 1/4 band (i.e., double bandwidth) at the highest spectral position would get distorted through 4-point DCT. This property plays an important role for "visual quality saturation effect" at further increasing a target bitrate.

The recent High Profile (and above) introduces 8x8 transform back to H.264. While the MB size remains at 16x16, these are divided up into 4x4 or 8x8 blocks. Then, a 4x4 or 8x8 block transform matrix is applied to every block of pixels in the MB. The new 8x8 transform closely approximates the 8x8 DCT, while its inverse transform is defined in the "transposed" manner. In case of 8x8 transform, the transform matrix is given by:

$$H_8 = \begin{bmatrix} 8 & 8 & 8 & 8 & 8 & 8 & 8 & 8 \\ 12 & 10 & 6 & 3 & -3 & -6 & -10 & -12 \\ 8 & 4 & -4 & -8 & -8 & -4 & 4 & 8 \\ 10 & -3 & -12 & -6 & 6 & 12 & 3 & -10 \\ 8 & -8 & -8 & 8 & 8 & -8 & -8 & 8 \\ 6 & -12 & 3 & 10 & -10 & -3 & 12 & -6 \\ 4 & -8 & 8 & -4 & -4 & 8 & -8 & 4 \\ 3 & -6 & 10 & -12 & 12 & -10 & 6 & -3 \end{bmatrix} \cdot 1/8$$

(4-59)

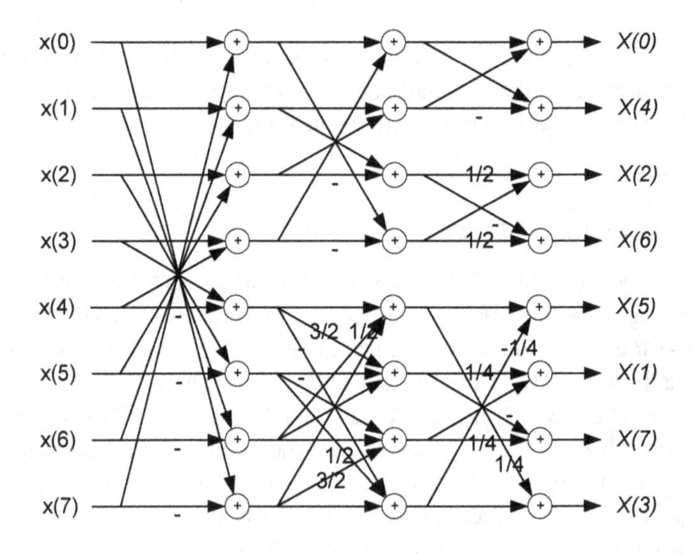


**Figure 4-23 H.264 8-point Fast Forward Transform**

A fast forward transform is shown in Figure 4-23, which was originally demonstrated in JVT [gordon:simplified]. Note that forward transform implementation is not defined in the standard. Figure 4-23 shows only one example of many possible implementations. From the standard perspective, it is only the inverse transform that matters. The inverse 8x8 transform is given by the transposed forward transform as follows:

$$H_{8inv} = \begin{bmatrix} 8 & 12 & 8 & 10 & 8 & 6 & 4 & 3 \\ 8 & 10 & 4 & -3 & -8 & -12 & -8 & -6 \\ 8 & 6 & -4 & -12 & -8 & 3 & 8 & 10 \\ 8 & 3 & -8 & -6 & 8 & 10 & -4 & -12 \\ 8 & -3 & -8 & 6 & 8 & -10 & -4 & 12 \\ 8 & -6 & -4 & 12 & -8 & -3 & 8 & -10 \\ 8 & -10 & 4 & 3 & -8 & 12 & -8 & 6 \\ 8 & -12 & 8 & -10 & 8 & -6 & 4 & -3 \end{bmatrix} \cdot 1/8$$

(4-60)

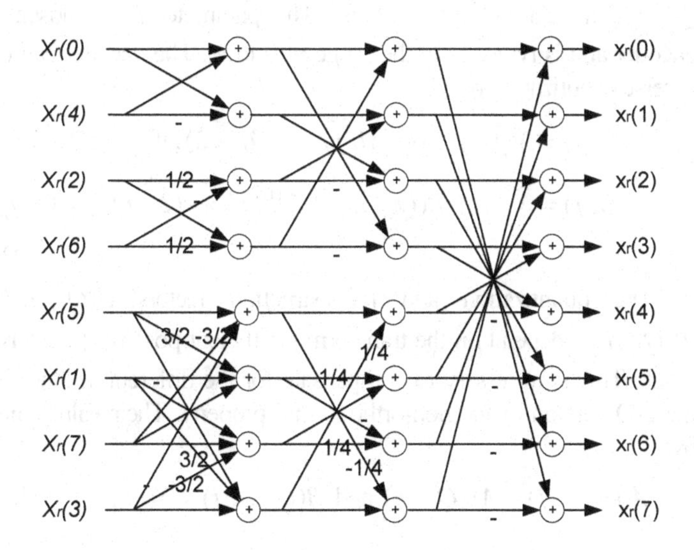


**Figure 4-24 H.264 8-point Fast Inverse Transform**

A fast inverse transform is shown in Figure 4-24, which was originally demonstrated in JVT. The forward and inverse transforms shown in Equation (4-59) and Equation (4-60) are only orthogonal, not orthonormal. Normality is achieved through properly chosen normalization factors as shown in next sections. Note that Figure 4-23 and Figure 4-24 are not transposed graphs since Figure 4-23 is shown as an implementation example. On top of that, notice that 8-point forward and Inverse transforms can still be implemented in the shift-operation domain as was the case for 4-point forward and Inverse transforms.

## Quantization of 8x8 Transform

For a given step size, the encoder can perform quantization by:

$$X_q(i,j) = sign\{X(i,j)\}[(|X(i,j)|A(Q_M,i,j) + f\cdot 2^{(16+Q_E)}) >> (16+Q_E)] \quad (4\text{-}61)$$

where i and j are the row and column indices and $Q_M \equiv Q \bmod 6$ with $Q_E \equiv Q/6$. The parameter $f$ is chosen by the encoder and is typically in the range of 0 to ½. The decoder can perform inverse quantization by:

$$X_r(i,j) = \widetilde{X}_q(i,j)B(Q_M,i,j) << (Q_E - 2), \text{ if } Q_E \ge 2, \quad (4\text{-}62)$$

$$X_r(i,j) = \widetilde{X}_q(i,j)B(Q_M,i,j) + 2^{(1-Q_E)} >> (2-Q_E) \text{ if } Q_E < 2 . \quad (4\text{-}63)$$

The quantization and reconstruction factors $A(Q_M,i,j)$ and $B(Q_M,i,j)$ depend on the transform coefficient position {i, j} inside the block. This is necessary to compensate for the different scaling factor in the 2-D version of the non-orthonormal property. Their values are given by:

$A(Q_M,i,j) = M(Q_M,r)$ and $B(Q_M,i,j) = S(Q_M,r)$ , where

$$A(Q_M, i, j) = \begin{cases} M(Q_M, 0) & \text{for} \quad i = [0,4], j = [0,4] \\ M(Q_M, 1) & \text{for} \quad i = [1,3,5,7], j = [1,3,5,7] \\ M(Q_M, 2) & \text{for} \quad i = [2,6], j = [2,6] \\ M(Q_M, 3) & \text{for} \quad (i = [0,4], j = [1,3,5,7]) \cap (i = [1,3,5,7], j = [0,4]) \\ M(Q_M, 4) & \text{for} \quad (i = [0,4], j = [2,6]) \cap (i = [2,6], j = [0,4]) \\ M(Q_M, 5) & \text{for} \quad (i = [2,6], j = [1,3,5,7]) \cap (i = [1,3,5,7], j = [2,6]) \end{cases} \tag{4-64}$$

and

$$B(Q_M, i, j) = \begin{cases} S(Q_M, 0) & \text{for} \quad i = [0,4], j = [0,4] \\ S(Q_M, 1) & \text{for} \quad i = [1,3,5,7], j = [1,3,5,7] \\ S(Q_M, 2) & \text{for} \quad i = [2,6], j = [2,6] \\ S(Q_M, 3) & \text{for} \quad (i = [0,4], j = [1,3,5,7]) \cap (i = [1,3,5,7], j = [0,4]) \\ S(Q_M, 4) & \text{for} \quad (i = [0,4], j = [2,6]) \cap (i = [2,6], j = [0,4]) \\ S(Q_M, 5) & \text{for} \quad (i = [2,6], j = [1,3,5,7]) \cap (i = [1,3,5,7], j = [2,6]) \end{cases} \tag{4-65}$$

with:

$$M = \begin{bmatrix} 13107 & 11428 & 20972 & 12222 & 16777 & 15481 \\ 11916 & 10826 & 19174 & 11058 & 14980 & 14290 \\ 10082 & 8943 & 15978 & 9675 & 12710 & 11985 \\ 9362 & 8228 & 14913 & 8931 & 11984 & 11259 \\ 8192 & 7346 & 13159 & 7740 & 10486 & 9777 \\ 7282 & 6428 & 11570 & 6830 & 9118 & 8640 \end{bmatrix} \tag{4-66}$$

$$\text{and } S = \begin{bmatrix} 20 & 18 & 32 & 19 & 25 & 24 \\ 22 & 19 & 35 & 21 & 28 & 26 \\ 26 & 23 & 42 & 24 & 33 & 31 \\ 28 & 25 & 45 & 26 & 35 & 33 \\ 32 & 28 & 51 & 30 & 40 & 38 \\ 36 & 32 & 58 & 34 & 46 & 43 \end{bmatrix}. \tag{4-67}$$

These matrices were designed to maximize the dynamic range of amplified transform coefficients and to satisfy a relationship of:

$$M(Q_M,r)S(Q_M,r)v(r) \cong 2^{24} \quad (4\text{-}68)$$

where $v(r) = \{8^2, \left(\frac{289}{32}\right)^2, 5^2, 8 \times \left(\frac{289}{32}\right), 8 \times 5, \left(\frac{289}{32}\right) \times 5\}$. (4-69)

The final scaling after reconstruction becomes:

$$x_r = (H_{inv} X_r + 2^5 e) >> 6 \quad (4\text{-}70)$$

where $e = [11111111]^T$.

Alternatively, total 24 right-shifts can be re-distributed in "three" serial modules without any problems – quantization, inverse quantization and final reconstruction, for example, with 12, 6 and 6, respectively. Then, the equations can be re-written as follows:

$$X_q(i,j) = sign\{X(i,j)\}[(|X(i,j)|A(Q_M,i,j) + f \cdot 2^{(12+Q_E)}) >> (12+Q_E)] \quad (4\text{-}71)$$

$$X_r(i,j) = \widetilde{X}_q(i,j)B(Q_M,i,j) << (Q_E - 6), \text{ if } Q_E \geq 6, \quad (4\text{-}72)$$

$$X_r(i,j) = \widetilde{X}_q(i,j)B(Q_M,i,j) + 2^{(5-Q_E)} >> (6 - Q_E) \text{ if } Q_E < 6 . \quad (4\text{-}73)$$

Since $Q_E \equiv Q/6$, the conditions, $Q_E \geq 6$ and $Q_E < 6$, mean that $Q \geq 36$ and $Q < 36$, respectively. Note that Equation (4-70) remains the same with this re-organization.

The quantization Equation (4-61) can still work with the inverse quantization Equation (4-72), Equation (4-73) and the reconstruction Equation (4-70). Since four more right-shifts are performed in Equation (4-61) compared with Equation (4-71), multiplication of input data by $2^4$ at any point in the signal path through the encoder can justify the use of Equation (4-61) with Equation (4-72), Equation (4-73) and Equation (4-70).

### 4x4 DC Transform of H.264

If the MB is encoded in 16x16 Intra prediction mode (i.e., the entire 16x16 luma component is predicted from neighboring samples) in YCbCr 4:4:4, each $4 \times 4$ residual block is first transformed using the "core" transform described in previous subsections, then Hadamard transformed on the DC values of each $4 \times 4$ block as shown in Figure 4-25 as follows:

$$X_2(i,j) = \left( \begin{bmatrix} 1 & 1 & 1 & 1 \\ 1 & 1 & -1 & -1 \\ 1 & -1 & -1 & 1 \\ 1 & -1 & 1 & -1 \end{bmatrix} \left[ X_1(k,l) \right] \begin{bmatrix} 1 & 1 & 1 & 1 \\ 1 & 1 & -1 & -1 \\ 1 & -1 & -1 & 1 \\ 1 & -1 & 1 & -1 \end{bmatrix} \right) / \Delta . \tag{4-74}$$

The inverse Hadamard Transform is performed at a decoder as follows:

$$X_{1r}(k,l) = \left( \begin{bmatrix} 1 & 1 & 1 & 1 \\ 1 & 1 & -1 & -1 \\ 1 & -1 & -1 & 1 \\ 1 & -1 & 1 & -1 \end{bmatrix} \left[ X_{2r}(i,j) \right] \begin{bmatrix} 1 & 1 & 1 & 1 \\ 1 & 1 & -1 & -1 \\ 1 & -1 & -1 & 1 \\ 1 & -1 & 1 & -1 \end{bmatrix} \right) . \tag{4-75}$$

The forward Hadamard Transform maps DC coefficients in the first transform domain to coefficients in the second transform domain. The standard defines only the inverse Transform in Equation (4-75). Since Hadamard Transform is orthonormal transform, the value of $\Delta$ is not position-dependent. The normalization value $\Delta$ in Equation (4-74) is to implement to provide an intended overall gain formula as shown in Equation (4-80) in H.264 encoders.

The reason of adopting hierarchical transform is to take advantage of the effect with long basis transform over a smooth area. It is reasonable to believe that the area is smooth when 16x16 intra prediction is selected over many viable prediction modes. When the second transform is applied to DC frequency components, this can emulate a kind of Wavelet transform that provides long basis functions over low frequency bands.

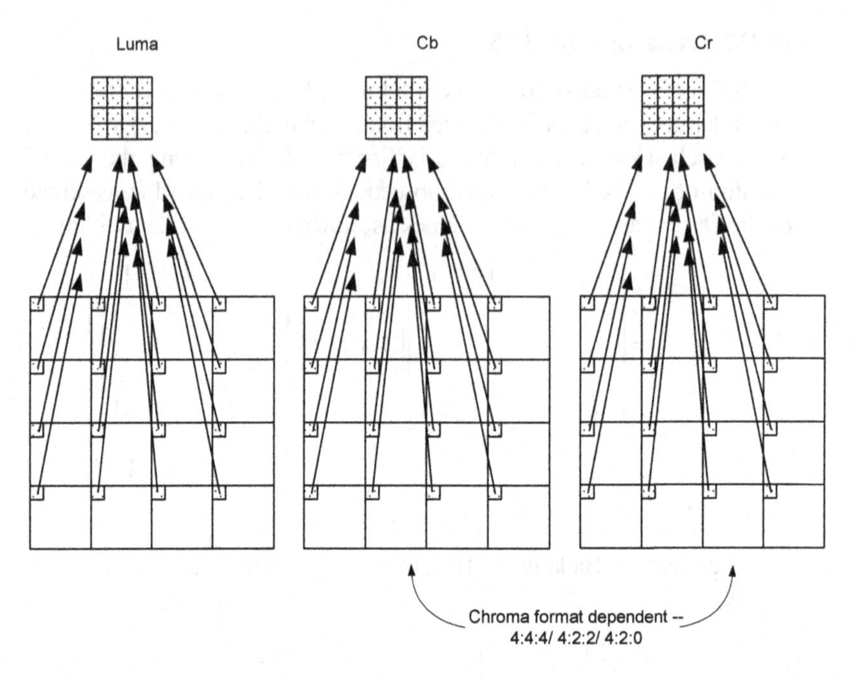


**Figure 4-25 Second Level Transform for 16x16 Intra-prediction Luma or Chroma**

### Quantization of 4x4 DC Transform

For a given step size, the encoder can perform quantization by:

$$X_{2q}(i,j) = sign\{X_2(i,j)\}[(|X_2(i,j)|A(Q'_M,0,0) + f \cdot 2^{(16+Q'_E)}) >> (16+Q'_E)] \quad (4\text{-}76)$$

where i and j are the row and column indices and $Q'_M \equiv Q' \bmod 6$ with $Q'_E \equiv Q'/6$, in case that no final scaling is assumed. The parameter $f$ is chosen by the encoder and is typically in the range of 0 to ½. The decoder can perform inverse quantization by:

$$X_{1r}(k,l) = X_{1q}(k,l)B(Q'_M,0,0) << (Q'_E - 2) \text{ if } Q'_E \geq 2, \quad (4\text{-}77)$$

$$X_{1r}(i,j) = X_{1q}(k,l)B(Q'_M,0,0) + 2^{(1-Q'_E)} >> (2 - Q'_E) \text{ if } Q'_E < 2. \quad (4\text{-}78)$$

The quantization and reconstruction factors are as follows: $A(Q'_M,0,0) = M(Q'_M,0)$ and $B(Q'_M,0,0) = S(Q'_M,0)$, with:

$$M = \begin{bmatrix} 13107 \\ 11916 \\ 10082 \\ 9362 \\ 8192 \\ 7282 \end{bmatrix} \text{ and } S = \begin{bmatrix} 10 \\ 11 \\ 13 \\ 14 \\ 16 \\ 18 \end{bmatrix}. \quad (4\text{-}79)$$

These matrices were designed to maximize the dynamic range of amplified transform coefficients and to satisfy a relationship of:

$$M(Q'_M,0)S(Q'_M,0)v(\Delta) \cong 2^{18} \quad (4\text{-}80)$$

where $v(\Delta) = 2$.

Alternatively, total 18 right-shifts can be re-distributed in "two" serial modules without any problems – quantization and inverse quantization, for example, with 12 and 6, respectively. Then, the equations can be re-written as follows:

$$X_{2q}(i,j) = sign\{X_2(i,j)\}[(|X_2(i,j)|A(Q'_M,i,j) + f \cdot 2^{(12+Q'_E)}) >> (12 + Q'_E)] \quad (4\text{-}81)$$

$$X_{1r}(k,l) = X_{1q}(k,l)B(Q'_M,0,0) << (Q'_E - 6) \text{ if } Q'_E \geq 6, \quad (4\text{-}82)$$

$$X_{1r}(i,j) = X_{1q}(k,l)B(Q'_M,0,0) + 2^{(5-Q'_E)} >> (6 - Q'_E) \text{ if } Q'_E < 6. \quad (4\text{-}83)$$

Since $Q'_E \equiv Q'/6$, the conditions, $Q'_E \geq 6$ and $Q'_E < 6$, mean that $Q' \geq 36$ and $Q' < 36$, respectively.

The quantization Equation (4-76) can still work with the inverse quantization Equation (4-82) and Equation (4-83). Since four more right-shifts are performed in Equation (4-76) compared with Equation (4-81), multiplication of input data by $2^4$ at any point in the signal path through the encoder can justify the use of Equation (4-76) with Equation (4-82) and Equation (4-83).

Note that amplification of DC values with scaling lists is still needed since the values are supposed to plug in with other transform coefficients that are amplified with scaling lists. In addition, the multiplication factors $A(Q'_M, 0,0)$ and $B(Q'_M, 0,0)$ are independent on position indices. Therefore, inverse transform and inverse quantization orders can be swapped. The H.264 standard defines the processing order to take the inverse transform prior to inverse quantization for the second layer Hadamard Transform.

### 2x2 DC Transform of U or V in YCbCr 4:2:0

If the MB is encoded in 16x16 Intra prediction mode (i.e., the entire 16x16 luma component is predicted from neighboring samples) in YCbCr 4:2:0, Y $4 \times 4$ residual block is first transformed using the "core" transform, followed by Hadamard Transform on the DC values of each $4 \times 4$ block as described in the previous subsection. Then, a CbCr $4 \times 4$ residual block is first transformed using the "core" transform, followed by Hadamard Transform on the DC values of each $2 \times 2$ block as follows:

$$X_2(i,j) = \left( \begin{bmatrix} 1 & 1 \\ 1 & -1 \end{bmatrix} [X_1(k,l)] \begin{bmatrix} 1 & 1 \\ 1 & -1 \end{bmatrix} \right) / \Delta . \tag{4-84}$$

The inverse Hadamard Transform is performed at the decoding stage as follows:

$$X_{1r}(k,l) = \left( \begin{bmatrix} 1 & 1 \\ 1 & -1 \end{bmatrix} [X_{2r}(i,j)] \begin{bmatrix} 1 & 1 \\ 1 & -1 \end{bmatrix} \right) . \tag{4-85}$$

The standard defines only the inverse transform in Equation (4-85). The normalization value $\Delta$ in Equation (4-84) is to implement to provide an intended overall gain formula as shown in Equation (4-90).

### Quantization of 2x2 DC Transform of U or V in YCbCr 4:2:0

For a given step size, the encoder can perform quantization by:

$$X_{2q}(i,j) = sign\{X_2(i,j)\}[(|X_2(i,j)|A(Q'_M,0,0) + f \cdot 2^{(16+Q'_E)}) >> (16+Q'_E)] \quad (4\text{-}86)$$

where i and j are the row and column indices and $Q'_M \equiv Q' \bmod 6$ with $Q'_E \equiv Q'/6$, in case that no final scaling is assumed. The parameter $f$ is chosen by the encoder and is typically in the range of 0 to ½. The decoder can perform inverse quantization by:

$$X_{1r}(k,l) = X_{1q}(k,l)B(Q'_M,0,0) << (Q'_E - 1) \text{ if } Q'_E \geq 1, \quad (4\text{-}87)$$

$$X_{1r}(k,l) = X_{1q}(k,l)B(Q'_M,0,0) >> 1 \text{ if } Q'_E < 1. \quad (4\text{-}88)$$

The quantization and reconstruction factors are as follows: $A(Q'_M,0,0) = M(Q'_M,0)$ and $B(Q'_M,0,0) = S(Q'_M,0)$ with:

$$M = \begin{bmatrix} 13107 \\ 11916 \\ 10082 \\ 9362 \\ 8192 \\ 7282 \end{bmatrix} \text{ and } S = \begin{bmatrix} 10 \\ 11 \\ 13 \\ 14 \\ 16 \\ 18 \end{bmatrix}. \quad (4\text{-}89)$$

These matrices were designed to maximize dynamic range of amplified transform coefficients and to satisfy a relationship of:

$$M(Q'_M,0)S(Q'_M,0)v(\Delta) \cong 2^{17} \quad (4\text{-}90)$$

where $v(\Delta) = 1$.

Alternatively, total 17 right-shifts can be re-distributed in "two" serial modules without any problems – quantization and inverse quantization, for example, with 12 and 5, respectively. Then, the equations can be re-written as follows:

$$X_{2q}(i,j) = sign\{X_2(i,j)\}[(|X_2(i,j)|A(Q'_M,i,j) + f \cdot 2^{(12+Q'_E)}) >> (12 + Q'_E)] \tag{4-91}$$

$$X_{1r}(k,l) = X_{1q}(k,l)B(Q'_M,0,0) << (Q'_E - 5) \text{ if } Q'_E \geq 5, \tag{4-92}$$

$$X_{1r}(k,l) = X_{1q}(k,l)B(Q'_M,0,0) + 2^{(4-Q'_E)} >> (5 - Q'_E) \text{ if } Q'_E < 5. \tag{4-93}$$

Since $Q'_E \equiv Q'/6$, the conditions, $Q'_E \geq 5$ and $Q'_E < 5$, mean that $Q' \geq 30$ and $Q' < 30$, respectively. The H.264 standard defines the processing order to take the inverse transform prior to inverse quantization for the second layer Hadamard Transform.

The quantization Equation (4-86) can still work with the inverse quantization Equation (4-92) and Equation (4-93). Since four more right-shifts are performed in Equation (4-86) compared with Equation (4-91), multiplication of input data by $2^4$ at any point in the signal path through the encoder can justify the use of Equation (4-76) with Equation (4-82) and Equation (4-83).

Note that one quantization implementation can actually handle all Transform cases when right-shift operations are carefully distributed – quantization Equations (4-56), (4-71), (4-81) and (4-91) are all the same from the implementation point of view. Even though Equation (4-56) looks slightly different, the quantization can be accommodated in the same implementation – one more right shift can be performed in Equation (4-56) as long as the input data are multiplied by 2 at any point in the signal path through the encoder. The same argument is true for the choice of quantization Equations (4-49), (4-61), (4-76) and (4-86).

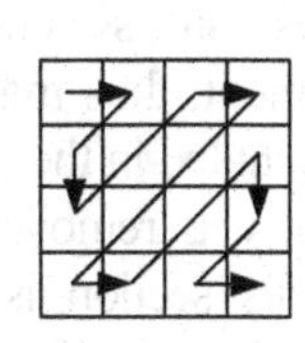

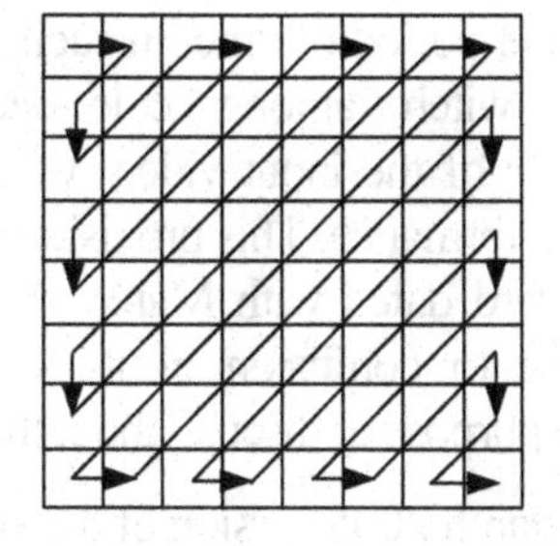

**Figure 4-26 Zig-zag Scan in H.264 Progressive Video**

### Inverse Zig-zag Scan in H.264

The inverse zig-zag scan interprets each (tailing one, zero-run, level) run for the block into 2-D transform coefficients. In H.264, triplet representation is not used. Figure 4-25 illustrates the H.264 scanning pattern for progressive video input. Note that the coding is actually taken in the reverse direction as explained in Chapter 9.

### Residual Color Transform and its Status in the standard

When a color decomposition is considered in video compression, it is generally chosen based on characteristics of human visual systems in response to color. For example, YCbCr is used since Y captures the more important black-and-white characteristics – as a result, Cb and Cr is down sampled by 2 in x and y axes to compress and represent in the YCbCr 4:2:0 domain. However, when we consider 4:4:4 representation in higher Profiles, we actually need to be more careful about color decomposition – no down sampling is performed in Cb and Cr to compress in x and y axes, even though there is clear human visual system characteristics to take advantage of. Therefore, an important question is whether YCbCr is still the best decomposition for 4:4:4 color spaces.

The idea of residual color transform in higher Profiles of H.264 is to allow a different selection of color decomposition in encoder and decoder pairs to adaptively maximize rate-distortion characteristics, meaning better quality in a certain color decomposition with less bits. For example, *A*-color-decomposition can seem to be better than B-color-decomposition during a certain period of time in terms of rate-distortion performance, while *B*-color-decomposition can be better than *A*-color decomposition during another period of time. In such a case, it is probably a good idea to adaptively switch among color-decomposition systems based on characteristics of the input video/ video segment, thus maximizing rate-distortion performance. This provision was included in the High Profile of H.264 standard dated with March, 2005, but was removed in the latest standard due to confusion of its value. This section is designed for informative purpose to discuss about the technology itself.

In the March 2005 version of the standard, the sequence parameter set has a "residual_colour_ transform_ flag" field. If an encoder uses this tool, then it has to choose a new color decomposition before its quantization. It is expected that in the new color domain, the quantization effect should be lessened. Otherwise, the encoder will not choose it. Also, the encoder tells decoders with the residual_colour_transform_flag field set that it is using this tool. Since the encoder changes the color space before quantization, a corresponding decoder has to change the color space back after inverse-quantization/ decompression as shown in Figure 4-27.

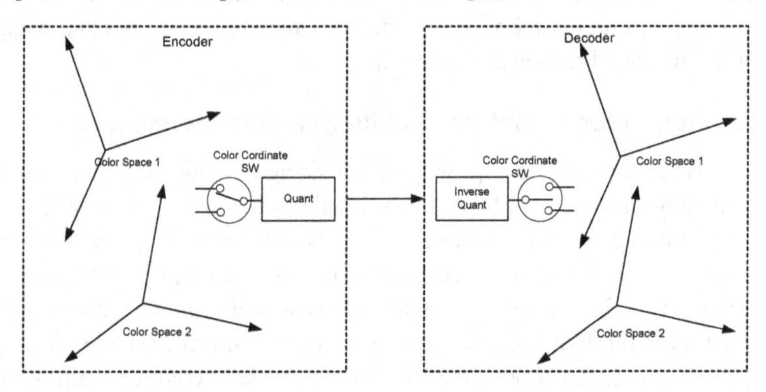


**Figure 4-27 Dynamic Color Space Change for a Better Rate-Distortion Characteristics**

In the current version of H.264, only one component (the first color component) has this option. That is why only $R_{Y,ij}$ is updated with $R_{G,ij}$ in the residual output. The other two components are left alone. For each i,j=0..ijMax, the residual color transform is computed as:

$$t = R_{Y,ij} - (R_{Cb,ij} >> 1) \qquad (4\text{-}94)$$

$$R_{G,ij} = t + R_{Cb,ij} \tag{4-95}$$

$$R_{B,ij} = t - (R_{Cr,ij} >> 1) \tag{4-96}$$

$$R_{R,ij} = R_{B,ij} + R_{Cr,ij} \tag{4-97}$$

where ijMax is set to 3 at transform_size_8x8_flag equal to 0, while it is set to 7 at transform_size_8x8_flag equal to 1.

The residual color transform is similar to the YCgCo transform specified in Appendix E of the H.264 standard. However, the residual color transform operates on the decoded residual difference data within the decoding process rather than operating as a post-processing step that is outside the decoding process specified in the H.264 standard.

# 5. Intra Prediction

## 5.1 Effect of Intra Prediction

### DCT Decomposition

The DCT can be considered as one of the filter bank structures shown in Figure 5-1. The delay term $z^{-1}$ is used to represent the actual sliding window operation we perform for DCT. All DCT basis signals shown in Chapter 4 are simply digital FIR filters in this re-organization. Each grouping of 8 signal points and the nullifying effect of such digital FIR filters other than grouping points can be justified with down-sampling by 8 after the DCT.

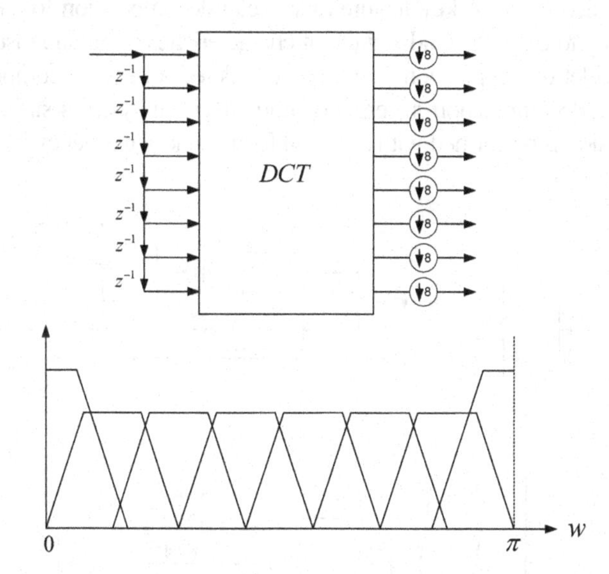


**Figure 5-1 A Bandwidth Decomposition with DCT**

The DCT decomposes a signal into 8 different spectrum bands as shown in Figure 5-1, and it is applied only one time unlike Wavelet

Transform. Each basis signal can lead to spectrum shifting in the frequency domain almost equidistantly. A key feature of DCT decomposition for signals is to take advantage of characteristics of human visual systems (e.g., coding bits can be allocated to low frequency bands after DCT).

## Wavelet Decomposition

An aspect of the wavelet transform is to decompose bandwidth repeatedly in the lower spectrum. The 3-stage tree structure is depicted in Figure 5-2, where $H_0$ is the low pass filter and $H_1$ is the high pass filter. Since 3 down-samplers by 2 are applied serially, the lowest band signal is downsampled by 8. Therefore, this is a reasonable structure to easily compare with 8-point DCT, where the lowest band signal is down sampled by 8. A key feature of wavelet decomposition for signals is the ability to easily take advantage of characteristics of human visual systems. Wavelet decomposition might be considered a better decomposition than DCT decomposition in terms of human visual systems since bandwidth division is performed with a special focus on low frequency bands.

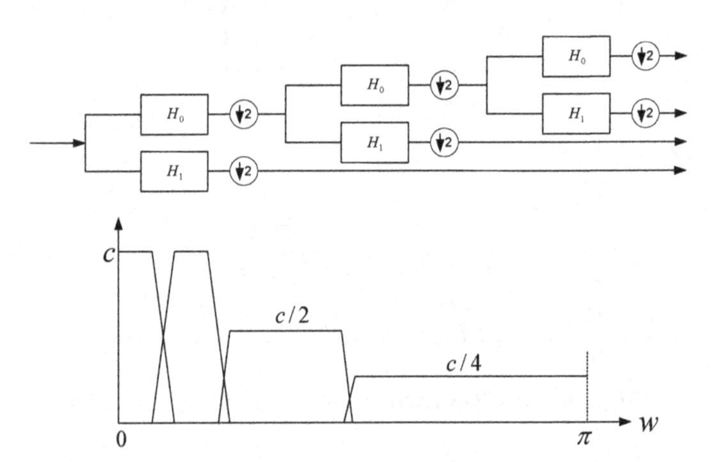


**Figure 5-2 A Bandwidth Decomposition with Wavelet**

A key disadvantage of Wavelet Transform is that it does not work very well for input images that contain only medium or high frequencies due to its biased frequency decomposition. In fact, DCT provides a better compression performance than Wavelet Transform for such types of input images. Therefore, Wavelet Transform has been mainly considered for Intra Picture compression, not for Inter Picture compression.

## Intra Prediction

Intra Prediction is a pre-processing operation before DCT to change/ tweak signal characteristics of input images for compression improvement by tuning them to a DCT basis. The main focus is to eliminate low frequency components in a predictable way that enables perfect reconstruction in the decoder as shown in Figure 5-3. As a result, the low frequency removed signals can work with DCT better than Wavelet Transform for compression since DCT provides better medium and high resolution in terms of signal decomposition. Intra Prediction has been present with almost all video compression standards such as MPEG-1, MPEG-2, MPEG-4 Part 2, VC-1 and H.264 [ISO:MPEG1, ITU:MPEG2, ISO:MPEG4, SMPTE:VC1, JVT:H.264]. However, actual techniques vary in each standard.

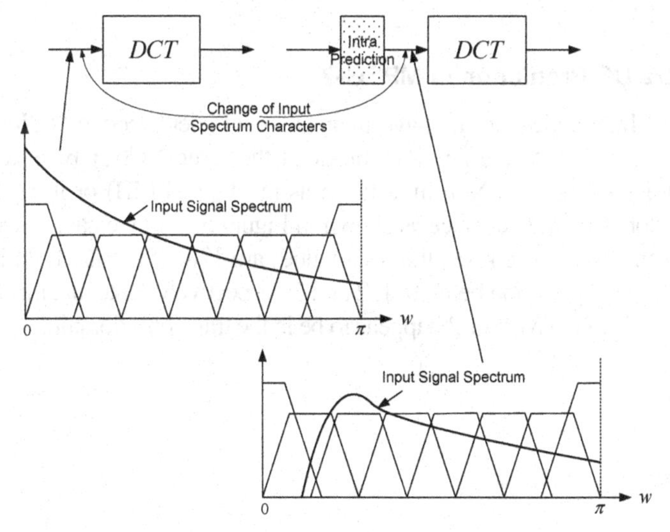


**Figure 5-3 Intra Prediction – Pre-processing to Tune Spectrum**

The shaped signal spectrum as shown in 5-3 just shows the concept. Generally, a better Intra Prediction suppresses low frequency parts more. For example, H.264 Intra Prediction introduces spatial domain processing with a lot of options to remove low frequency components. Since many viable options are developed, popular spatial patterns can most likely be accommodated into one of many Intra Prediction modes.

## Adaptive Intra Prediction

Given an image, removal of only the DC component for the entire frame might not be enough since image is typically non-stationary and the statistical property of it changes from one local area to another. Therefore, devising localized and adaptive methods is an important task. For example, a different DC value shall be removed from each 8x8 block of a MB since each local area would have different DC values. In addition, different low frequency components shall be eliminated from each 8x8 block of a MB. For example, prediction for horizontal low frequency components shall be performed for certain 8x8 blocks, while prediction for vertical low frequency components shall be considered for some other area "adaptively."

## Intra DC Prediction in MPEG-2

Intra-coded blocks have their DC coefficients coded "differentially" with respect to the previous block of the same YCbCr type, unless the previous block is Non-Intra (such as in Skipped MB) or in an MB that belongs to another Slice as shown in Figure 5-4. In the case where a new predictor is to be reset, the reset values are 128, 256, 512, 1024 based on intra_dc_precision being 0, 1, 2 and 3, respectively. Note that this Intra DC Prediction in MPEG-2 happens to be in the transform domain.

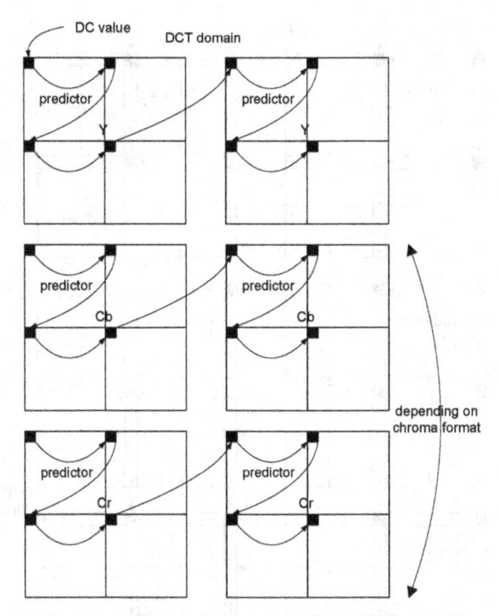


**Figure 5-4 Intra DC Prediction of 2 MBs in MPEG-2**

## 5.2 VC-1 Intra Prediction

### DC/ AC Prediction

Intra MBs are in I frames and optionally the P/B frames of the VC-1 standard. Recent video compression standards take advantage of Intra-prediction methods such as DC/ AC Prediction from adjacent blocks. The VC-1 has DC/ AC Prediction in the transform domain like in MPEG-4 part 2. DC Prediction is always mandatory, while AC Prediction is optional with AC_pred_flag. Luma and Chroma data perform independent Intra Prediction as shown in Figure 5-5.

One interesting point of differential DC values in Advanced Profile of VC-1 is that Huffman-encoding/ decoding of them is treated differently with TRANSDCTAB value. If TRANSDCTAB=0, then the low motion luma DC differential and the low motion Color-differential DC Tables shall be used to decoded the DC differential for luma and luma components respectively.

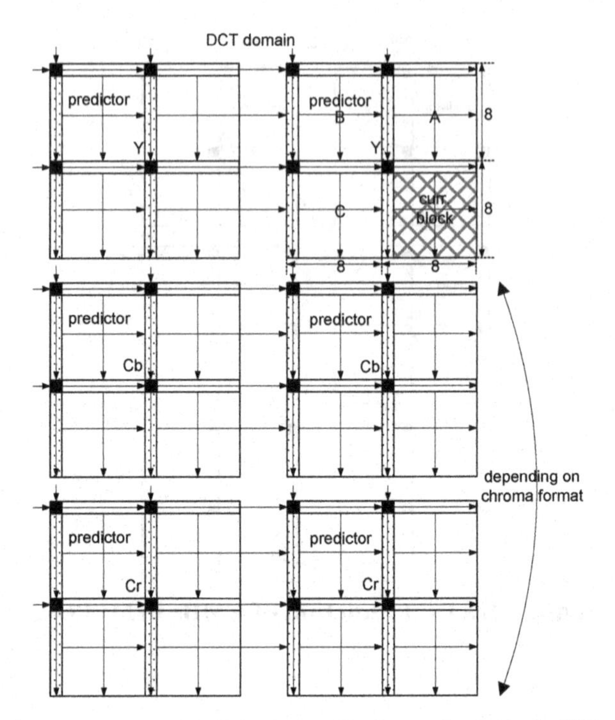


**Figure 5-5 Intra DC/AC Prediction of 2 MBs in VC-1**

To improve coding efficiency, there are 8 AC coding sets. The 8 coding sets are divided into two groups of four and are nominally called "intra" and "inter" coding sets. For Y blocks, one of four intra coding sets shall be used. For Cb and Cr blocks, one of four inter coding set shall be used. One important note is that coding set index is considered differently for $PQINDEX \leq 8$ or for $PQINDEX > 8$. In other words, different Huffman Tables are used based on PQINDEX values. In the sense, this can be thought of as Context-Adaptive VLC.

A new feature in VC-1 is to introduce 8x8 Intra blocks in an Inter MB. In other words, an Inter MB with 4MV mode can have more than one 8x8 Intra block. The initial setting where no adjacent Intra block is available should be considered carefully. Such argument should also go to Slice boundary.

When Quantization Parameters (Qps) between two adjacent MBs are different, such difference should be factored into inverse scaling for inverse Intra Prediction.

The quantized DC value for the current block shall be obtained by adding the DC predictor to the DC differential value, where the direction of predictor is depicted in Figure 5-5. For legacy reasons, the direction of prediction is different with different Profiles. To encode/ decode the DC of the current block as shown in Figure 5-5 for Advanced Profile (AP), either horizontal directional predictor or vertical directional predictor is selected. When predictors of A and C are available, the direction of prediction is determined by:

If $(|DC_B - DC_A| \le |DC_B - DC_C|)$ (5-1)

Predict from block C

Else

Predict from block A.

When predictor C is only available, predictor is taken from block C. Otherwise, the predictor for DC is chosen to be 0 and variable prediction_direction is given with LEFT.

The DC differential value is generally a VLC code, but is specially treated when ESCAPECODE is parsed from the position. When ESCAPECODE is parsed, DCDifferential is described in Fixed Length Code (FLC). When the QUANT value is 1, 2 or others, DCDifferntial is described in 10 bits, 9 bits or 8 bits in FLC, respectively. If ESCAPECODE is not parsed from the stream, the second level interpretation is performed by:

If (QUANT==1)

DCDifferential=DCDifferential $\times$ 4+FLC_decode(2)-3

(5-2)

Else if (QUANT==2)

DCDifferential=DCDifferential $\times$ 2+FLC_decode(1)-1.

(5-4)

The next bit from the stream determines the sign of DC value. In summary, the DC differential is represented in combination of VLC and FLC in the bitstream.

If the ACPRED syntax element specifies that AC prediction is used for the blocks, then AC coefficient prediction is carried out in a similar way for the first row or column of AC coefficients predicted in the direction determined for the DC coefficient as shown in Figure 5-5. If there is no intra block in this direction, the AC predictor shall be set to zero. Figure 5-5 shows how to use the first row of AC coefficients in the block immediately above or immediately left.

There is an additional coefficient scaling step if the MB quantizers of the neighboring blocks are different than that of the current block. The scaling process is described below:

$$\overline{DC_p} = (DC_p \times DCSTEP_p \times DQScale[DCSTEP_c] + 0x20000) >> 18, \quad (5\text{-}4)$$

$$\overline{AC_p} = (AC_p \times STEP_p \times DQScale[STEP_c] + 0x20000) >> 18 \quad (5\text{-}5)$$

where $\overline{DC_p}$ is the scaled DC coefficient in the predictor block, $DC_p$ is the original DC coefficient in the predictor block, $DCSTEP_p$ is the DCStepSize (in Table 4-3) of the predictor block, $DCSTEP_c$ is the DCStepSize (in Table 4-3) in the current block, $\overline{AC_p}$ is the scaled AC coefficient in the predictor block, $AC_p$ is the original AC coefficient in the predictor block, $STEP_p$ is the double_quant-1 in the predictor block, $STEP_c$ is the double_quant-1 in the current block and $DQScale$ is a 63 element integer array as provided in Table 5-1 (IS: Table 74). This approach fundamentally prevents the encoder and decoder pair from possible "drift."

**Table 5-1 DQScale**

| Index | DQScale[Index] |
|---|---|
| 1 | 262144 |
| 2 | 131072 |
| 3 | 87381 |
| 4 | 65536 |

| 5 | 52429 |
|---|---|
| …. | …. |
| 60 | 4369 |
| 61 | 4297 |
| 62 | 4228 |
| 63 | 4161 |

## 5.3 H.264 Intra Prediction

### Luma Prediction

Intra frames/slices generally do not refer to other reference pictures. This means that temporal de-correlation in the pixel domain cannot be performed for Intra frames/slices. Therefore, compression efficiency generally suffers for Intra frames/slices. In H.264, Intra prediction in the pixel domain of Intra frames/slices is devised to reduce spatial correlation emulating temporal de-correlation in the pixel domain of Inter frames/slices.

There are three types of Intra prediction as shown in Figure 5-6, Figure 5-7 and Figure 5-8 – Intra 4x4 prediction, Intra 16x16 prediction and Intra 8x8 prediction. In particular, Intra 8x8 prediction has been recently introduced for High Profile. Different Intra prediction modes define different flows of transform handling. For example, Intra 16x16 prediction additionally defines the second layer Haar transform on DC values of the first layer 4x4 Integer DCT transforms in a MB.

There are nine luma options in 4x4 intra prediction for each 4x4 luma block while there are four chroma options for each 8x8 chroma block. Figure 5-6 depicts how a MB is broken down for 4x4 blocks with luma options.

There are four luma options in 16x16 intra prediction for each 16x16 luma block, while there are four chroma options for each 8x8 chroma block. Figure 5-7 depicts how a MB undergoes Intra 16x16 prediction with luma options.

There are nine luma options in 8x8 intra prediction for each 8x8 luma block, while there are four chroma options for each 8x8 chroma block. Figure 5-8 depicts about how a MB is broken down for 8x8 blocks with luma options.

In terms of chroma data handling in intra prediction, there is no difference under any given luma intra prediction mode as shown in Figure 5-6, Figure 5-7 and Figure 5-8. However, there are some differences in handling chroma intra prediction when input videos differ in chroma formats such as 4:2:0, 4:2:2 and 4:4:4 as shown in Figure 5-9.

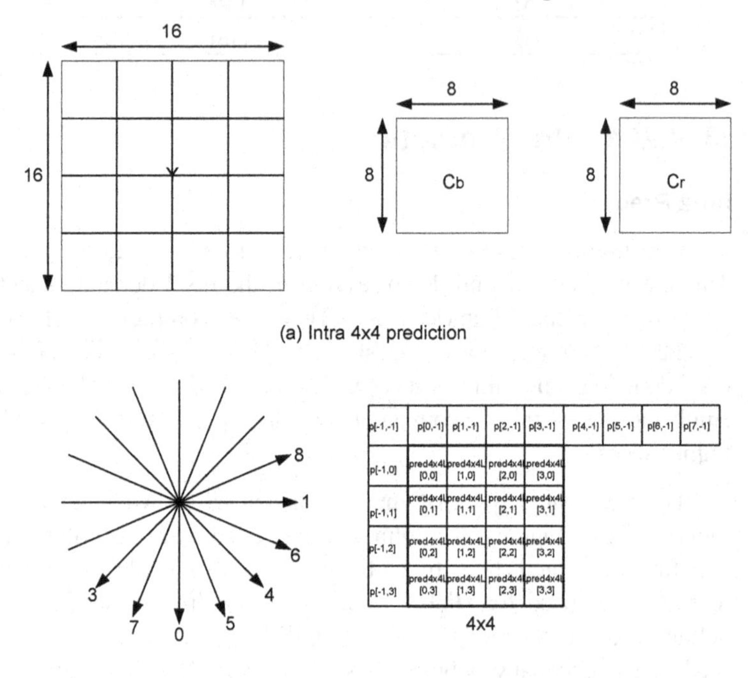


**Figure 5-6 4x4 Intra Prediction**

The algorithm in an encoder is composed of two steps – the mode decision about which mode of intra prediction works best, and predictor computation based on a selected mode. The factors of mode decision are the size of prediction (16x16/8x8/4x4) and the direction (9 or 4 different options). The decision mode taken in the encoder is transmitted to a decoder to compute the same predictor in the decoding process. Currently there are many more luma options than those of chroma.

A key assumption is that there is significant spatial correlation in relatively small areas (16x16/ 8x8/ 4x4) with directional pattern. This is generally a good assumption for typical still image and video. If there is no directional pattern, the DC value is supposed to be a good candidate for the predictor. When predictors are computed, neighborhood pixels serve this purpose. Un-filtered data in the scanning order of MBs are used from neighborhood MBs for the computation of predictors. Note that in-loop filtering is performed on a decoded frame after all MB decoding process for the frame is finished.

The modes in Intra4x4PredMode consist of 0 (Intra 4x4 Vertical), 1 (Intra 4x4 Horizontal), 2 (Intra 4x4 DC), 3 (Intra 4x4 Diagonal Down Left), 4 (Intra 4x4 Diagonal Down Right), 5 (Intra 4x4 Vertical Right), 6 (Intra 4x4 Horizontal Down), 7 (Intra 4x4 Vertical Left) and 8 (Intra 4x4 Horizontal Up). Each mode can be matched with the direction of a particular predictor as shown in Figure 5-6. The most computational complexities lie with mode 5 and mode 6, where 4 conditional branches involve actual computations other than just copying over some data.

The Intra 4x4 prediction algorithm is as follows:

1. Intra_4x4_Vertical prediction mode shall be used when Intra4x4PredMode[luma4x4BlkIdx] is equal to 0.

   This mode shall be used only when the samples p[x,-1] with x=0..3 are marked as "available for Intra_4x4 prediction."

   The values of prediction samples $pred4x4_L[x,y]$ with x,y=0..3 are derived by

   $pred4x4_L[x,y]=p[x,-1]$ with x, y=0..3. (5-6)

2. Intra_4x4_Horizontal prediction mode shall be used when Intra4x4PredMode[luma4x4BlkIdx] is equal to 1.

   This mode shall be used only when the samples p[-1,y] with y=0..3 are marked as "available for Intra_4x4 prediction."

   The values of prediction samples $pred4x4_L[x,y]$ with x,y=0..3 are derived by

   $pred4x4_L[x,y]=p[-1,y]$ with x, y=0..3. (5-7)

3. Intra_4x4_DC prediction mode shall be used when Intra4x4PredMode[luma4x4BlkIdx] is equal to 2.

   - If all samples p[x,-1] with x=0..3 and p[-1,y] with y=0..3 are marked as "available for Intra_4x4 prediction," the values of prediction samples pred4x4L[x,y] with x,y=0..3 are derived by:

     $$pred4x4_L[x,y]=(p[0,-1]+p[1,-1]+p[2,-1]+p[3,-1]+p[-1,0]+p[-1,1]+p[-1,2]+p[-1,3]+4)>>3 . \quad (5\text{-}8)$$

   - If any samples p[x,-1] with x, y=0..3. are marked as "not available for Intra_4x4 prediction" and all samples p[-1,y] with y=..3 are marked as "available for Intra_4x4 prediction," the values of the prediction samples pred4x4L[x,y] with x,y=0..3 are derived by:

     $$pred4x4_L[x,y]=(p[-1,0]+p[-1,1]+p[-1,2]+p[-1,3]+2)>>2. \quad (5\text{-}9)$$

   - If all samples p[-1,y] with y=0..3 are marked as "not available for Intra_4x4 prediction" and all samples p[x,-1] with x=0..3 are marked as "available for Intra_4x4 prediction," the values of prediction samples pred4x4L[x,y] with x,y=0..3 are derived by:

     $$pred4x4_L[x,y]=(p[0,-1]+p[1,-1]+p[2,-1]+p[3,-1]+2)>>2 . \quad (5\text{-}10)$$

   - Some samples p[x,-1] with x=0..3 and some samples p[-1,y] with y=0..3 are marked as "not available for Intra_4x4 prediction," the values of prediction samples pred4x4L[x,y] with x,y=0..3 are derived by:

     $$pred4x4_L[x,y]=(1<<(BitDepth_Y-1)). \quad (5\text{-}11)$$

4. Intra_4x4_Diagonal_Down_Left prediction mode shall be used when Intra4x4PredMode[luma4x4BlkIdx] is equal to 3. This mode shall be used only when the samples p[x,-1] with x=0..7 are marked as "available for Intra_4x4 prediction."

- If x is equal to 3 and y is equal to 3, the values of prediction samples pred4x4L[x,y] with x,y=3 are derived by:

  $$pred4x4_L[x,y]=(p[6,-1]+3\times p[7,-1]+2)>>2\,. \quad (5\text{-}12)$$

- If x is not equal to 3 and y is not equal to 3, the values of the prediction samples pred4x4L[x,y] with x,y=0..3 are derived by:

  $$pred4x4_L[x,y]=(p[x+y,-1]+2\times p[x+y+1,-1]+p[x+y+2,-1]+2)>>2. \quad (5\text{-}13)$$

5. Intra_4x4_Diagonal_Down_Right prediction mode shall be used when Intra4x4PredMode[luma4x4BlkIdx] is equal to 4. This mode shall be used only when the samples p[x,-1] with x=0..3 and p[-1,y] with y=-1..3 are marked as "available for Intra_4x4 prediction."

   - If x is greater than y, the values of prediction samples pred4x4L[x,y] are derived by:

     $$pred4x4_L[x,y]=(p[x-y-2,-1]+2\times p[x-y-1,-1]+p[x-y,-1]+2)>>2\,. \quad (5\text{-}14)$$

   - If x is less than y, the values of prediction samples pred4x4L[x,y] are derived by:

     $$pred4x4_L[x,y]=(p[-1,y-x-2]+2\times p[-1,y-x-1]+p[-1,y-x]+2)>>2\,. \quad (5\text{-}15)$$

   - If x is equal to y, the values of prediction samples pred4x4L[x,y] are derived by:

     $$pred4x4_L[x,y]=(p[0,-1]+2\times p[-1,-1]+p[-1,0]+2)>>2\,. \quad (5\text{-}16)$$

6. Intra_4x4_Vertical_Right prediction mode shall be used when Intra4x4PredMode[luma4x4BlkIdx] is equal to 5. This mode shall be used only when the samples p[x,-1] with x=0..3 and p[-1,y] with y=-1..3 are marked as "available for Intra_4x4 prediction."

- If $(2 \times x-y)$ is equal to 0, 2, 4 or 6, the values of prediction samples $pred4x4_L[x,y]$ are derived by:

$$pred4x4_L[x,y]=(p[x-(y>>1)-1,-1]+p[x-(y>>1),-1]+1)>>1 . \quad (5-17)$$

- If $(2 \times x-y)$ is equal to 1, 3 or 5, the values of prediction samples $pred4x4_L[x,y]$ are derived by:

$$pred4x4_L[x,y]=(p[x-(y>>1)-2,-1]+2 \times p[x-(y>>1)-1,-1]+p[x-(y>>1),-1]+ 2)>>2. \quad (5-18)$$

- If $(2 \times x-y)$ is equal to -1, the values of prediction samples $pred4x4_L[x,y]$ are derived by:

$$pred4x4_L[x,y]=(p[-1,0]+2 \times p[-1,-1]+p[0,-1]+ 2)>>2 . \quad (5-19)$$

- If $(2 \times x-y)$ is equal to –2 or -3, the values of prediction samples $pred4x4_L[x,y]$ are derived by:

$$pred4x4_L[x,y]=(p[-1,y-1]+2 \times p[-1,y-2]+p[-1,y-3]+2)>>2 . \quad (5-20)$$

7. Intra_4x4_Horizontal_Down prediction mode shall be used when Intra4x4PredMode[luma4x4BlkIdx] is equal to 6. This mode shall be used only when the samples p[x,-1] with x=0..3 and p[-1,y] with y=-1..3 are marked as "available for Intra_4x4 prediction."

   - If $(2 \times y-x)$ is equal to 0, 2, 4 or 6, the values of prediction samples $pred4x4_L[x,y]$ are derived by:

   $$pred4x4_L[x,y]=(p[-1,y-(x>>1)-1]+p[-1,y-(x>>1)]+1)>>1 . \quad (5-21)$$

   - If $(2 \times y-x)$ is equal to 1, 3 or 5, the values of prediction samples $pred4x4_L[x,y]$ are derived by:

   $$pred4x4_L[x,y]=(p[-1,y-(x>>1)-2]+2 \times p[-1,y-(x>>1)-1]+p[-1,y-(x>>1)]+ 2)>>2. \quad (5-22)$$

   - If $(2 \times y-x)$ is equal to -1, the values of prediction samples $pred4x4_L[x,y]$ are derived by:

   $$pred4x4_L[x,y]=(p[-1,0]+2 \times p[-1,-1]+p[0,-1]+ 2)>>2 .$$

(5-23)

- If $(2 \times y-x)$ is equal to –2 or -3, the values of prediction samples $pred4x4_L[x,y]$ are derived by:

  $$pred4x4_L[x,y]=(p[x-1,-1]+2\times p[x-2,-1]+p[x-3,-1]+2)>>2 . \quad (5\text{-}24)$$

8. Intra_4x4_Vertical_Left prediction mode shall be used when Intra4x4PredMode[luma4x4BlkIdx] is equal to 7. This mode shall be used only when the samples p[x,-1] with x=0..3 are marked as "available for Intra_4x4 prediction."

   - If y is equal to 0 or 2, the values of prediction samples $pred4x4_L[x,y]$ are derived by:

     $$pred4x4_L[x,y]=(p[x+(y>>1),-1]+p[x+(y>>1)+1,-1]+1)>>1 . \quad (5\text{-}25)$$

   - If y is equal to 1 or 3, the values of prediction samples $pred4x4_L[x,y]$ are derived by:

     $$pred4x4_L[x,y]=(p[x+(y>>1),-1]+2\times p[x+(y>>1)+1,-1]+p[x+(y>>1)+2,-1]+2)>>2. \quad (5\text{-}26)$$

9. Intra_4x4_Horizontal_Up prediction mode shall be used when Intra4x4PredMode[luma4x4BlkIdx] is equal to 8. This mode shall be used only when the samples p[-1,y] with y=0..3 are marked as "available for Intra_4x4 prediction."

   - If $(x+2\times y)$ is equal to 0, 2 or 4, the values of prediction samples $pred4x4_L[x,y]$ are derived by:

     $$pred4x4_L[x,y]=(p[-1,y+(x>>1)]+p[-1,y+(x>>1)+1]+1)>>1 . \quad (5\text{-}27)$$

   - If $(x+2\times y)$ is equal to 1 or 3, the values of prediction samples $pred4x4_L[x,y]$ are derived by:

     $$pred4x4_L[x,y]=(p[-1,y+(x>>1)]+2\times p[-1,y+(x>>1)+1]+p[-1,y+(x>>1)+2]+2)>>2. \quad (5\text{-}28)$$

   - If $(x+2\times y)$ is equal to 5, the values of prediction samples $pred4x4_L[x,y]$ are derived by:

$$\text{pred4x4}_L[x,y]=(p[-1,2]+3\times p[-1,3]+2)>>1\,. \quad (5\text{-}29)$$

- If $(x+2\times y)$ is greater than 5, the values of prediction samples $\text{pred4x4}_L[x,y]$ are derived by:

$$\text{pred4x4}_L[x,y]=p[-1,3]. \quad (5\text{-}30)$$

The modes in Intra16x16PredMode consist of 0 (Intra 16x16 Vertical), 1 (Intra 16x16 Horizontal), 2 (Intra 16x16 DC) and 3 (Intra 16x16 Plane). Each mode can be matched with the direction of a predictor shown in Figure 5-7.

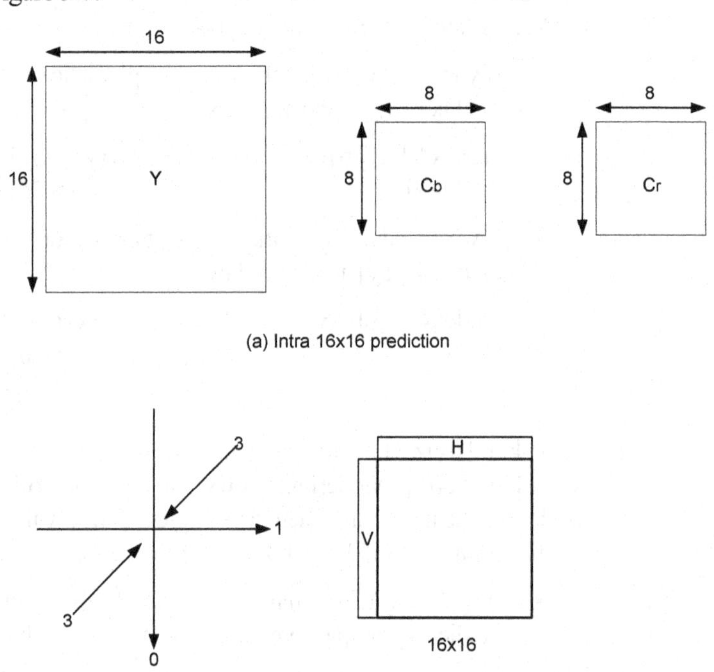


**Figure 5-7 16x16 Intra Prediction**

The most computational complexities lie with mode 3 where the computation involves non-trivial multiplications.

The Intra 16x16 prediction algorithm is as follows:

1. Intra_16x16_Vertical prediction mode shall be used when Intra16x16PredMode is equal to 0.

   This mode shall be used only when the samples p[x,-1] with x=0..15 are marked as "available for Intra_16x16 prediction."

   The values of prediction samples $pred16x16_L[x,y]$ with x,y=0..15 are derived by

   $$pred16x16_L[x,y]=p[x,-1] \text{ with } x, y=0..15. \tag{5-31}$$

2. Intra_16x16_Horizontal prediction mode shall be used when Intra16x16PredMode is equal to 1.

   This mode shall be used only when the samples p[-1,y] with y=0..15 are marked as "available for Intra_16x16 prediction."

   The values of prediction samples $pred16x16_L[x,y]$ with x,y=0..15 are derived by

   $$pred16x16_L[x,y]=p[-1,y] \text{ with } x, y=0..15. \tag{5-32}$$

3. Intra_16x16_DC prediction mode shall be used when Intra16x16PredMode is equal to 2.

   - If all samples p[x,-1] with x=0..15 and p[-1,y] with y=0..15 are marked as "available for Intra_16x16 prediction," the prediction for all luma samples in the MB are derived with x, y=0..15 by:

     $$pred16x16_L[x,y]=\left(\sum_{i=0}^{15} p[i,-1]+\sum_{j=0}^{15} p[-1,j]+16\right)>>5. \tag{5-33}$$

   - If any of samples p[x,-1] with x=0..15 are marked as "not available for Intra_16x16 prediction" and all samples p[-1,y] with y=0..15 are marked as "available for Intra_16x16 prediction," the prediction for all luma samples in the MB are derived with x, y=0..15 by:

$$\text{pred16x16}_L[x,y]=(\sum_{j=0}^{15} p[-1,j]+8)>>4. \quad (5\text{-}34)$$

- If any of samples p[-1,y] with y=0..15 are marked as "not available for Intra_16x16 prediction" and all samples p[x,-1] with x=0..15 are marked as "available for Intra_16x16 prediction," the prediction for all luma samples in the MB are derived with x, y=0..15 by:

$$\text{pred16x16}_L[x,y]=(\sum_{i=0}^{15} p[i,-1]+8)>>4. \quad (5\text{-}35)$$

- Some samples p[x,-1] with x=0..15 and some samples p[-1,y] with y=0..15 are marked as "not available for Intra_16x16 prediction," the values of prediction samples $\text{pred16x16}_L[x,y]$ with x,y=0..15 are derived by:

$$\text{pred16x16}_L[x,y]=(1<<(\text{BitDepth}_Y-1)). \quad (5\text{-}36)$$

4. Intra_16x16_Plane prediction mode shall be used when Intra16x16PredMode is equal to 3. This mode shall be used only when the samples p[x,-1] with x=-1..15 and p[-1,y] with y=0..15 are marked as "available for Intra_16x16 prediction."

   The values of prediction samples $\text{pred16x16}_L[x,y]$ with x,y=0..15 are derived by

$$\text{pred16x16}_L[x,y]=\text{Clip1}_Y(a+b\times(x-7)+c\times(y-7)+16)>>5), \quad (5\text{-}37)$$

   where $a=16\times(p[-1,15]+p[15,-1])$, $b=(5\times H+32)>>6$ and $c=(5\times V+32)>>6$. Here, H and V are proposed as follows:

$$H=(\sum_{i=0}^{7}(i+1)\times(p[8+i,-1]-p[6-i,-1]), \quad (5\text{-}38)$$

$$V=(\sum_{j=0}^{7}(j+1)\times(p[-1,8+j]-p[-1,6-j]). \quad (5\text{-}39)$$

The modes in Intra8x8PredMode consist of 0 (Intra 8x8 Vertical), 1 (Intra 8x8 Horizontal), 2 (Intra 8x8 DC), 3 (Intra 8x8 Diagonal Down Left), 4 (Intra 8x8 Diagonal Down Right), 5 (Intra 8x8 Vertical Right), 6 (Intra 8x8 Horizontal Down), 7 (Intra 8x8 Vertical Left) and 8 (Intra 8x8 Horizontal Up). Each mode can be matched with the direction of a particular predictor shown in Figure 5-8. The most computational complexities lie with mode 5 and mode 6, where there are four conditional branches involving actual computations.

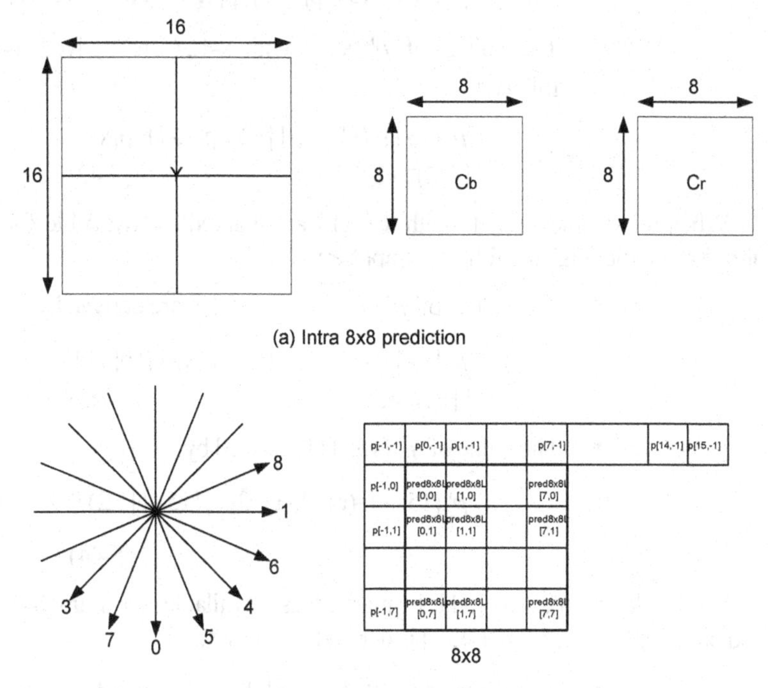


**Figure 5-8 8x8 Intra Prediction**

The Intra 8x8 prediction algorithm is as follows:

When all samples p[x,-1] with x=0..7 are marked as "available for Intra_8x8 prediction," the following applies:

- The value of $p'[0,-1]$ is derived as follows:

- If p[-1,-1] is marked as "available for Intra_8x8 prediction,"

  $p'[0,-1]=(p[-1,-1]+2\times p[0,-1]+p[1,-1]+2)>>2.$ (5-40)

- If p[-1,-1] is marked as "not available for Intra_8x8 prediction,"

  $p'[0,-1]=(3\times p[0,-1]+p[1,-1]+2)>>2.$ (5-41)

- The values of $p'[x,-1]$ with x-1..7 are derived as follows:
  - $p'[x,-1]=(p[x-1,-1]+2\times p[x,-1]+p[x+1,-1]+2)>>2.$ (5-42)

When all samples p[x,-1] with x=7..15 are marked as "available for Intra_8x8 prediction," the following applies:

- The values of $p'[x,-1]$ with x=8..14 are derived by:
  - $p'[x,-1]=(p[x-1,-1]+2\times p[x,-1]+p[x+1,-1]+2)>>2.$ (5-43)
- The value of $p'[15,-1]$ is derived by:
  - $p'[15,-1]=(p[14,-1]+3\times p[15,-1]+2)>>2.$

    (5-44)

When all samples p[-1,-1] is marked as "available for Intra_8x8 prediction," the value of $p'[-1,-1]$ is derived as follows:

- If the sample p[0,-1] or p[-1,0] is marked as "not available for Intra_8x8 prediction," the following applies:
  - If p[0,-1] is marked as "available for Intra_8x8 prediction,"

    $p'[-1,-1]=(3\times p[-1,-1]+p[0,-1]+2)>>2.$

    (5-45)

- If p[0,-1] is marked as "not available for Intra_8x8 prediction" and the sample p[-1,0] is marked as "available for Intra_8x8 prediction,"

  $p'[-1,-1]=(3\times p[-1,-1]+p[-1,0]+2)>>2.$ (5-46)

- The sample p[0,-1] and the sample p[-1,0] are marked as "available for Intra_8x8 prediction," $p'[-1,-1]$ is derived by:
  - $p'[-1,-1]=(p[0,-1]+2\times p[-1,-1]+p[-1,0]+2)>>2.$ (5-47)

When all samples p[-1,y] with y=0..7 are marked as "available for Intra_8x8 prediction," the following applies:

- The value of $p'[-1,0]$ is derived as follows:
  - If p[-1,-1] is marked as "available for Intra_8x8 prediction,"

    $p'[-1,0]=(p[-1,-1]+2\times p[-1,0]+p[-1,1]+2)>>2.$ (5-48)
  - If p[-1,-1] is marked as "not available for Intra_8x8 prediction"

    $p'[-1,0]=(3\times p[-1,0]+p[-1,-1]+2)>>2.$ (5-49)

- The values of $p'[-1,y]$ with y=1..6 are derived by:
  - $p'[-1,y]=(p[-1,y-1]+2\times p[-1,y]+p[-1,y+1]+2)>>2.$ (5-50)

- The values of $p'[-1,7]$ is derived by:
  - $p'[-1,7]=(p[-1,6]+3\times p[-1,7]+2)>>2.$ (5-51)

1. Intra_8x8_Vertical prediction mode shall be used when Intra8x8PredMode [luma8x8BlkIdx] is equal to 0. This mode shall be used only when the samples p[x,-1] with x=0..7 are marked as "available for Intra_8x8 prediction."

   The values of prediction samples $pred8x8_L[x,y]$ with x,y=0..7 are derived by

   $$pred8x8_L[x,y] = p'[x,-1] \text{ with } x, y=0..7. \quad (5\text{-}52)$$

2. Intra_8x8_Horizontal prediction mode shall be used when Intra8x8PredMode [luma8x8BlkIdx] is equal to 1. This mode shall be used only when the samples p[-1,y] with x=0..7 are marked as "available for Intra_8x8 prediction."

   The values of prediction samples $pred8x8_L[x,y]$ with x,y=0..7 are derived by $pred8x8_L[x,y] = p'[-1,y]$ with x, y=0..7. (5-53)

3. Intra_8x8_DC prediction mode shall be used when Intra8x8PredMode [luma8x8BlkIdx] is equal to 2. The values of prediction samples $pred8x8_L[x,y]$ with x,y=0..7 are derived by:

   - If all samples p[x,-1] with x=0..7 and p[-1,y] with y=0..7 are marked as "available for Intra_8x8 prediction," the values of the prediction samples $pred8x8_L[x,y]$ with x,y=0..7 are derived by:

     $$pred8x8_L[x,y] = \left(\sum_{i=0}^{7} p'[i,-1] + \sum_{j=0}^{7} p'[-1,j] + 8\right) >> 4. \quad (5\text{-}54)$$

   - If any samples p[x,-1] with x=0..7 are marked as "not available for Intra_8x8 prediction" and all samples p[-1,y] with y=0..7 are marked as "available for Intra_8x8 prediction," the values of the prediction samples $pred8x8_L[x,y]$ with x,y=0..7 are derived by:

     $$pred8x8_L[x,y] = \left(\sum_{j=0}^{7} p'[-1,j] + 4\right) >> 3. \quad (5\text{-}55)$$

- If any samples p[-1,y] with y=0..7 are marked as "not available for Intra_8x8 prediction" and all samples p[x,-1] with x=0..7 are marked as "available for Intra_8x8 prediction," the values of the prediction samples $pred8x8_L[x,y]$ with x,y=0..7 are derived by:

$$pred8x8_L[x,y]=(\sum_{i=0}^{7} p'[i,-1]+4)>>3 . \quad (5\text{-}56)$$

- If any samples p[x,-1] with x=0..7 and some samples p[-1,y] are marked as "not available for Intra_8x8 prediction," the values of the prediction samples $pred8x8_L[x,y]$ with x,y=0..7 are derived by:

$$pred8x8_L[x,y]=(1<<(BitDepth_Y-1)) . \quad (5\text{-}57)$$

4. Intra_8x8_Diagonal_Down_Left prediction mode shall be used when Intra8x8PredMode [luma8x8BlkIdx] is equal to 3. This mode shall be used only when the samples p[x,-1] with x=0..15 are marked as "available for Intra_8x8 prediction."

   The values of prediction samples $pred8x8_L[x,y]$ with x,y=0..7 are derived by:

   - If x is equal to 7 and any y is equal to 7,

$$pred8x8_L[x,y]=( p'[14,-1]+3\times p'[15,-1]+2)>>2. \quad (5\text{-}58)$$

   - If x is not equal to 7 or any y is not equal to 7,

$$pred8x8_L[x,y]=( p'[x+y,-1]+2\times p'[x+y+1,-1]+ p'[x+y+2,-1]+2)>>2. \quad (5\text{-}59)$$

5. Intra_8x8_Diagonal_Down_Right prediction mode shall be used when Intra8x8PredMode [luma8x8BlkIdx] is equal to 4. This mode shall be used only when the samples p[x,-1] with x=0..7 and p[-1,y] with y=-1..7 are marked as "available for Intra_8x8 prediction."

   The values of prediction samples $pred8x8_L[x,y]$ with x,y=0..7 are derived by:

- If x is greater than y,

  $pred8x8_L[x,y]=( p'[x-y-2,-1]+2\times p'[x-y-1,-1]+ p'[x-y,-1]+2)>>2.$ (5-60)

- If x is less than y,

  $pred8x8_L[x,y]=( p'[-1,y-x-2]+2\times p'[-1,y-x-1]+ p'[-1,y-x]+2)>>2.$ (5-61)

- If x is equal to y,

  $pred8x8_L[x,y]=( p'[0,-1]+2\times p'[-1,-1]+ p'[-1,0]+2)>>2.$ (5-62)

6. Intra_8x8_Vertical_Right prediction mode shall be used when Intra8x8PredMode [luma8x8BlkIdx] is equal to 5. This mode shall be used only when the samples p[x,-1] with x=0..7 and p[-1,y] with y=-1..7 are marked as "available for Intra_8x8 prediction."

   The values of prediction samples $pred8x8_L[x,y]$ with x,y=0..7 are derived by:

   - If (2× x-y) is equal to 0, 2, 4, 6, 8, 10, 12 or 14,

     $pred8x8_L[x,y]=( p'[x-(y>>1)-1,-1]+ p'[x-(y>>1),-1]+1)>>1.$ (5-63)

   - If (2× x-y) is equal to 1, 3, 5, 7, 9, 11 or 13,

     $pred8x8_L[x,y]=( p'[x-(y>>1)-2,-1]+2\times p'[x-(y>>1)-1,-1]+ p'[x-(y>>1),-1]+2)>>2.$ (5-64)

   - If (2× x-y) is equal to -1,

     $pred8x8_L[x,y]=( p'[-1,0]+2\times p'[-1,-1]+ p'[0,-1]+2)>>2.$ (5-65)

   - If (2× x-y) is equal to -2, -3, -4, -5, -6 or -7,

     $pred8x8_L[x,y]=( p'[-1,y-2\times x-1]+2\times p'[-1,y-2\times x-2]+ p'[-1,y-2\times x-3]+2)>>2.$ (5-66)

7. Intra_8x8_Horizontal_Down prediction mode shall be used when Intra8x8PredMode [luma8x8BlkIdx] is equal to 6. This mode shall be used only when the samples p[x,-1] with x=0..7 and p[-1,y] with y=-1..7 are marked as "available for Intra_8x8 prediction."

   The values of prediction samples $pred8x8_L[x,y]$ with x,y=0..7 are derived by:

   - If $(2\times y-x)$ is equal to 0, 2, 4, 6, 8, 10, 12 or 14,

     $pred8x8_L[x,y]=(p'[-1,y-(x>>1)-1]+p'[-1,y-(x>>1)]+1)>>1.$ (5-67)

   - If $(2\times y-x)$ is equal to 1, 3, 5, 7, 9, 11 or 13,

     $pred8x8_L[x,y]=(p'[-1,y-(x>>1)-2]+2\times p'[-1,y-(x>>1)-1]+p'[-1,y-(x>>1)]+2)>>2.$ (5-68)

   - If $(2\times y-x)$ is equal to -1,

     $pred8x8_L[x,y]=(p'[-1,0]+2\times p'[-1,-1]+p'[0,-1]+2)>>2.$ (5-69)

   - If $(2\times y-x)$ is equal to -2, -3, -4, -5, -6 or -7,

     $pred8x8_L[x,y]=(p'[x-2\times y-1,-1]+2\times p'[x-2\times y-2,-1]+p'[x-2\times y-3,-1]+2)>>2.$ (5-70)

8. Intra_8x8_Vertical_Left prediction mode shall be used when Intra8x8PredMode [luma8x8BlkIdx] is equal to 7. This mode shall be used only when the samples p[x,-1] with x=0..15 are marked as "available for Intra_8x8 prediction."

   The values of prediction samples $pred8x8_L[x,y]$ with x,y=0..7 are derived by:

   - If y is equal to 0, 2, 4 or 6,

     $pred8x8_L[x,y]=(p'[x+(y>>1),-1]+p'[x+(y>>1)+1,-1]+1)>>1.$ (5-71)

   - If y is equal to 1, 3, 5 or 7,

$$pred8x8_L[x,y]=(p'[x+(y>>1),-1]+2 \times p'[x+(y>>1)+1,-1]+p'[x+(y>>1)+2,-1]+2)>>2. \quad (5\text{-}72)$$

9. Intra_8x8_Horizontal_Up prediction mode shall be used when Intra8x8PredMode [luma8x8BlkIdx] is equal to 8. This mode shall be used only when the samples p[-1,y] with y=0..15 are marked as "available for Intra_8x8 prediction."

   The values of prediction samples $pred8x8_L[x,y]$ with x,y=0..7 are derived by:

   - If $(x+2\times y)$ is equal to 0, 2, 4, 6, 8, 10 or 12,

     $$pred8x8_L[x,y]=(p'[-1,y+(x>>1)]+p'[-1,y+(x>>1)+1]+1)>>1. \quad (5\text{-}73)$$

   - If $(x+2\times y)$ is equal to 1, 3, 5, 7, 9 or 11,

     $$pred8x8_L[x,y]=(p'[-1,y+(x>>1)]+2\times p'[-1,y+(x>>1)+1]+p'[-1,y+(x>>1)+2]+2)>>2. \quad (5\text{-}74)$$

   - If $(x+2\times y)$ is equal to 13,

     $$pred8x8_L[x,y]=(p'[-1,6]+3\times p'[-1,7]+2)>>2. \quad (5\text{-}75)$$

   - If $(x+2\times y)$ is greater than 13,

     $$pred8x8_L[x,y]=p'[-1,7]. \quad (5\text{-}76)$$

## Chroma Prediction

The modes in IntraChromaPredMode consist of 0 (Intra Chroma DC), 1 (Intra Chroma Horizontal), 2 (Intra Chroma Vertical) and 3 (Intra Chroma Plane). Each mode can be matched with the direction of a predictor shown in Figure 5-9. The most computational complexities lie with mode 3, where the computation involves non-trivial multiplications. Note that the definition for DC predictor for Chroma is slightly different than those of luma DC predictors. In addition, the computation of DC predictor is taken slightly differently based on the Chroma format of input video as shown in Figure 5-9 (a). It can be considered as a natural extension of the derivation method for 4:2:0 format, though.

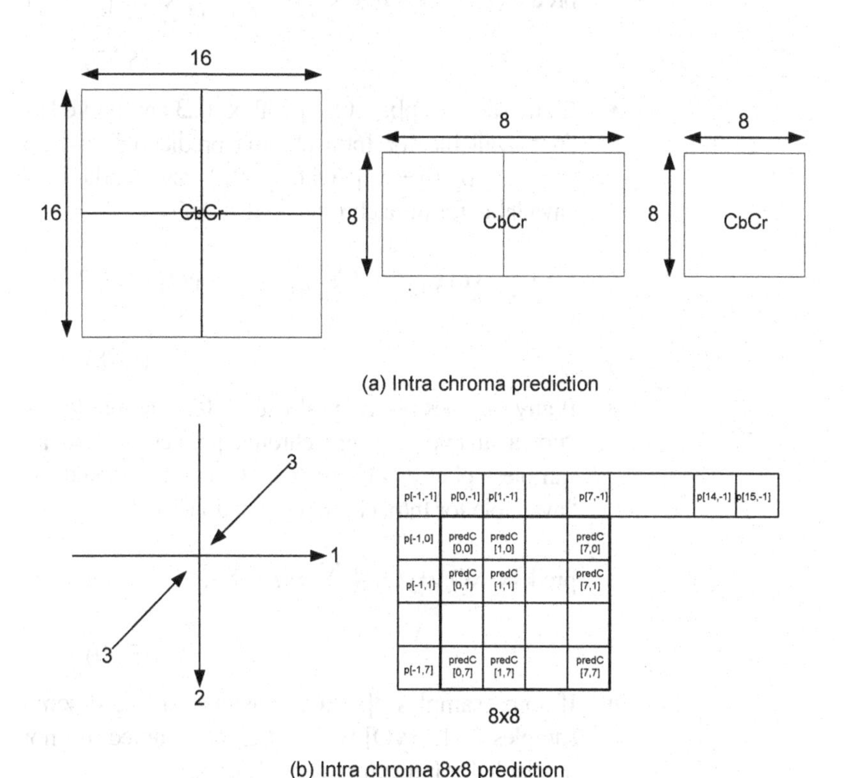


(a) Intra chroma prediction

(b) Intra chroma 8x8 prediction

**Figure 5-9 Chroma Intra Prediction**

The Intra chroma prediction algorithm is as follows:

1. Intra_Chroma_DC prediction mode shall be used when IntraChromaPredMode is equal to 0. Let (xO,yO) be the position of the upper-left sample of a 4x4 chroma block with index chorma4x4BlkIdx. If (xO,yO) is equal to (0,0) or xO and yO are greater than 0, the values of prediction samples predc[x,y] with x,y=0..3 are derived as follows:

   - If all samples p[x+xO,y+yO] with x,y=0..3 and p[-1,y+yO] with y=0..3 are marked as "available for Intra chroma prediction,"

     $$predc[x+xO,y+yO]=\left(\sum_{i=0}^{3} p[i+xO,-1]+\sum_{j=0}^{3} p[-1,j+yO]+4\right)>>3. \quad (5\text{-}77)$$

   - If any samples p[x+xO,-1] with x=0..3 are marked as "not available for Intra chroma prediction" and all samples p[-1,y+yO] with y=0..3 are marked as "available for Intra chroma prediction,"

     $$predc[x+xO,y+yO]=\left(\sum_{j=0}^{3} p[-1,j+yO]+2\right)>>2. \quad (5\text{-}78)$$

   - If any samples p[-1,y+yO] with y=0..3 are marked as "not available for Intra chroma prediction" and all samples p[x+xO,-1] with x=0..3 are marked as "available for Intra chroma prediction,"

     $$predc[x+xO,y+yO]=\left(\sum_{i=0}^{3} p[i+xO,-1]+2\right)>>2. \quad (5\text{-}79)$$

   - If some samples p[x+xO,-1] with x=0..3 and some samples p[-1,y+yO] with y=0..3 are marked as "not available for Intra chroma prediction,"

     $$predc[x+xO,y+yO]=(1<<(BitDepthc-1)). \quad (5\text{-}80)$$

If xO is greater than 0 and yO is equal to 0, the value of prediction samples predc[x,y] with x,y=0..3 are derived as follows:

- If all samples p[x+xO,-1] with x=0..3 are marked as "available for Intra chroma prediction,"

$$\text{predc}[x+xO,y+yO]=(\sum_{i=0}^{3} p[i+xO,-1]+2)>>2. \quad (5\text{-}81)$$

- If any samples p[-1,y+yO] with y=0..3 are marked as "available for Intra chroma prediction" and all samples p[x+xO,-1] with x=0..3 are marked as "available for Intra chroma prediction,"

$$\text{predc}[x+xO,y+yO]=(\sum_{j=0}^{3} p[-1,j+yO]+2)>>2. \quad (5\text{-}82)$$

- If some samples p[x+xO,-1] with x=0..3 and some samples p[-1,y+yO] with y=0..3 are marked as "not available for Intra chroma prediction,"

$$\text{predc}[x+xO,y+yO]=(1<<(\text{BitDepthc}-1)). \quad (5\text{-}83)$$

If xO is equal to 0 and yO is greater than 0, the value of prediction samples predc[x,y] with x,y=0..3 are derived as follows:

- If all samples p[-1,y+yO] with y=0..3 are marked as "available for Intra chroma prediction,"

$$\text{predc}[x+xO,y+yO]=(\sum_{j=0}^{3} p[-1,j+yO]+2)>>2. \quad (5\text{-}84)$$

- If all samples p[x+xO,-1] with x=0..3 are marked as "available for Intra chroma prediction,"

$$\text{predc}[x+xO,y+yO]=(\sum_{i=0}^{3} p[i+xO,-1]+2)>>2\,.$$

(5-85)

- If some samples p[x+xO,-1] with x=0..3 and some samples p[-1,y+yO] with y=0..3 are marked as "not available for Intra chroma prediction,"

  predc[x+xO,y+yO]=(1<<(BitDepthc-1)) . (5-86)

2. Intra_Chroma_Horizontal prediction mode shall be used when IntraChromaPredMode is equal to 1.

   This mode shall be used only when the samples p[-1,y] with y=0..MbHeightC-1 are marked as "available for Intra chroma prediction."

   The values of prediction samples predc[x,y] are derived by

   predc[x,y]=p[-1,y] with x=0..MbWidthC-1,
   y=0..MbHeightC-1. (5-87)

3. Intra_Chroma_Vertical prediction mode shall be used when IntraChromaPredMode is equal to 2.

   This mode shall be used only when the samples p[x,-1] with x=0..MbWidthC-1 are marked as "available for Intra chroma prediction."

   The values of prediction samples predc[x,y] are derived by

   predc[x,y]=p[x,-1] with x=0..MbWidthC-1,
   y=0..MbHeightC-1. (5-88)

4. Intra_Chroma_Vertical prediction mode shall be used when IntraChromaPredMode is equal to 3.

   This mode shall be used only when the samples p[x,-1] with x=0..MbWidthC-1 and p[-1,y] with y=0..MbHeightC-1 are marked as "available for Intra chroma prediction."

The values of prediction samples predc[x,y] are derived as follows: Let the variable xCF be set equal to $4\times$(chroma_format_idc==3) and let the variable yCF be set equal to $4\times$(chroma_format_idc!=1).

$$\text{predc}[x,y]=\text{Clip1c}((a+b\times(x-3-xCF)+c\times(y-3-yCF)+16)>>5), \tag{5-89}$$

where a=16 $\times$ (p[-1,MbHeightC-1]+p[MbWidthC-1,-1]), b=((34-29 $\times$ (chorma_format_idc==3)) $\times$ H+32)>>6 and b=((34-29 $\times$ (chorma_format_idc!=1)) $\times$ V+32)>>6. Here, H and V are proposed as follows:

$$H=(\sum_{i=0}^{3+xCF}(i+1)\times(p[4+xCF+i,-1]-p[2+xCF-i,-1]), \tag{5-90}$$

$$V=(\sum_{j=0}^{3+yCF}(j+1)\times(p[-1,4+yCF+j]-p[-1,2+yCF-j])\cdot \tag{5-91}$$

# 6. Inter Prediction

## 6.1 Inter Prediction

### Inter Prediction and Temporal Masking Effect

Inter Prediction is the heart of moving image compression, whereas the Transform is the heart of still image compression. Fundamentally, Inter Prediction utilizes "semantic" information (i.e., Motion Vectors – MV) to represent a group of pixel data (i.e., MB) whose texture is the closest pattern from a past picture (i.e., reference picture). This was a significant jump from straightforward texture representation. Figure 6-1 illustrates this concept. A positive value for the horizontal or vertical component of a MV indicates that the prediction is made from samples in the reference pictures that lie to the right or below the samples being predicted, respectively. Interestingly, a few bits of two MV components in the x and y axes can replace 3072 bits and 6144 bits of one MB data in YCbCr 4:2:0 and YCbCr 4:4:4, respectively, when only transitional movement happens in the part of a scene that the current MB belongs to. One extreme case is "Skip" mode. Skip implies that there is no MV differential data or zero MV data, based on where the MB is, and that there is no residual data required to represent the current MB. Certain standards such as H.263 assign only "1" bit to represent this scenario (a.k.a., COD_flag in the standard), while some other standards such as MPEG-2 assign no VLC code for this, and handle it with MB address increment. When this mode is used for certain scenarios, the compression ratio becomes 3072 or 6144 to 1, depending on Chroma formats for those MBs. When this kind of efficient representation happens more in the compression, the compression ratio goes up much higher. It is important to understand that the occurring rate of such representations can be controlled by an encoder to the improve compression ratio.

The "temporal masking effect" means that human visual systems cannot detect texture details or errors in fast moving scenes. In other words, the visual threshold increases for a given stimulus in the presence of signals in the mask regions. The "temporal masking effect" is adopted in two levels of compression systems. The low level adoption is as follows: since a given stimulus is hard to see in moving images, the aspect of "Contrast Sensitivity" can be also somewhat ignored in moving scenes. In other words, visual weighting factors along with different frequency components are not necessary. Therefore, "one" large fixed size quantization is applied to the components throughout different frequencies. The high level adoption is as follows: due to the "temporal masking effect," not all picture qualities need to be equally good. Some pictures in between two relatively good quality pictures can contain deteriorated contents. Therefore, some rate control algorithms force usage of "degraded" quantization in B pictures to save bits as shown in Chapter 3. This is the main reason why generated bits from B pictures are the smallest of all among different picture types.

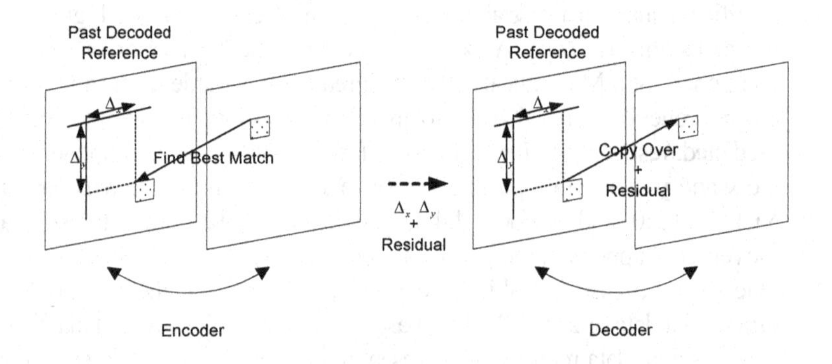


**Figure 6-1 Inter Prediction and Motion Compensation**

## Fractional-Pel Motion Estimation and Compensation

Fractional-pel resolution is introduced in Motion Estimation (ME) and Motion Compensation (MC) for recent video standards such as MPEG-2, VC-1, H.264, etc. Recent research surveys that broadcast video in NTSC or PAL can represent resolutions of 0.25 pel to reasonably perform temporal decorrelation. Figure 6-2 depicts an example about artificial refinement in the ¼-pel domain interpolation for enhanced prediction performmance [bhaskaran:image]. Generally, two applications work better with higher resolution (i.e., fractional-pel) ME/ MC – high quality video (HD applications) and small size video (mobile applications). It makes sense to use delicate motion prediction methods for high quality video.

However, it might not be so intuitive to utilize fractional-pel ME for small size video. The logic is as follows: the portion of one pixel in small size video is more significant in terms of visual information compared with that in HD size video. It might contain the "right eye" in a human face in QCIF resolution video, while it might be a small spot in an iris of the same eye in HD resolution video. Therefore, introducing fractional resolution for small size video would help significantly in improving video compression ratio. MPEG-4 Part 2 or H.263 fully exploit fractional-pel ME/ MC when they are applied even to mobile video applications.

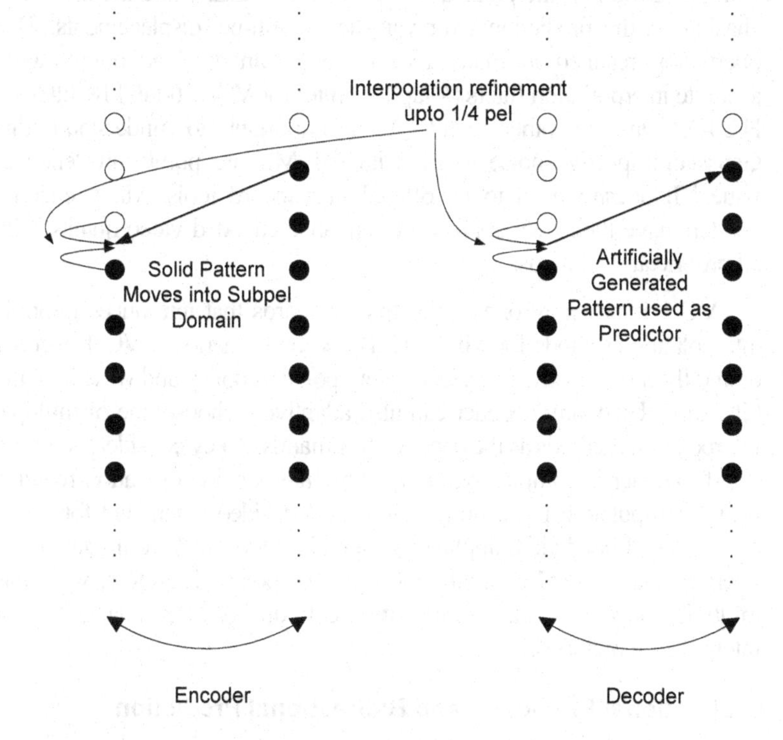


**Figure 6-2 A Case In Need Of Sub-Pel ME/ MC**

### Interpolation Filters and Adaptation

Digital cameras introduce two main negative effects at video capturing – spatial aliasing and Gaussian/Impulsive noise. When an image is captured in the digital domain from analog (i.e., space-continuous) domain, aliasing terms are naturally introduced. To eliminate the aliasing defect, a dedicated analog low pass filter can be applied before obtaining a digital picture. However, such processing cannot be performed since digital cameras acquire images directly on discretely-spaced photo sensors. In addition, photo sensors are noise-prone due to device physics. This noise typically can be seen as RGB "flecks," especially noticeable in the shadow areas of an image. These negative effects obtained at capturing time become more serious with fractional-pel ME and MC since such noise can be propogated to interpolated pixels in the fractional resolution domain. Recent researches, for example, show that aliasing introduces the worst distortion at interpolated half-pel domain data, while the impact of aliasing on the prediction error vanishes at full-pel displacements. This obervation resulted in many recent video standards' adopting highly accurate interporlation filters (4-tap FIR filter for VC-1/ 6-tap FIR filter for H.264). On the other hand, it is important to understand that Gaussian/Impulsive noise issues with ME/MC are purely an "encoder issue." It doesn't need to be official in standard tools. Many encoder venders have their own approach to enhance encoded video quality with all proprietary solutions.

VC-1 is probably one of the first standards that introduces multiple interpolation methods for ME/ MC. Based on scenarios, a VC-1 encoder on the fly can "adpatively" select its interpolation policy and write it on the bitstream. Then, any decoder can also adaptively choose one of multiple interpolation methods as the encoder commands. A key consideration for a VC-1 encoder is complexity and quality. Interpolation operation requires heavy computation in general. If an encoded video is targeted for a low quality-small bandwidth application, spending too much computation for interpolation at ME/ MC might not be a good policy. Therefere, VC-1 has multiple options w.r.t. complexity and quality for adaptation of interpolation methods.

## Unidirectional Prediction and Bidirectional Prediction

Figure 6-1 shows unidirectional prediction, while Figure 6-3 illustrates bidirectional prediction in time. Unidirectional predicted pictures, denoted as P, take advantage of translational movement of contents whose representation is captured with a MV as shown in Figure 6-1. Sometimes a uniform area for the same movement can be even smaller than that of a MB such as 4x4 or 8x8. Bidirectional predicted pictures, denoted as B, come in between two reference I/P pictures or P/P pictures, in general, as shown in Figure 6-3. Typically compression efficiency is better in B pictures than in P pictures, as noted in Chapter 3. However, this doesn't mean that constructing more B pictures between two reference I/P pictures or P/P pictures produces a higher compression ratio. This is because constructing more B pictures implies the distance between P pictures or I/P pictures becomes large, thus making their temporal correlation decrease. In other words, more B pictures would get higher compression efficiency over all, but the compression efficiency of P pictures would get worse. B pictures perform well when there is a big scene change between two reference I/P pictures or P/P pictures. To this end, B pictures have generally two or four options to get a reasonable predictor with "forward" MV, "backward" MV, "interpolative" MV or "direct" MV modes. The first two options (forward MV, backward MV) work well with a big scene change. For example, if the B picture comes before a big scene change, reference 0 would be better suited for prediction as shown in Figure 6-3. If the B picture comes after a big scene change, reference 1 would be better suited for prediction as shown in Figure 6-3. If either reference is not close enough due to high speed motion, the average of the references could be a good predictor candidate. This is the third option, called "interpolative." The fourth "direct" mode can be a good candidate when solid translational movement happens in this period. The next section explains this idea. If any of these prediction modes cannot produce a reasonable predictor for a MB in B pictures, the MB can be coded as Intra mode.

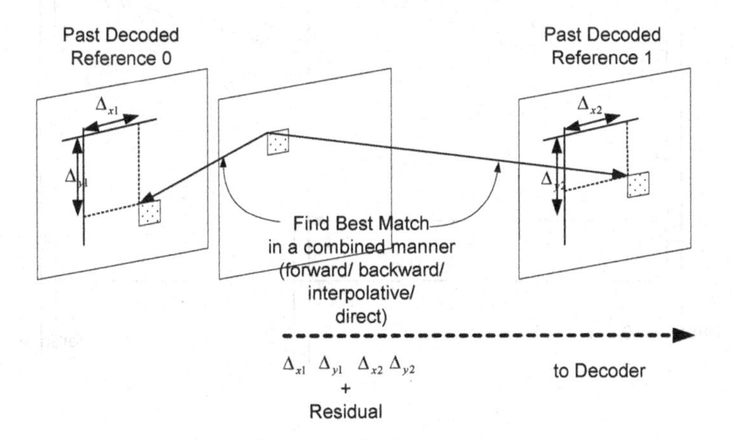


**Figure 6-3 B Prediction**

### Direct Mode in Bidirectional Prediction

The key idea of "direct" mode is to represent a MB without any MV data for a MB in B pictures. The missing MV data is derived from the co-located MB in a P picture as shown in Figure 6-4. This derivation holds only when solid translation motion happens in the scene. Let's say the object in reference 0 is translationally moving with $MV_C$ in reference 1. If the object is solid and its speed is constant, neighborhood areas are also moving with the same direction and speed. To compute MVs for a MB in a specific B picture, $MV_C$ can be proportionally divided as shown in Figure 6-4. Here *TD* means temporal distance. The forward/ backward MVs of a "direct mode" declared MB in B pictures are computed from those at co-located MB in the reference P picture. The MVs are obtained as follows:

$$MV_0 = \frac{TD_B}{TD_D} \times MV_C \text{ , } MV_1 = \frac{TD_B - TD_D}{TD_D} \times MV_C \qquad (6\text{-}1)$$

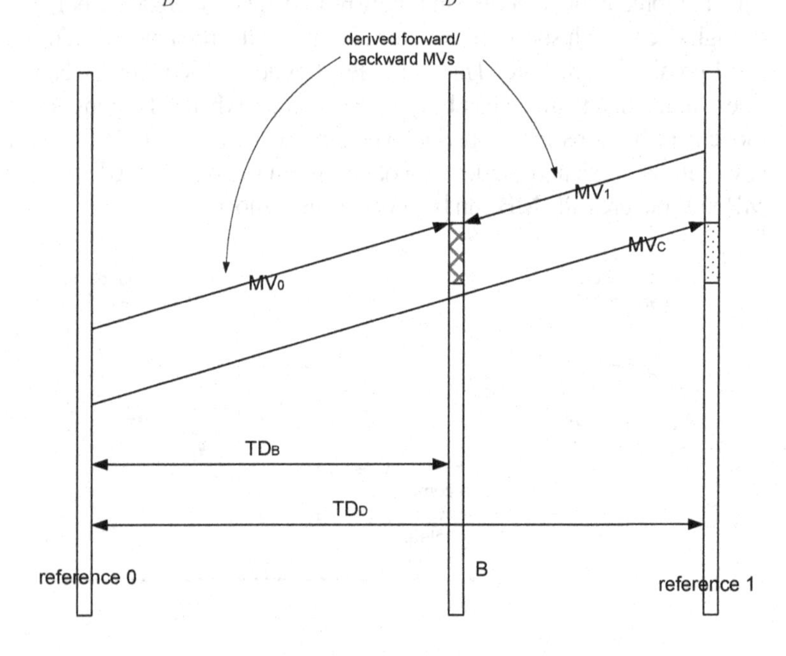


**Figure 6-4 Direct Mode based on MV of Co-located MB**

## Display Order and Coding Order

Since B pictures utilize past decoded reference pictures, future reference P pictures should be coded before B pictures. In other words, display order and coding order are different, in general, as shown in Figure 6-3. This aspect introduces certain re-ordering delays after decoding. One of important aspect of H.264 B slices is that they no longer need to be in between two reference pictures. In other words, conventional display order doesn't hold in H.264 as shown in Figure 6-5. In addition, pictures containing B slices can be used as reference pictures in the standard as shown in Figure 6-5. Chapter 3 explains about this in detail.

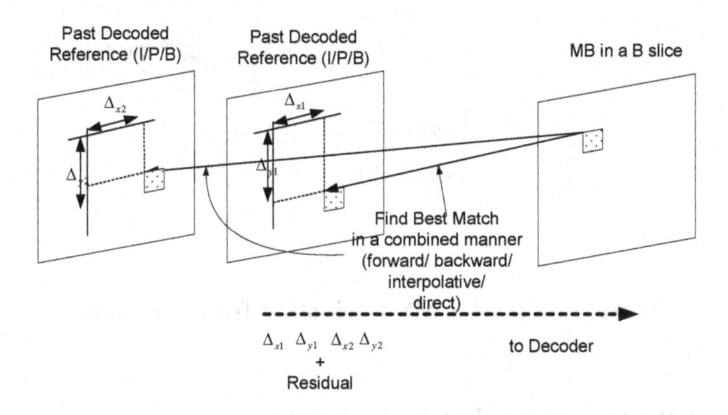


**Figure 6-5 H.264 B Prediction**

## Chroma Motion Vectors

The term "MV" generally means displacement information for Luma data, not for Chroma data. In compression systems, no dedicated Chroma MV is inscribed in the bitstream. Instead, the Luma MV is used to derive the Chroma MV. This is basically an issue of cost and performance in Inter prediction. If separate Luma/ Chroma MVs were used, prediction results might give better performance. However, such systems might be pretty expensive due to extra bits in terms of performing Inter predicted compression. The general observation is that the location of the best match in the Chroma domain is very close to that of Luma data. Therefore, "derivation" of chroma MVs from luma MVs works satisfyingly. The interpretation of position should be performed based on Chroma format, though, as shown in Figure 6-6.

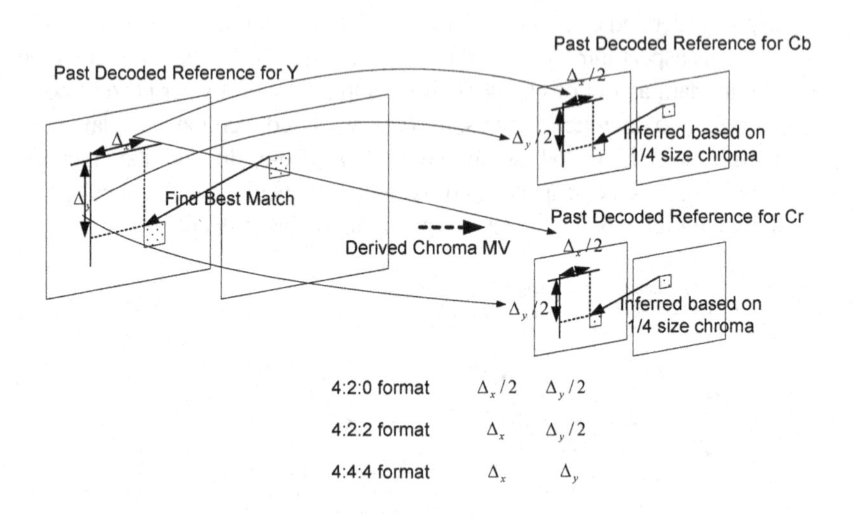


**Figure 6-6 Chroma MV Derivation from Luma MV**

## Transform Coding of Residual Signals

With MVs, residual data is also sent to decoders. Typically, residual data is compressed with Transform coding. The Transformed residual data is zig-zag scanned to form as many zero runs as possible. Then, statistical representation of such codes is used for compact expression. Huffman codes are the most popular one. When a decoder receives MV and residual data, it tries to copy the best match from past decoded reference pictures based on MV data. Then, the decoded residual is combined with data from the best match.

## Visual Weighting and Quantization

When residual data is compressed, quantization and visual weighting are applied. As discussed in the previous section, visual weighting for Inter predicted MBs is generally fixed and uniform to incorporate the aspect of temporal masking effects. Quantization is controlled by a rate control policy that, again, tries to utilize contrast sensitivity and temporal masking effects in various parts of the rate control algorithm.

### Motion Vector Predictor and Motion Vector Differential

Coding the MVs for current the MB or the current 8x8 block (when 8x8 block partition is allowed) can be efficient with a MV Predictor (MVP). Most recent video standards take advantage of locality of MVs in adjacent blocks. One popular candidate is selected from blocks in left (A), top (B) and top-right (C) positions as shown in Figure 6-7. Typically the MV Predictor MVP is taken as median value of three predictor candidates as follows:

$$MVP.x = median(MV_A.x, MV_B.x, MV_C.x) \quad (6\text{-}2)$$

$$MVP.y = median(MV_A.y, MV_B.y, MV_C.y). \quad (6\text{-}3)$$

Note that median operator is applied independently on each x or y component of the three candidate MVs. The MV differential (MVD) is obtained based on difference values between the MVs of the current block and *MVP* s.

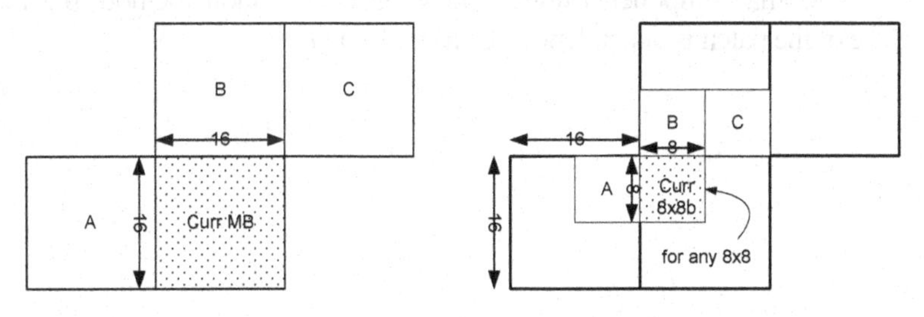


**Figure 6-7 MV Predictor Example for MVD with 16x16, 8x8**

### Inter Prediction in MPEG-2

The size of ME/ MC in MPEG-2 is 16x16 in Luma for progressive video. The size can be halved for interlaced video as explained in Chapter 8. The interpolation method of MPEG-2 video is a simple average into the half-pel domain. All MVs are specified to an accuracy of a ½-sample. If a component of a MV is odd, the samples are read from mid-way between the actual samples in the reference pictures, where half-pels are computed by simple linear interpolation from the actual samples. Due to the data requirement for complete interpolation, 17x17 size Luma data should be fetched from reference memory for a hardware MC unit as shown in Figure 6-9.

Full-pel domain data are defined as *A*, *B*, *C* and *D*, while half-pel domain data are defined as *a*, *b*, *c* and *d* as shown in Figure 6-8. The interpolation method for Luma component is proposed in MPEG-2 as follows:

$$a = round((A+B)/2)\ ,\ b = round((A+C)/2), \qquad (6\text{-}4)$$

$$c = round((A+B+C+D)/4)\ ,\ d = round((B+D)/2),$$

$$e = round((C+D)/2).$$

Chroma components utilize the same interpolation method, but the size of the patch is dependent on Chroma format.

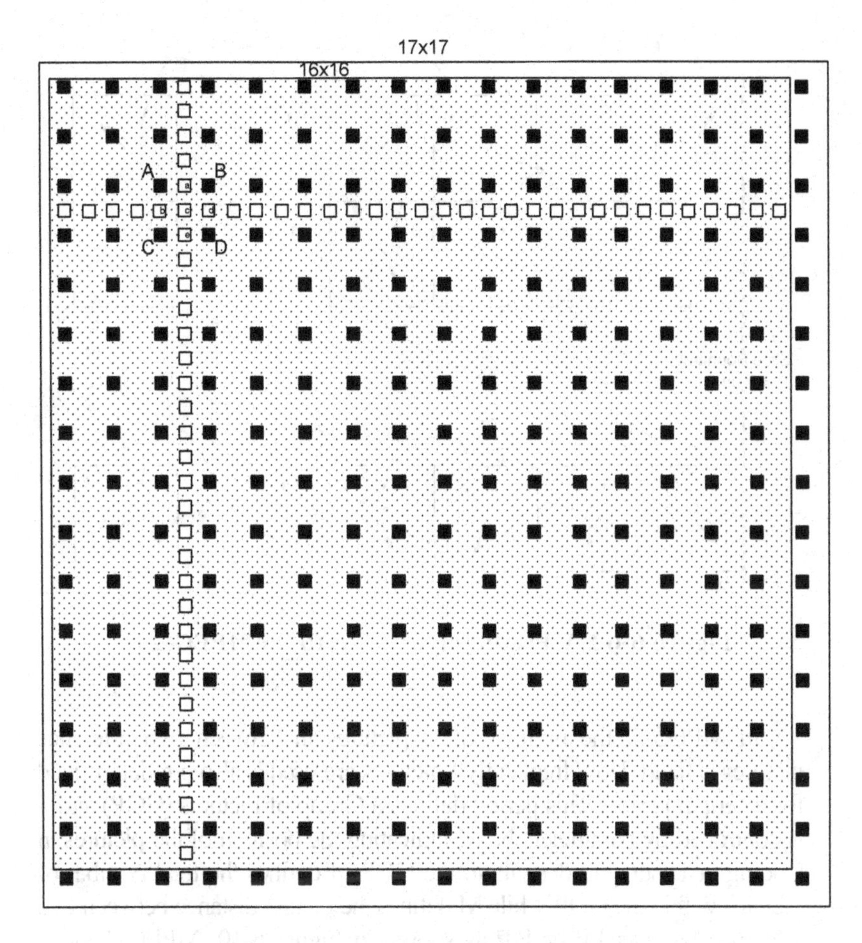


**Figure 6-8 Inter Predictor with 16x16 for MPEG-2**

The MVP of MPEG-2 video is not taken as median values of candidates from adjacent MBs. For I/P pictures, MVP is the MV of the previous MB, if it has one, as shown in Figure 6-9. Otherwise, $MVP = 0$.

For B pictures, the forward MVs are predicted from forward $MVP$, while the backward MVs are predicted from backward $MVP$. In other words, predictions are performed separately.

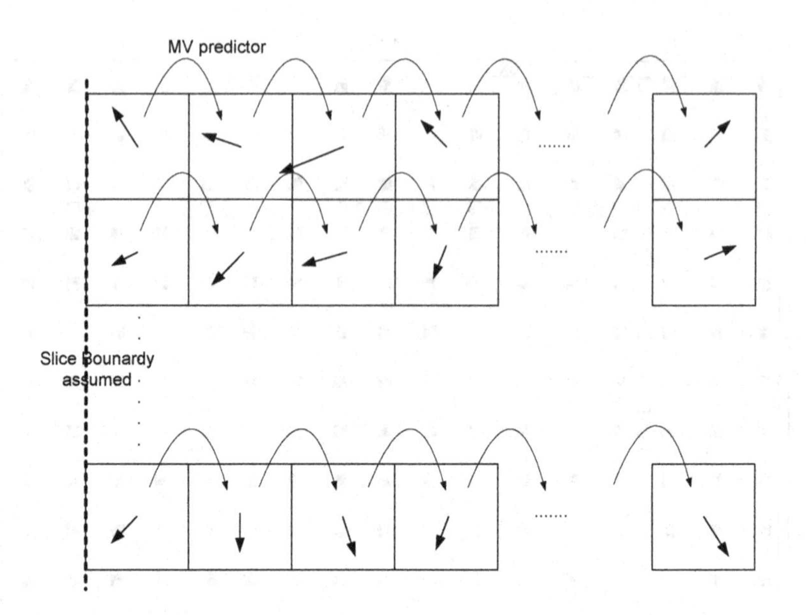


**Figure 6-9 MV Predictor with 16x16 for MPEG-2**

In MPEG-2, the number of B pictures in between two reference pictures is fixed. How frequently I pictures are used is also pre-determined for a certain period. This period defines a Group of Pictures (GOP) and it is controlled by parameters N, M indicating the position of the pictures in Coding or Display Order within the GOP. N defines the total number of pictures within the GOP, while M defines the picture distance between two reference pictures I/P or P/P as shown in Figure 6-10. MPEG-2 video standard allows GOPs to be of arbitrary structure and length. In other words, it is legal not to have any GOP headers in a MPEG-2 stream so that no pre-determined period is used in the stream. In this case, any number of decoded B pictures could be positioned in the most recently decoded reference pictures for rearrangement in the Display Order. Note that distance of two reference pictures does not matter too much in decoding process since there is no Direct mode in MPEG-2. The GOP header may be used as the basic unit for editing an MPEG-2 video stream.

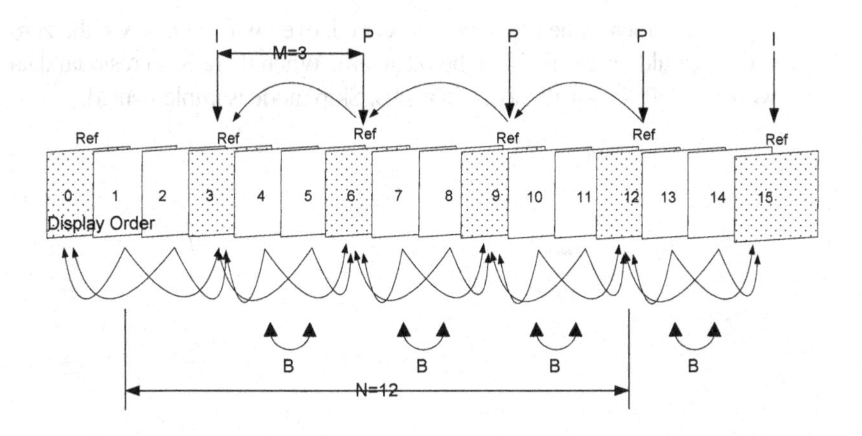


**Figure 6-10 GOP Example with N=12 and M=3 for MPEG-2**

The MBs in B pictures have only three MV mode options – forward MV, backward MV and interpolative MV. There is no Direct mode in MPEG-2. Coding modes for P pictures and B pictures in MPEG-2 video can be summarized as shown in Figure 6-11. Three MB coding types are defined in this book for convenience – "coded," "not coded" and "skipped." The "coded" type means to contain any Transform residual data, while the "not coded" type means to contain only MV data without any residual data. In contrast, Skip mode doesn't contain any MV data or any Transform residual data. Note that skipped MBs in B pictures have the same MVs and the same MB types as the previous MB, while skipped MBs in P pictures have "zero" MVs. The reason to treat skipped MBs differently in P and B pictures is that B-MB has forward/ backward/ interpolative MV modes. When the MVD becomes zero, the information about which MV mode that induces the MVD being zero is necessary for the decoder to infer. In contrast, P-MB only has "forward directional" MV modes for both Progressive and Interlace videos (Interlace options are covered in Chapter 8). Therefore, interpretation for the MVD has no confusion about which MV the inference should be performed from in terms of MV mode. The interpretation for skipped MB is based on MB address increment in MPEG-2, thus making no assignment of VLC for this mode.

When at least one block is to be coded, even with zero MVs, the zero vector should be inscribed in the bitstream. When there is no residual data with zero MVs with all blocks in a MB, Skip mode is implemented.

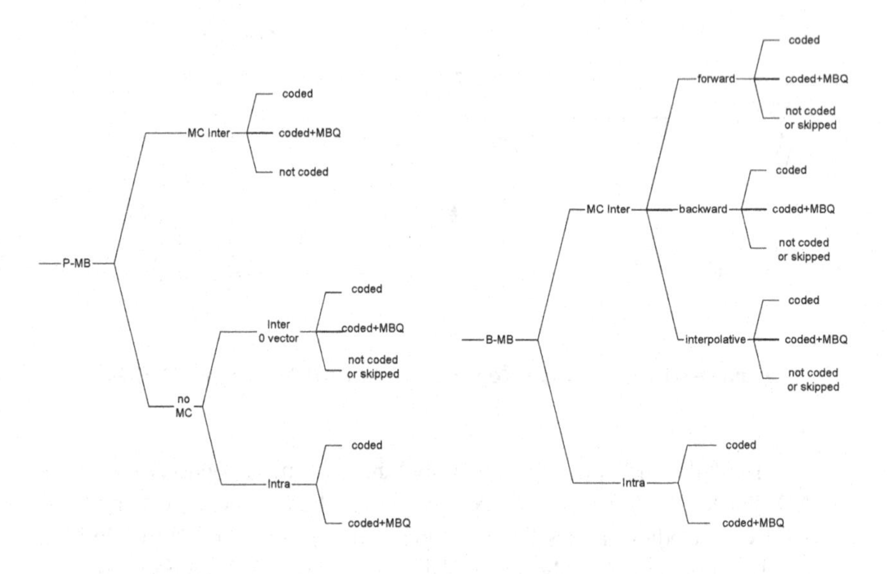


**Figure 6-11 MB Coding Mode in P and B Pictures for MPEG-2**

## 6.2 VC-1 Inter Prediction

### MC Block Partitions

As shown in Figure 6-12, a 16x16 luma can be broken into four 8x8 areas for individual MC. The 8x8 size MC is pretty effective when the area to cover is not uniform in motion of texture such as in object boundaries. For example, one of four 8x8 blocks in an object boundary MB can move in different direction when it falls in the background with three other blocks occupied in the foreground. In such cases where motion is not uniformly formed, smaller regions for MC can provide better performance in compression. This has been adopted in previous standards such as MPEG-4 Part 2. VC-1 moves forward one more step. Any of the 8x8 blocks in a MB can be coded as Intra mode as shown in Figure 6-12. For example, when a new pattern reveals from the background (previously hidden) as foreground content is moving, the Intra coding option can be considered for the area.

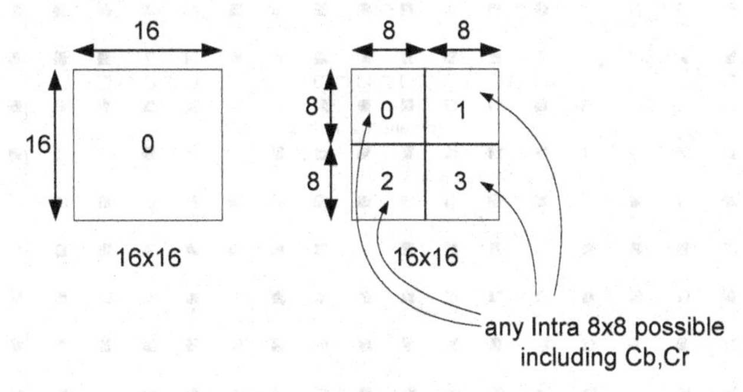


**Figure 6-12 Partition for Motion Compensation in VC-1**

## Luma Interpolation

The size of ME/ MC in VC-1 is 16x16 or 8x8 in luma for progressive video. The size can be halved for interlace video as explained in Chapter 8. Multiple interpolation methods of VC-1 video are proposed and one of the methods is selected by the encoder. The key idea is to prioritize MC methods based on a combination of size of MC/ interpolation methods/ pel-accuracy. VC-1 proposes four MC methods as follows: 1. 16x16 block size (1MV) ½-pel bilinear, 2. 16x16 block size (1MV) ½-pel bi-cubic, 3. 16x16 block (1MV) ¼-pel bi-cubic, 4. 8x8 block (4MV) ¼-pel bi-cubic MC methods. Note that the order of arrangement is in order of complexity and quality. Each application scenario can select a MC method. For example, mobile handheld devices have relatively low computational power and typically do not need high quality reproduction of video. In such a case, 16x16 block size ½-pel bilinear interpolation-MC must be a good candidate for an encoder to select since this doesn't require high computational power and produces good quality. For HD-DVD, 8x8 block ¼-pel bi-cubic interpolation-MC might be a good choice since this would produce excellent quality through sophisticated predictors. Due to data requirement for complete interpolation, 19x19 size Luma data should be fetched from reference memory for a hardware MC unit as shown in Figure 6-13. When only ½-pel resolution is required, interpolation up to ½-pel is performed. When ¼-pel resolution is demanded, interpolation up to ¼-pel is performed. The size to fetch for an 8x8 block is 11x11 as shown in Figure 6-15.

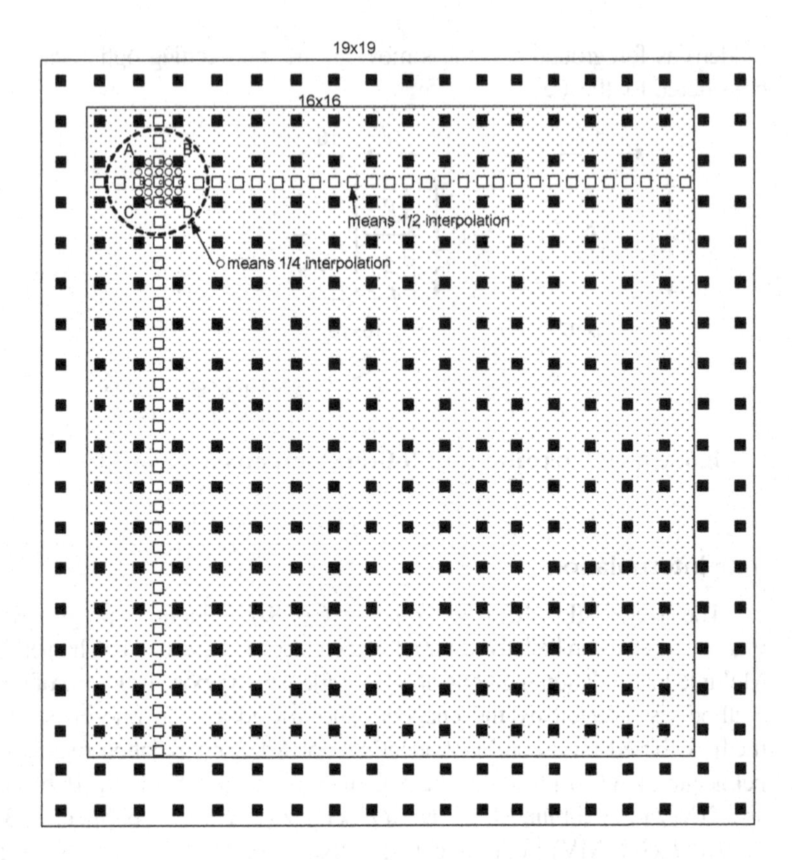


**Figure 6-13 19x19 Fetch Out Area for 16x16 Luma Block**

Note that interpolation methods proposed for Luma and Chroma are the same as shown in Figure 6-13 and Figure 6-15, respectively. Three extra neighboring pixel rows/ columns need to be considered for Luma and Chroma interpolation. There are two interpolation methods defined in VC-1 – bi-linear and bi-cubic. Applied orders are not important for bi-linear interpolation since outcomes are defined to compute from the four closest integer pixels in a 2D manner. However, the applied orders of bi-cubic interpolation are critical and defined to be vertical before horizontal since the interpolation method is defined to compute from four consecutive integer pixels in a 1D manner. And, rounding operation is parameterized with a *RND* value. Bi-linear interpolation in Figure 6-13 is computed as follows:

$$\text{Given } F[] = \{4,3,2,1,0\} \text{ and } G[] = \{0,1,2,3,4\}, \tag{6-5}$$

$$p = (F[x]F[y]C + F[x]G[y]A + G[x]G[y]B + G[x]F[y]D + 8 - RND) >> 4$$

where $x$ and $y$ are the sub-pixel shifts in the horizontal (left to right) and vertical (bottom to top) direction multiplied by four.

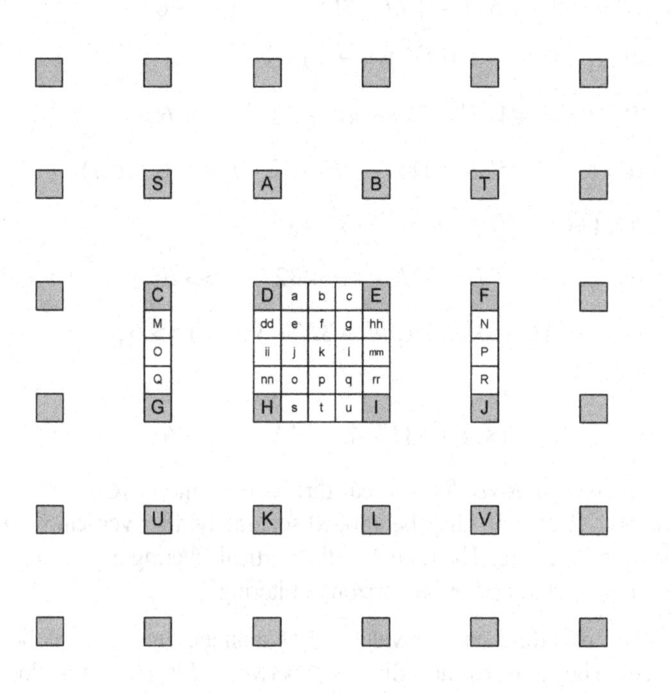


**Figure 6-14 Luma Interpolation for VC-1**

Bi-cubic interpolation in Figure 6-14 is computed as follows:

$$dd = (-4A + 53D + 18H - 3K + 32 - r) >> 6, \tag{6-6}$$

$$ii = (-A + 9D + 9H - K + 8 - r) >> 4,$$

$$nn = (-3A + 18D + 53H - 4K + 32 - r) >> 6,$$

$$hh = (-4B + 53E + 18I - 3L + 32 - r) >> 6,$$

$$mm = (-B + 9E + 9I - L + 8 - r) >> 4,$$

$$rr = (-3B + 18E + 53I - 4L + 32 - r) >> 6,$$

$$M = (-4S + 53C + 18G - 3U + 32 - r) >> 6,$$

$$O = (-S + 9C + 9G - U + 8 - r) >> 4,$$

$$Q = (-3S + 18C + 53G - 4U + 32 - r) >> 6,$$

$$N = (-4T + 53F + 18J - 3V + 32 - r) >> 6,$$

$$P = (-T + 9F + 9J - V + 8 - r) >> 4,$$

$$R = (-3T + 18F + 53J - 4V + 32 - r) >> 6,$$

$$a = (-4C + 53D + 18E - 3F + 32 - r) >> 6, \quad (6\text{-}7)$$

$$b = (-C + 9D + 9E - F + 8 - r) >> 4,$$

$$c = (-3C + 18D + 53E - 4F + 32 - r) >> 6,$$

$$e = (-4M + 53dd + 18hh - 3N + 32 - r) >> 6,$$

$$\ldots\ldots\ldots,$$

$$u = (-3G + 18H + 53I - 4J + 32 - r) >> 6,$$

where *r=1-RND* for vertical direction, while *r=RND* for horizontal direction. Rounding shall be applied separately after vertical filtering and horizontal filtering. The output of the vertical filtering after rounding shall be used as the input for the horizontal filtering.

The rounding control value *RND* manages rounding policy in the codec. The policies are different between Simple/Main Profile and Advanced Profile. For Simple/Main Profile, at each I and BI picture, the value *RND* is reset to 1. The *RND* toggles between 0 and 1 at each P picture. And, the *RND* for B pictures remain the same as prior anchor pictures (I or P). For Advanced Profile, *RND* is derived from RNDCTRL syntax elements in the Picture header.

### Chroma Interpolation

Chroma components always utilize the bi-linear interpolation method, and only one chroma format, YCbCr 4:2:0, is currently defined for VC-1. Due to the data requirement for complete interpolation, 11x11 size

Chroma data should be fetched from reference memory for a hardware MC unit as shown in Figure 6-15.

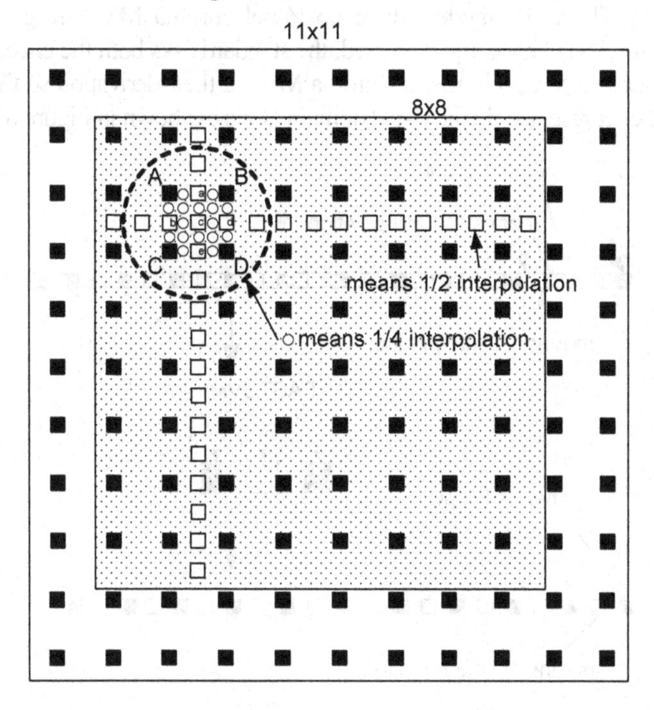


**Figure 6-15 11x11 Fetch Out Area for 8x8 Luma/Chroma Block**

As was described in the previous section, a Luma MV generally has a resolution of ¼-pel. Straightforward derivation of Chroma MV naturally has resolution of 1/8-pel. However, as shown in Figure 6-15, Chroma data is interpolated up to the ¼-pel domain only and not up to the 1/8-pel domain as in VC-1. The method suggested in VC-1 to restrict Chroma MVs to the ¼-pel domain is to "biased" round Luma MV/2 into the ¼-pel domain. The fundamental rule is to look at the last two bits of a Luma MV to process differently. The MVs whose last two-bit values are "0," "1" or "2" are simply divided by two. The MVs whose last 2-bit values are "3" are incremented by one and then divided by two.

There is a mode called "FASTUVMC" designed for software decoders to reduce computational complexity for the interpolation operation because the interpolation unit is one of the most complicated parts in software decoders in terms of computation. VC-1 encoders can use FASTUVMC mode, where no ¼-pel chroma MVs are generated. When FASTUVMC mode is used, the standard asks both the encoder and the decoder to deeply round Chroma MVs at their derivation so that they can be represented in the ½-pel domain alone as shown in Figure 6-16.

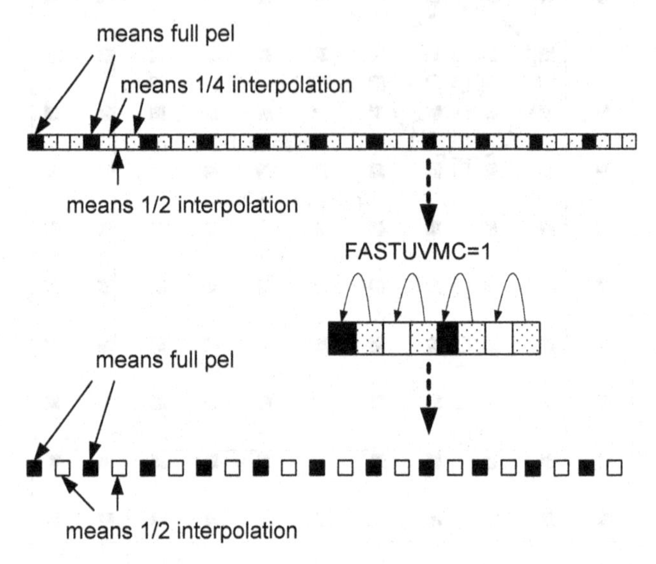


**Figure 6-16 Chroma Resolution Restriction with FASTUVMC**

## Extended Padding and Motion Vector Pullback

When a solid object moves into a scene or a camera is panning to generate a global motion as shown in Figure 6-17, inter prediction suffers at picture edges due to reference loss or newly coming pattern. Therefore, picture boundaries produce too much Transform coefficient data and MV differential (MVD). To effectively represent Transform coefficients in picture boundaries, extended padding techniques have been recently introduced in video coding such as in MPEG-4 Part 2. The idea is to extrapolate edge pixels of the picture so that artificially generated texture can be used as virtual reference data as shown in Figure 6-17.

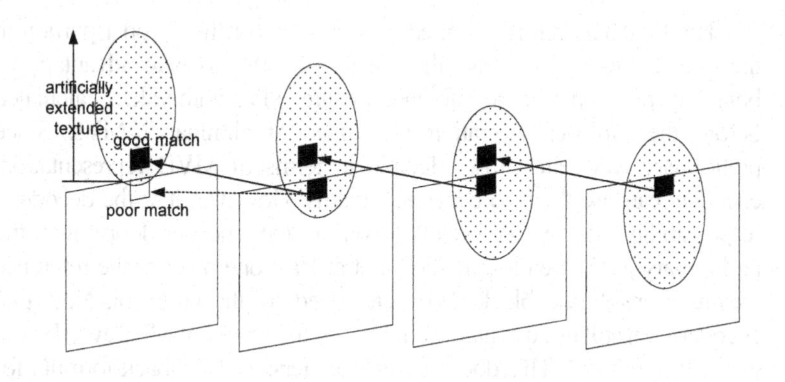


**Figure 6-17 Effectiveness of Extended Padding**

To effectively represent MVD in picture boundaries, infinitely (i.e., in actuality, its maximum is MV range maximum) extended padding techniques are recently introduced in video coding such as H.264. The idea is to extrapolate edge pixels of the picture large enough so that a big global motion vector designated by a single MV (or few MVs) doesn't need to be refracted on picture boundaries as shown in Figure 6-17 and Figure 6-18.

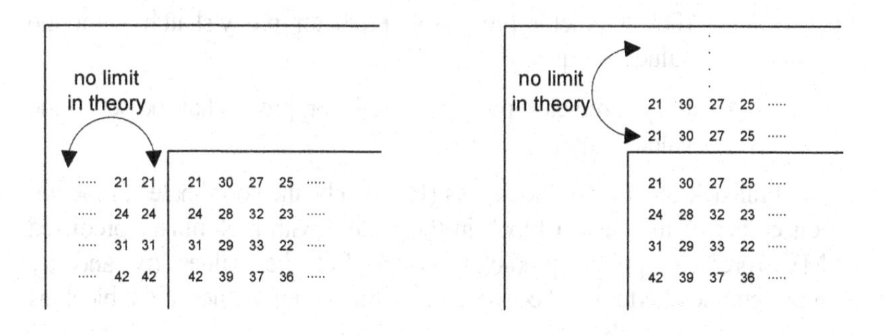


**Figure 6-18 Extended Padding before MV Pullback**

The Pullback unit is designed for decoders (or the decoder portion in the encoder loop) to rectify the values of MVs to take advantage of boundary pixel data in the reference picture. MVs with values calculated before the Pullback operation are called "preliminary" MVs. Since preliminary MVs are most effective in terms of MVD representation, encoders can use them to generate streams. However, it is the decoder's responsibility (or the decoder portion in the encoder loop) that the preliminary MVs are clipped such that at least one pixel of the reference picture is inside the block/ MB referenced by the predictor. Note that encoders can still use this algorithm to limit the values of MVs even before generating streams. This doesn't hurt decoders. The Pullback formula for the operation are defined as follows:

Let $X=16\times((\text{CodedWidth}+15)/16)\times 4-4$ and $Y=16\times((\text{CodedHeight}+15)/16)\times 4-4$.

Pullback of MVP of MB: Let (MBx, MBy) be the current MB in the picture with preliminary predicted MV (predictor_pre_x, predictor_pre_y). Let the values qx and qy represent the quarter pixel coordinates of the top left corner of the MB as $qx=MBx\times 16\times 4$, $qy=MBy\times 16\times 4$.

- If qx+predictor_pre_x<-60, predictor_pre_x shall be set to the value (-60-qx).
- If qx+predictor_pre_x>X, predictor_pre_x shall be set to the value (X-qx).
- If qy+predictor_pre_y<-60, predictor_pre_y shall be set to the value (-60-qx).
- If qy+predictor_pre_y>Y, predictor_pre_y shall be set to the value (Y-qx).

Pullback of MVP of block: Let (Bx, By) be the coordinates of the top left corner of the current block in the picture with preliminary predicted MV (predictor_pre_x, predictor_pre_y). Let the values qx and qy represent the quarter pixel coordinates of the top left corner of the block as $qx=Bx\times 8\times 4$, $qy=By\times 8\times 4$.

- If qx+predictor_pre_x<-28, predictor_pre_x shall be set to the value (-28-qx).
- If qx+predictor_pre_x>X, predictor_pre_x shall be set to the value (X-qx).

- If qy+predictor_pre_y<-28, predictor_pre_y shall be set to the value (-28-qy).
- If qy+predictor_pre_y>Y, predictor_pre_y shall be set to the value (Y-qy).

### Hybrid Motion Vector Prediction

When a solid content is moving in a scene, more than two MBs or more than two blocks are moving rigidly. In such a case, the MVP value might be quite different from that of the actual MV. Let's say, the area of the current MB is rigidly moving together with MB-C as shown in Figure 6-19. As depicted, let's consider that MB-A and MB-B are moving in the same direction. Then, the MVP, which is median value of three MBs, could be very different from the direction of the actual MV. VC-1 allows specification of this situation in a syntax element called HYBRIDMV. The syntax element explicitly declares which MV is used as the predictor among MVs of MB-A and MB-C. Note that MB-B is excluded in this representation as shown in Figure 6-19. This is only valid when one of those MVs is too different (i.e., difference measure more than 32) from the regularly predicted MVP as shown in Figure 6-19. Note that Hybrid Motion Vector Prediction is only allowed for P pictures, not for B pictures.

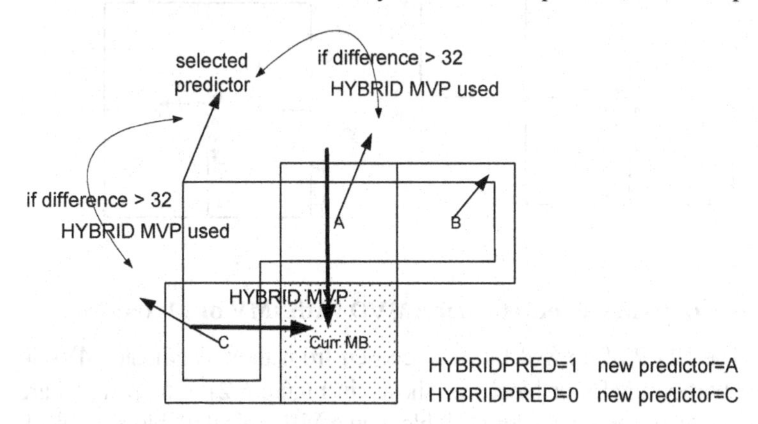


**Figure 6-19 Hybrid Motion Vector Prediction**

## Motion Vector Predictors

The MVP of a MB is chosen from three candidate MVs in neighborhood MBs or blocks as shown in Figure 6-20. The upper two patterns are for 16x16 block size MC mode. In typical cases, the first pattern is used for selecting candidates. If the position of current MB is the last one in a row, the MB-B is not available. In such a case, the second pattern is used for selecting candidates. The lower two patterns are for mixed size MC mode, where the current MB is on 16x16 block size MC and neighborhood MBs are on 8x8 size MC mode. They are similar patterns for selecting candidates from neighborhood MBs except that closest 8x8 blocks are selected out of four 8x8 blocks in each MB.

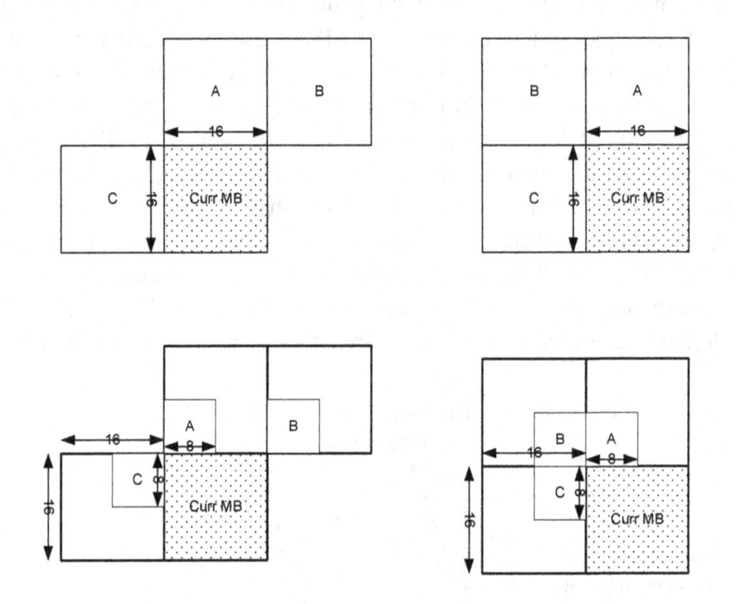


**Figure 6-20 MV Predictors for MVD with 1MV or Mixed-MV**

The MVP of an 8x8 block is chosen from three candidate MVs in neighborhood MBs or blocks as shown in Figure 6-21. The patterns are depicted with candidates for each block in a MB – block 0, block 1, block 2 or block 3. For some special cases, certain different patterns are chosen mainly due to unavailability of MV data in the neighborhood as shown in Figure 6-21.

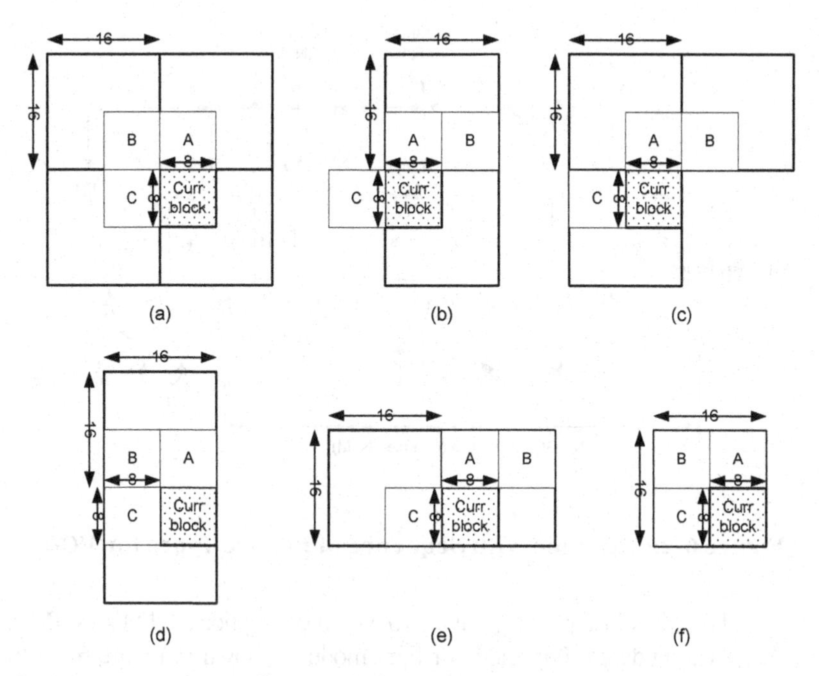


**Figure 6-21 MV Predictors for MVD for 4MV: (a) Predictors for block 0 if not the first MB in row, (b) Predictors for block 0 if first MB in row, (c) Predictors for block 1 if not last MB in row, (d) Predictors for block 1 if last MB in row, (e) Predictors for block 2, (f) Predictors for block 3**

## Sequence of Picture Types

In VC-1, there is no fixed GOP structure as shown in Figure 6-22. I and P pictures can be reference pictures. However, selection of either I and P entirely depends on the encoder's choice. Therefore, there is no fixed N parameter. In addition, the distance between two reference pictures is not fixed. However, there is a maximum consecutive number allowed for B pictures – it is seven. More than seven B pictures cannot be in between two references so that a tolerable compression can be achieved due to reasonable correlation between two references.

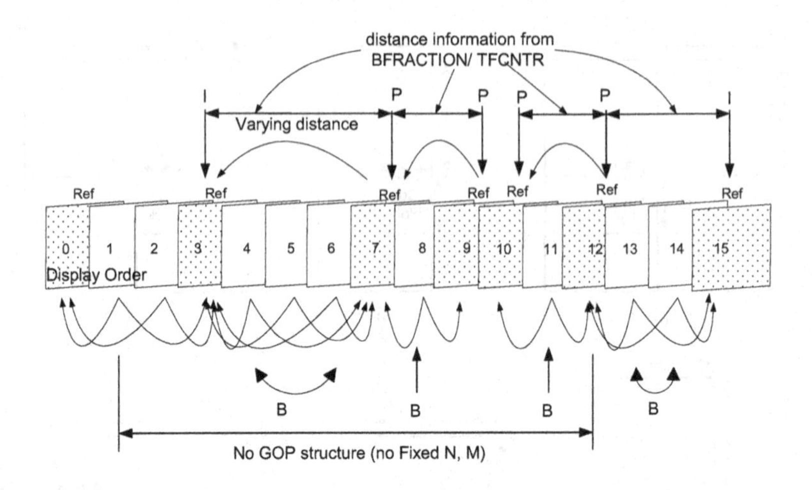


**Figure 6-22 Example with Sequence of Picture Types for VC-1**

The MBs in P pictures have two MV mode options – 1MV or 4MV. MBs can undergo Skip mode or Intra mode as shown in Figure 6-23. In addition, some 8x8 blocks can be coded as Intra mode, too. Skip mode for the 1MV case means that there is no MVD and no residual data for corresponding 1 MV. Skip mode for 4MV case implies that there are no MVDs and no residual data for corresponding 4 MVs. The MBs in B pictures have four MV mode options – forward MV, backward MV, interpolative MV and direct MB. Coding modes for P pictures and B pictures in VC-1 video can be summarized as shown in Figure 6-24. Note that skipped MBs in B pictures have the same MVs as that of the predictor and the same MB types as the previous MB while skipped MBs in P pictures have only the same MVs as that of the predictor. The reason to treat skipped MBs differently in P and B pictures is that B-MB has forward/ backward/ interpolative/ direct MV modes. When MVD becomes zero, the decoder must infer which MV mode induces MVD to be zero. In contrast, P-MB only has "forward" MV mode. Therefore, interpretation for MVD has no confusion about which MV to use. When at least one block is to be coded zero MVD, the zero MV should be inscribed in the bitstream. When there is no residual data with zero MVDs in all blocks in a MB, the Skip mode is implemented. Figure 6-24 applies both for 1MV and 4MV cases.

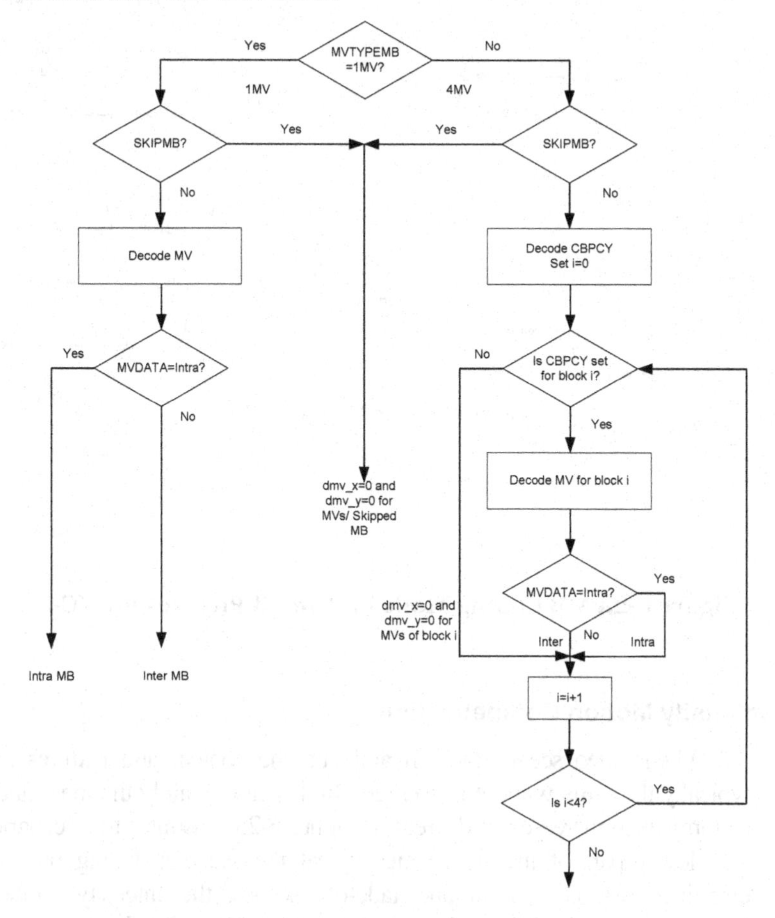


**Figure 6-23 MV Logic in P Pictures for VC-1**

The Skip mode is signaled by the SKIPMB syntax element and it is extended to contain Hybrid MV prediction mode.

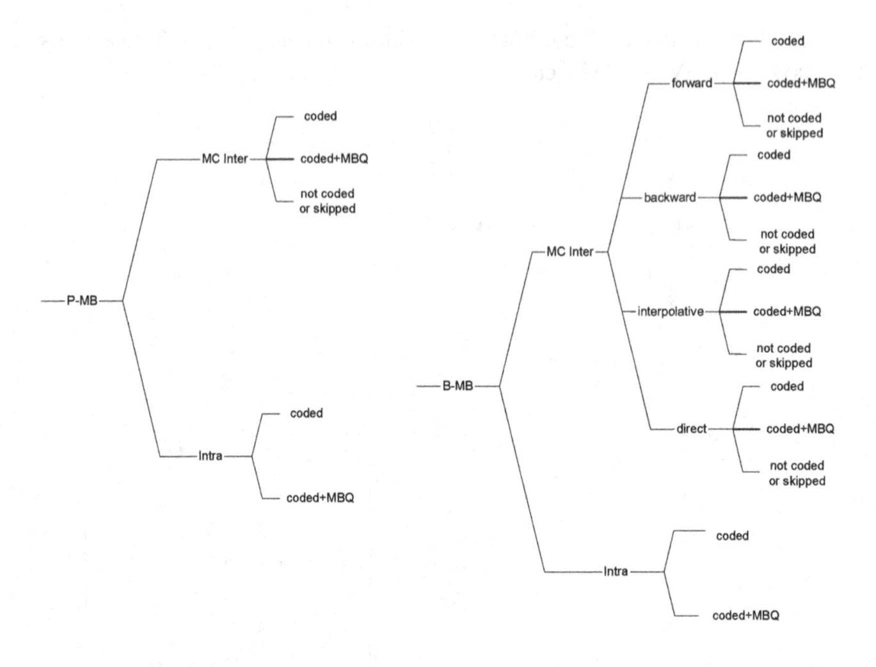


**Figure 6-24 MB Coding Mode in P and B Pictures for VC-1**

## Intensity Motion Compensation

When video scenes fade in and out, the content and textures are typically the same without movement. In this case, mainly the magnitude of luminance increases or decreases. Figure 6-25 illustrates this scenario. The lower part of the figure means that the scene is fading out. To effectively handle fade-in and fade-out scenes, the Intensity Motion Compensation (IMC) technique is devised in VC-1. The key idea is to map Luma and Chroma data of the reference into an intensity-decreased or intensity-increased domain of the reference data as shown in Figure 6-25. If typical MC is applied, pretty big residual data would be produced as shown in Figure 6-25. However, if the same MC is applied for a remapped reference picture, almost minimal amounts of residual data are produced. This technique is only defined for P pictures in VC-1. B pictures are not considered for this tool.

To remap reference data, a linear mapping is suggested in VC-1. To define a linear line, parameters LUMSCALE and LUMSHIFT are introduced. When a VC-1 encoder uses IMC, the encoder should determine parameters LUMSCALE and LUMSHIFT for the best mapping in order that minimal residual data is to produce in the upcoming MC process.

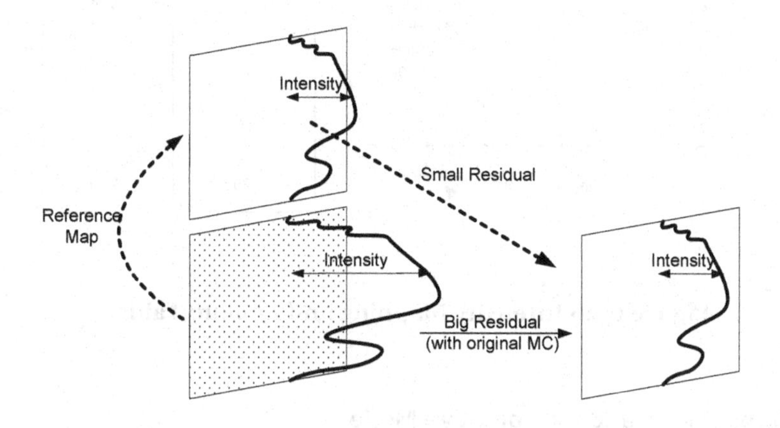


**Figure 6-25 Intensity Compensation**

When a decoder performs IMC, the decoder first builds up a Lookup Table based on LUMSCALE and LUMSHIFT parameters that a VC-1 encoder inscribed in the bitstream. The Lookup Table is a pre-computed mapping table between original intensity and mapped intensity as shown in Figure 6-26. This saves a lot of computational complexity in advance. Otherwise, each mapped data must be computed on the fly at decoders.

Such a mapping shown in Figure 6-26 applies to color components, too. Therefore, a decoder has to prepare two mapping tables as follows:

*mapped_luma_value =LUTY[original_luma_value],* (6-8)

*mapped_chroma_value =LUTUV[original_chroma_value].* (6-9)

The mapped reference picture data shall also be used for prediction of all subsequent B pictures.

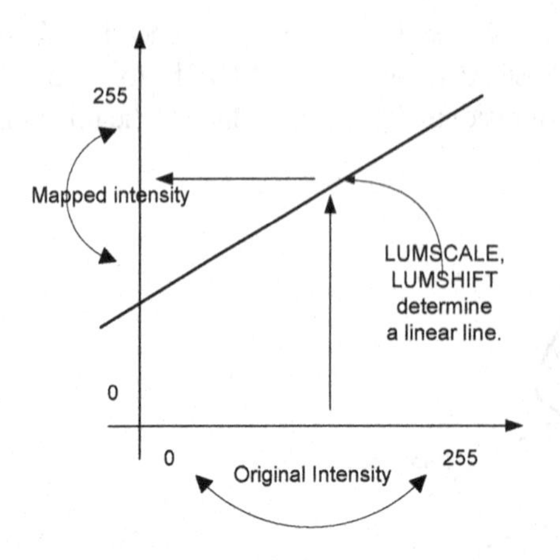


| Original Intensity | Mapped Intensity |
|---|---|
| 0 | 123 |
| 1 | 123 |
| 2 | 124 |
| . | . |
| . | . |
| . | . |
| . | . |
| 255 | 255 |

**Figure 6-26 Intensity Mapping and Lookup Table**

## Direct Mode and Interpolative Mode

MBs with Direct or Interpolative modes use both reference pictures for prediction. They use two sets of MVs for forward and backward MVs. In both cases the pixels shall be interpolated from two reference pictures, which shall be followed by a pixel average operation with round-up to compute the pixels in the inter predictor.

*pixel value in the predictor = (interpolated value from reference 0 + interpolated valued from reference 1 + 1)>>1* (6-10)

In interpolative mode, the forward and backward MVs are explicitly coded within the bitstream. In direct mode, the forward and backward MVs are derived by scaling the co-located MVs from the backward reference picture. These scaling operations are performed in the ¼-pel domain.

When the co-located MB is Intra coded, the direct mode MVs shall be set to (0, 0). When reference P picture's co-located MV is 1MV, that MV will be buffered for the next B pictures to be coded for direct mode. If it is 4MV, the following process is performed:

- median4( ) of the 4MVs shall be used if none of the blocks is Intra coded.

- median3( ) of the 3MVs shall be used if one of the blocks is Intra coded.
- (MVa+MVb)/2 shall be used if 2 of the blocks are Intra coded. Here, Mva and MVb are the MVs of the 2 Inter coded blocks.
- (0, 0) shall be used if 3 or 4 of the blocks are Intra coded.

There are three steps defined to derive direct MVs in the VC-1 standard. The first step is to determine "ScaleFactor" based on BFRACTION information in the Picture Level. The second step is to actually derive forward and backward MVs ( $MV_F.x$ , $MV_F.y$ , $MV_B.x$ and $MV_B.y$ ) based on the determined ScaleFactor. The third step is to apply the Pullback algorithm.

The first step defines several variables of NumShortLVC[ ] = {1,1,2,1,3,1,2}, DenShortVLC[ ] = {2,3,3,4,4,5,5}, NumLongVLC[ ] = { 3,4,1,5,1,2,3,4,5,6,1,3,7}, DenLongVLC[ ] = {5,5,6,6,7,7,7,7,7,7,8,8,8,8} and Inverse[ ] = {256, 128, 85, 64, 51, 43, 37, 32}. The code words for BRACTION are classified into two categories – "short" and "long" code words. Codewords 000b through 110b are considered as "short" code words, while the rest are considered "long." Briefly, short code words cover cases where the number of B pictures is less than five, while long code words cover cases where the number of B pictures is more than five. These variables are all devised to compute Equation (6-1) with the introduction of Lookup Tables only.

For long words, two important variables, FrameReciprocal and ScaleFactor, are derived as follows:

$$Numerator = NumLongVLC[code\ word - 112], \qquad (6\text{-}11)$$

$$Denominator = DenLongVLC[code\ word - 112],$$

$$FrameReciprocal = Inverse[Denominator\text{-}1]\ and$$

$$ScaleFactor = Numerator \times FrameReciprocal.$$

For short words, two important variables, FrameReciprocal and ScaleFactor, are derived as follows:

$$Numerator = NumShortVLC[code\ word], \qquad (6\text{-}12)$$

*Denominator = DenShortVLC[code word],*

*FrameReciprocal = Inverse[Denominator-1] and*

*ScaleFactor=Numerator × FrameReciprocal.*

The Inverse[ ] values are designed to preserve high precision in MV computation through Equation (6-13) and Equation (6-14).

The second step derives $MV_F.x$, $MV_F.y$, $MV_B.x$ and $MV_B.y$ in all quarter-pel units.

For MV modes being 1MV ½-pel or 1MV ½-pel bilinear, $MV_F.x$, $MV_F.y$, $MV_B.x$ and $MV_B.y$ are derived as follows:

$$MV_F.x = 2\times (MV.x\times ScaleFactor+255)>>9), \quad (6\text{-}13)$$

$$MV_F.y = 2\times (MV.y\times ScaleFactor+255)>>9),$$

$$MV_B.x = 2\times (MV.x\times (ScaleFactor-256)+255)>>9),$$

$$\text{and } MV_B.y = 2\times (MV.y\times (ScaleFactor-256)+255)>>9).$$

For all other ¼-pel MV modes, $MV_F.x$, $MV_F.y$, $MV_B.x$ and $MV_B.y$ are derived as follows:

$$MV_F.x = (MV.x\times ScaleFactor+128)>>8), \quad (6\text{-}14)$$

$$MV_F.y = (MV.y\times ScaleFactor+128)>>8),$$

$$MV_B.x = (MV.x\times (ScaleFactor-256)+128)>>8),$$

$$\text{and } MV_B.y = (MV.y\times (ScaleFactor-256)+128)>>8).$$

The third step applies a Pullback algorithm on $MV_F.x$, $MV_F.y$, $MV_B.x$ and $MV_B.y$.

Note that the distance between two reference pictures is forced to 256 to allocate high precision computation in conjunction with Figure 6-4. This approach fundamentally prevents the encoder and decoder pair from possible "drift." The division of 256 can be implemented with >>8. To preserve high precision, rounding operation (i.e., +128 before right-shift 8) is performed through Equation (6-13) and Equation (6-14). The direct mode is signaled by the DIRECTBIT syntax element. If not direct, it is signaled by BMVTYPE.

## 6.3 H.264 Inter Prediction

### MC Block Partitions

H.264 MC block partition has many options as shown in Figure 6-27. The upper illustration represents partitions for the 16x16 area, while the lower part shows partitions for the 8x8 area. These can be combined to represent any MC partition for a MB. For an extreme scenario, a MB can be broken into 16 4x4 areas for individual MC.

Table 2-7 indicates the constraints on Levels of H.264. Note that there is an important provision in MaxMVsPerMb parameter that limits the number of MVs in two consecutive MBs. In other words, the maximum number of MVs in a MB is averaged in two consecutive MBs. Level 3 handles 32 MVs in two consecutive MBs and has the largest number in all Levels.

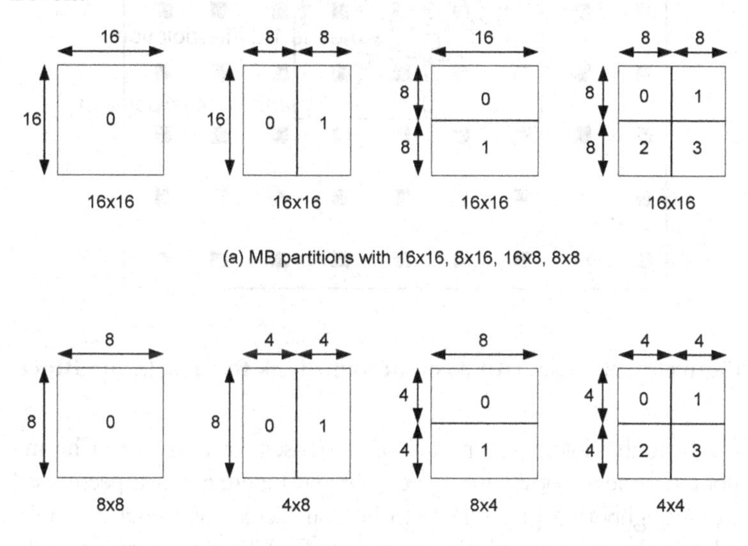


**Figure 6-27 Partition for Motion Compensation in H.264**

### Luma Interpolation

Due to interpolation, an extended 9x9 block data is required for Luma when a 4x4 block is used for MC. To interpolate the right-most columns of a 4x4 block, three pixel columns out of the next block are used as shown in Figure 6-28. Also, three pixel rows out of the block are required to interpolate the lowest rows of the block. Due to the data requirement for complete interpolation, 9x9 size Luma data should be fetched from reference memory for hardware MC unit as shown in Figure 6-28.

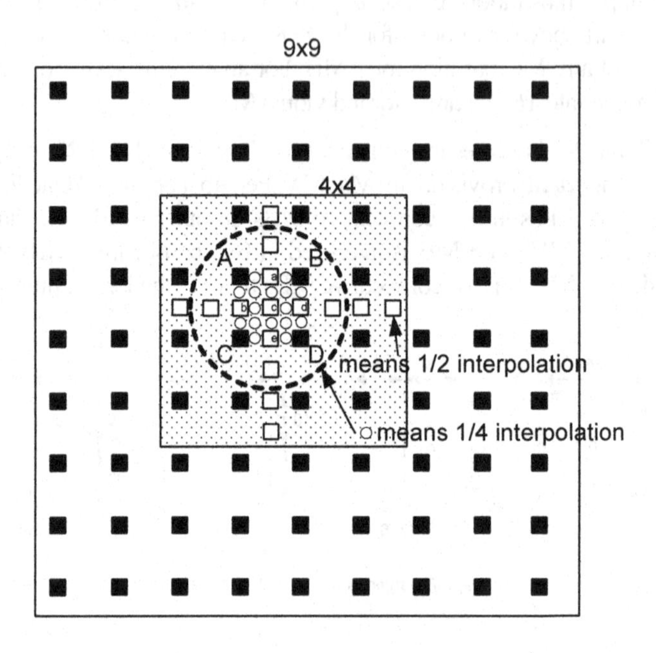


**Figure 6-28 Fetch Out Area of 9x9 Block for 4x4 Luma Block**

Note that interpolation methods proposed for Luma and Chroma are not the same as shown in Figure 6-28 and Figure 6-30, respectively. Five extra neighboring pixels need to be considered for Luma interpolation, while only one extra pixel is necessary for Chroma interpolation. Luma interpolation is performed in two steps – the first step is to apply 6-tap filtering to generate half-resolution pixels, while the second step is to generate quarter-resolution pixels with linear interpolation. The detailed procedures (see Figure 6-29) are as follows:

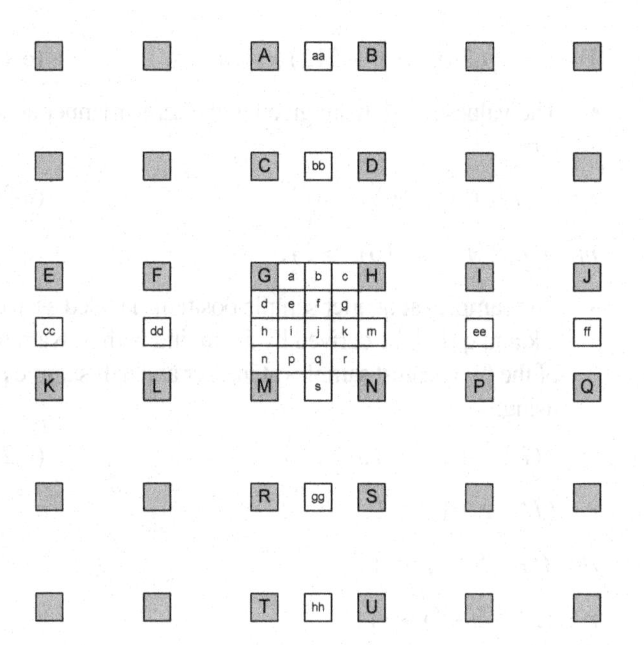


**Figure 6-29 Luma Interpolation for H.264**

- The samples at half position labeled as b and h shall be derived by applying the 6-tap filter to the nearest integer position samples as:

$$b_1 = (E - 5 \times F + 20 \times G + 20 \times H - 5 \times I + J), \quad (6\text{-}14)$$

$$h_1 = (A - 5 \times C + 20 \times G + 20 \times M - 5 \times R + T). \quad (6\text{-}15)$$

Then, $b = Clip1_Y((b_1 + 16) >> 5)$, (6-16)

$$h = Clip1_Y((h_1 + 16) >> 5). \quad (6\text{-}17)$$

Here, $Clip1_Y(x)$ means "clipping $x$ in [0..255]."

- The samples at half position labeled as j shall be derived by applying the 6-tap filter to the nearest integer position samples as:

$$j_1 = (cc - 5 \times dd + 20 \times h_1 + 20 \times m_1 - 5 \times ee + ff), \quad (6\text{-}18)$$

or $j_1 = (aa - 5 \times bb + 20 \times b_1 + 20 \times s_1 - 5 \times gg + hh)$.

Then, $j = Clip1_Y((j_1 + 512) >> 10)$. (6-19)

- The values s and m are given in the same manner as for $b$ and $h$ by:

$$s = Clip1_Y((s_1 + 16) >> 5), \quad (6\text{-}20)$$

$$m = Clip1_Y((m_1 + 16) >> 5).$$

- The samples at quarter sample positions labeled as a, c, d, n, f, i, k and q shall be derived by averaging with upward rounding of the two nearest samples at integer and half sample positions using:

$$a = (G + b + 1) >> 1, \quad (6\text{-}21)$$

$$c = (H + b + 1) >> 1,$$

$$d = (G + h + 1) >> 1,$$

$$n = (M + h + 1) >> 1,$$

$$f = (b + j + 1) >> 1,$$

$$i = (h + j + 1) >> 1,$$

$$k = (j + m + 1) >> 1,$$

$$q = (j + s + 1) >> 1.$$

- The samples at quarter sample positions labeled as e, g, p and r shall be derived by averaging with upward rounding of the two nearest samples at half sample positions using:

$$e = (b + h + 1) >> 1, \quad (6\text{-}22)$$

$$g = (b + m + 1) >> 1,$$

$$p = (h + s + 1) >> 1,$$

$$r = (m + s + 1) >> 1.$$

### Chroma Interpolation

Due to interpolation, an extended 5x5 block data is required for Chroma when a 4x4 block is used for MC. To interpolate the right-most columns of 4x4 block, one extra neighboring pixel column needs to be considered as shown in Figure 6-30. Also, one pixel row out of the block is required to interpolate the lowest rows of the block. Basically, Chroma interpolation is performed with linear interpolation. Note that the interpolated resolution for chroma is 1/8-pel, which is different from that of VC-1.

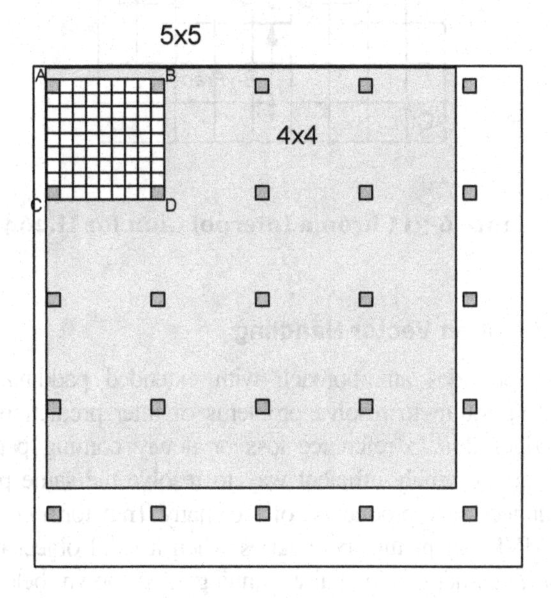


**Figure 6-30 Fetch Out Area of 5x5 Block for 4x4 Chroma Block**

In Figure 6-31, the positions labeled with A, B, C and D represent chroma samples at full-sample locations inside the given 2D component image for chroma samples. The interpolated chroma sample value $predPartLX_C[x_C,y_C]$ is derived as follows:

$$predPartLX_c[x_c, y_c] = ((8 - xFrac_c) \times (8 - yFrac_c) \times A + xFrac_c \times (8 - yFrac_c) \times B + (8 - xFrac_c) \times yFrac_c \times C + xFrac_c \times yFrac_c \times D + 32) >> 6.$$

(6-23)

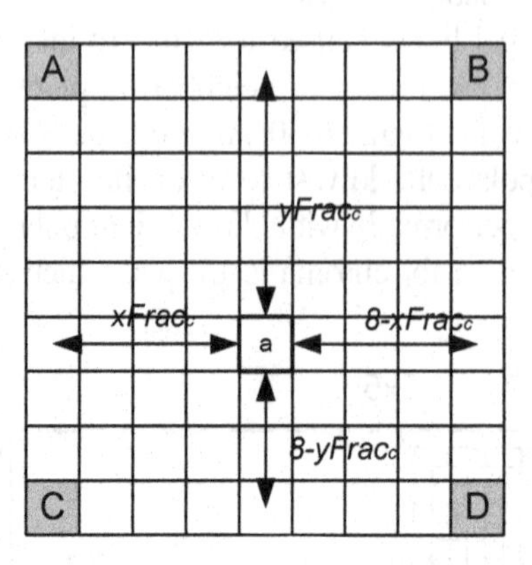


**Figure 6-31 Chroma Interpolation for H.264**

## Extended Motion Vector Handling

VC-1 provides an approach with extended padding and a MV Pullback algorithm to resolve problems of inter prediction suffering at picture edges due to reference loss or newly coming patterns. H.264 proposes an extremely efficient way to resolve the same problems that lead to unnecessary production of too many Transform coefficients and too big MVDs at picture boundaries when a solid object moves into a scene or a camera is globally panning. As shown below, the main difference between the approaches of VC-1 and H.264 is that VC-1 tries to define pixel values outside the picture boundaries, while H.264 designates a "way" to obtain pixel values outside the picture boundaries in terms of a formula. In theory, both methods could provide unlimited number of padded pixels. However, in reality, the range of MV is restricted by Profile/ Level limits defined in Table 2-7.

Let's define a Luma location in full-sample units be ($xInt_L$, $yInt_L$). In other words, this set can be the full-pel part of MV data that can be pointing to either inside or outside the picture boundaries. The Luma sample arrays are taken from the selected reference picture $refPicLX_L$.

In Figure 6-29, the positions labeled with upper-case letters within shaded blocks represent Luma samples at full-sample locations inside the given 2D array *refPicLX*L of Luma samples. These samples may be used for generating the interpolated Luma sample values. The location for each of the corresponding Luma samples Z (i.e., Z may be A, B, C, D, E, F, G, H, I, J, K, L, M, N, P, Q, R, S, T or U) inside the given array *refPicLX*L are derived as follows:

$$xZ_L = Clip3(0, PicWidthInSamples_L - 1, xInt_L + xDZ_L) \quad (6\text{-}24)$$

$$yZ_L = Clip3(0, PicHeightInSamples_L - 1, yInt_L + yDZ_L).$$

Here, $Clip3(0, \max, x)$ means to be "clipping $x$ in [0..max]." And, the following Table 6-1 specifies $(xDZ_L, yDZ_L)$ for different replacements of Z.

| *Z* | *A* | *B* | *C* | *D* | *E* | *F* | *G* | *H* | *I* | *J* |
|---|---|---|---|---|---|---|---|---|---|---|
| *xDZ*L | 0 | 1 | 0 | 1 | -2 | -1 | 0 | 1 | 2 | 3 |
| *yDZ*L | -2 | -2 | -1 | -1 | 0 | 0 | 0 | 0 | 0 | 0 |
| *Z* | *K* | *L* | *M* | *N* | *P* | *Q* | *R* | *S* | *T* | *U* |
| *xDZ*L | -2 | -1 | 0 | 1 | 2 | 3 | 0 | 1 | 0 | 1 |
| *yDZ*L | 1 | 1 | 1 | 1 | 1 | 1 | 2 | 2 | 3 | 3 |

**Table 6-1 Differential Full-sample Luma Locations**

Let's define a Chroma location in full-sample units be (*xInt*c, *yInt*c). In other words, this set can be the full-pel part of MV data that can be pointing to either inside or outside the picture boundaries. The Chroma sample arrays are taken from the selected reference picture *refPicLXc*.

In Figure 6-31, the positions labeled with A, B, C and D represent chroma samples at full-sample locations inside the given 2D array *refPicLX*c of chroma samples are derived as follows:

$$xA_c = Clip3(0, PicWidthInSamples_c - 1, xInt_c), \quad (6\text{-}25)$$

$$yA_c = Clip3(0, PicHeightInSamples_c - 1, yInt_c).$$

$$xB_c = Clip3(0, PicWidthInSamples_c - 1, xInt_c + 1), \quad (6\text{-}26)$$

$$yB_c = Clip3(0, PicHeightInSamples_c - 1, yInt_c).$$

$$xC_c = Clip3(0, PicWidthInSamples_c - 1, xInt_c), \quad (6\text{-}27)$$

$$yC_c = Clip3(0, PicHeightInSamples_c - 1, yInt_c + 1).$$

$$xD_c = Clip3(0, PicWidthInSamples_c - 1, xInt_c + 1), \quad (6\text{-}28)$$

$$yD_c = Clip3(0, PicHeightInSamples_c - 1, yInt_c + 1).$$

Since the formulas given for Luma and Chroma sample selection are valid for MVs that point to outside of the picture boundaries, they virtually have extended padding effects. Note that the H.264 approach for extended MV handling does not require any Pullback algorithm as was used for VC-1.

### Motion Vector Predictors

The MVP of a MB is chosen from three candidate MVs in neighborhood MBs, blocks or sub-partitions as shown in Figure 6-32 and Figure 6-33. Basically, there are two kinds of MVPs – predictors based on non-median operations and predictors based on median operations.

Figure 6-32 illustrates non-median prediction. This applies when the current MB is taking either 16x8 or 8x16 size MC. When 16x8 size is used as shown in the upper part of Figure 6-32, MVP is set as the candidate vector from either MB-A or MB-B. For the upper 16x8 partition, the vector from MB-B is used. For the lower 16x8 partition, the vector from MB-A is used. When 8x16 size is used as shown in the lower part of Figure 6-32, MVP is set as the candidate vector from either MB-A or MB-C. For the left 8x16 partition, the vector from MB-A is used. For the right 8x16 partition, the vector from MB-C is used. Note that the notation for A, B and C are different from those of VC-1 as shown in Figure 6-20 and Figure 6-21.

Figure 6-33 illustrates median prediction. This applies except for cases mentioned in Figure 6-32. In other words, predictors are derived when sizes of MC are 8x8, 8x4, 4x8, 4x4 and 16x16. Note that for 16x16 size MC, median prediction is utilized for MB-A, MB-B and MB-C. If there is more than one block/ sub-partition immediately to the left of the current MB/block/sub-partition, the topmost of these partitions is chosen as A as illustrated in Figure 6-33. If there is more than one block/ sub-partition immediately above the current MB/block/sub-partition, the leftmost of these partitions is chosen as B as illustrated in Figure 6-33. If there is more than one block/ sub-partition immediately to the above-right of the current MB/block/sub-partition, the leftmost of these partitions is chosen as C as illustrated in Figure 6-33.

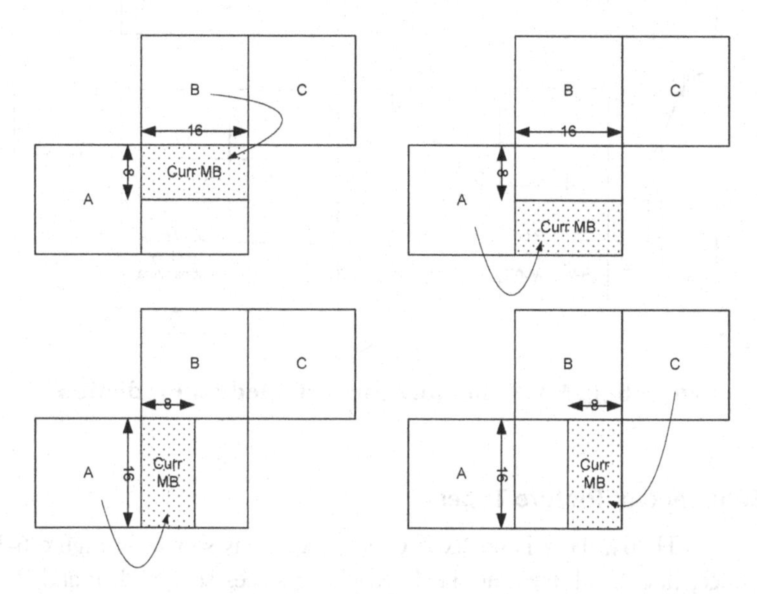


**Figure 6-32 MV Predictors through Non-median Prediction**

In the upper left illustration in Figure 6-33, it is shown that the candidate A is propagated to B and C when the candidates of B and C are not available. In the upper right illustration in Figure 6-33, it is shown that the candidate is chosen as the predictor when one and only one of the reference indices is equal to the reference index of the current partition. In the lower left illustration in Figure 6-33, a typical median predictor is shown, while the lower right illustration depicts that the candidate D is chosen when the candidate of C is not available. Note that if none of candidates of A, B and C are available or all of them are Intra coded, MVP is set to zero.

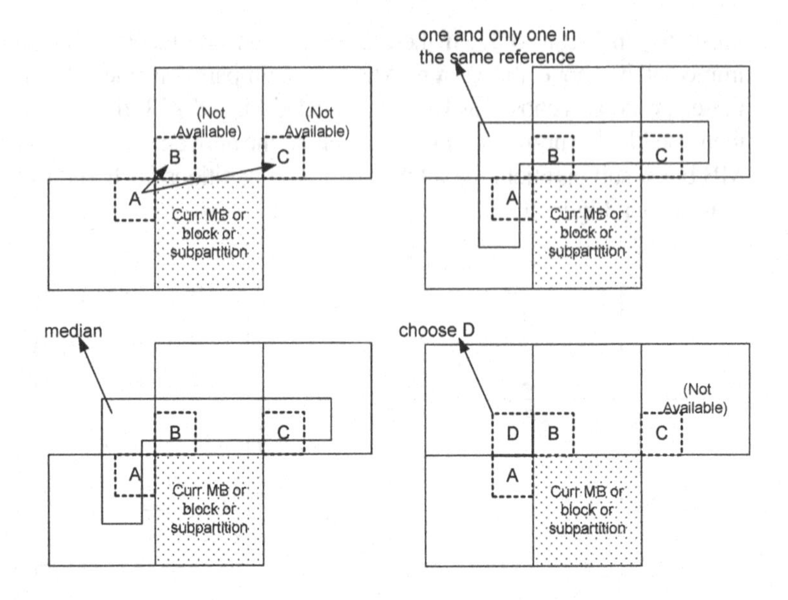


**Figure 6-33 MV Predictors through Median Prediction**

## Sequence of Picture Types

In H.264, there is no fixed GOP structure as shown in Figure 6-34 since, first of all, there are no fixed picture types such as I, P and B. A picture can be composed of many different slice types. The only matter is to select reference pictures. Pictures at any position can be reference pictures if the encoder wants to use them as references. However, such a selection entirely depends on the encoder. The maximum number of reference pictures can be, in theory, as many as "num_ref_frames," where the nun_ref_frames shall be in the range of 0 and MaxDpbSize. The number is practically limited with $DPB_{max}$ in Table 2-7 in Chapter 2.

MBs in I slices are all Intra coded, while MBs in P or B slices are either Inter coded or Intra coded. When Inter mode is used, MBs can use any reference pictures as shown in Figure 6-34. For example, P-MBs in Picture 6 can use Picture 3 as the reference and B-MBs in Picture 5 can use Picture 3 and Picture 7 as references. Note that B-MBs in H.264 do not need to take forward and backward directional references always. As seen in Figure 6-34, B-MBs in Picture 10 can use Picture 0 and Picture 7 as references, both of which give "forward" MVs.

The maximum number of MVs with P-MBs can be 16 for Luma since 16 4x4 blocks can be the MC mode. In contrast, the maximum number of MVs with B-MBs can be 32 for Luma since 16 4x4 blocks can take both forward and backward MVs simultaneously. It is important to understand that the maximum number of reference pictures used for each MB is four for unidirectional prediction. Each 8x8 block can take one reference picture for unidirectional prediction. However, all sub-partitions need to use the same reference picture no matter how many partitions are generated from the same 8x8 block. Note that the maximum number of reference pictures for bidirectional prediction is eight due to double references in each 8x8 block.

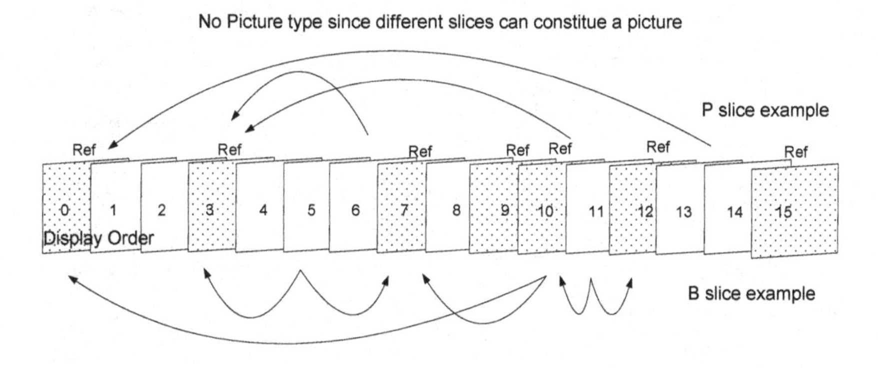


**Figure 6-34 Example with Sequence of Picture Types for H.264**

The MBs in P pictures have two coding mode options – Intra or Inter modes. MBs can also use Skip mode. In addition, I_PCM mode is included for Intra mode, too. Skip mode means that there is no MVD for corresponding MVs and no residual data. The MBs in B pictures have four MV mode options – forward MV, backward MV, interpolative MV and direct MB. Coding modes for P pictures and B pictures in H.264 video can be summarized as shown in Figure 6-35. Note that skipped MBs in B pictures have the same MVs as that of the predictor and the same MB types as the previous MB, while skipped MBs in P pictures have only the same MVs as that of the predictor. The interpretation for skipped MB is based on dedicated flags P_Skip and B_Skip in H.264. The P_Skip can be interpreted straightforwardly as aforementioned. However, B_Skip can mean both skipped MB or direct MB based on the interpretation of CBP. If CBP is zero, then it is really "Skip" mode, where no MVD and no residual data are allowed. If CBP is not zero, then it is inferred as "Direct" mode, where residual data can be non-zero. When at least one block is to be coded with zero MVD, the zero MV should be inscribed in the bitstream.

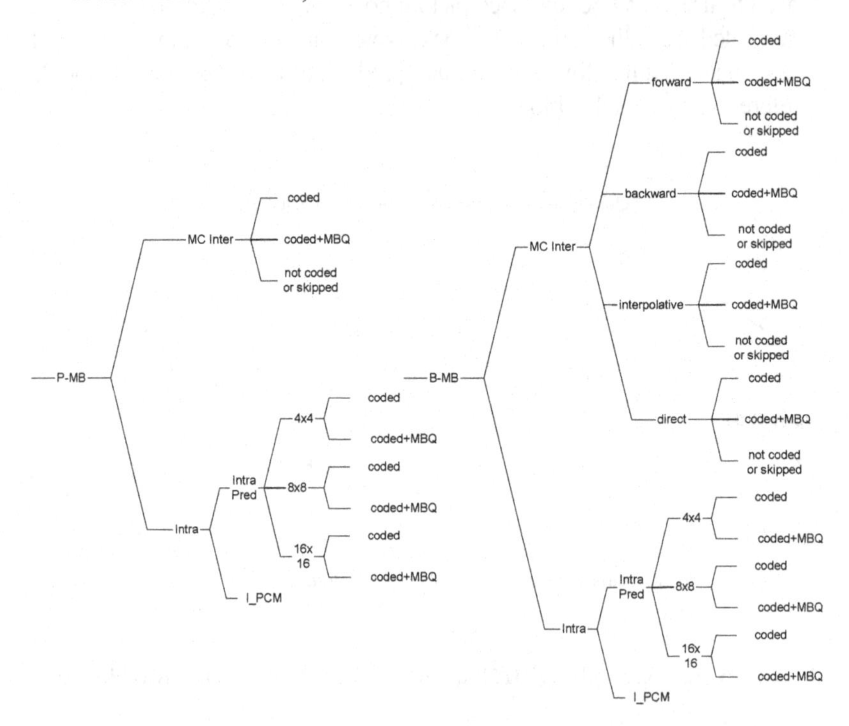


**Figure 6-35 MB Coding Mode in P and B Pictures for H.264**

In the bitstreams conforming to the Main, High, High 10, High 4:2:2, High 4:4:4 Predictive or Extended Profiles, relatively small size "interpolative" MC mode is not allowed for B for certain levels – the value of sub_mb_type in B-MB shall not be equal to B_Bi_8x4, B_Bi_4x8, B_Bi_4x4 for the levels above Level 3.1. Similarly, in the bitstreams conforming to Main, High, High 10, High 4:2:2, High 4:4:4 Predictive or Extended Profiles, only frame MB mode is allowed for certain levels – sequence parameter sets shall have frame_mbs_only_flag equal to 1 for the levels above Level 4.2. Note that frame_mbs_only_flag is equal to 1 for all levels of the Baseline Profile. In the bitstreams conforming to the Main, High, High 10, High 4:2:2 or High 4:4:4 Predictive Profiles, sequence parameter sets shall have direct_8x8_inference_flag equal to 1 for the levels above Level 3. Enabling this flag would result in motion vectors being computed at the 8x8 block level instead of 4x4. The condition here to avoid is when the current MB is a Field MB, co-located MB is Frame MB, and the direct_8x8_inference_flag=0. By doing so, any confusion at interpreting references of Direct mode for 4x4 blocks can be eliminated. In this mode, MVs of 8x8 blocks in the co-located MB can be obtained to consider the MV of outer 4x4 block in each 8x8 block as representative.

### Temporal Direct Mode and Weighted Prediction

Direct mode uses bidirectional prediction based on derived forward and backward MVs. In the main coding representation, only residual data is described. Figure 6-4 depicts the derivation process of forward/ backward MVs of direct mode MB from the MV of a co-located MB, where two MVs are obtained in Equation (6-1). Enhanced precision can be achieved based on the following procedures introduced in H.264 to prevent the encoder and decoder pair from possible "drift":

$$\text{with } X = \frac{(16384 + abs(\frac{TD_D(POC)}{2}))}{TD_D(POC)} \text{ and} \tag{6-29}$$

$$ScaleFactor = clip3(-1024, 1023, (TD_B(POC) \times X + 32) >> 6) \quad , \tag{6-30}$$

$$MV_0 = (ScaleFactor \times MV_C + 128) >> 8 \ , \tag{6-31}$$

$$MV_1 = MV_0 - MV_C \, . \tag{6-32}$$

where +32 or +128 are intended for rounding operation. Note that the current MB uses the same partition as the co-located MB. In typical cases, selection of references is to take the 0-indexed element from List1 as colPic and the lowest valued reference indexed element from List0 that references refPicCol. Here, refPicCol is the picture that was referred by the co-located MB inside the picture colPic. List0 shall contain refPicCol. This conventional Direct mode is called "Temporal direct" in the context of H.264, as opposed to "Spatial direct" mode introduced in the next section.

The direct mode in H.264 can be improved by weighted blending of the prediction signal. There are two weighted prediction modes indicated in the SPS for P and SP slices using "weighted_pred_flag" and for B slices using the "weighted_bipred_idc" field. Weights can be represented either explicitly (i.e., Explicit mode) or implicitly (i.e., Implicit mode). Explicit mode is supported in P, SP and B slices, while Implicit mode is supported only in B slices and is used only for bi-directionally predicted MBs. The weighted_bipred_idc field being set to 2 means Implicit mode and weighting factors are not explicitly transmitted under this mode in the slice header. This is somewhat similar to the Intensity Motion Compensation in VC-1 in a sense that it would work best for a fade-in/ fade-out scenario of the video. The Implicit weighted blending technique uses bidirectional mode MVs including direct mode MVs and weights for the calculation of the predicted block based on this distance. The Implicit weighted blending technique calculates the prediction block $c$ for Direct mode coded MB according to:

$$c = \frac{c_0(TD_D - TD_B) + c_1 TD_B}{TD_D} \qquad (6\text{-}33)$$

where $c_0$ and $c_1$ is the predicted blocks from the List0 and List1. And, $c$ is the weighted predictor. Note that Explicit mode is indicated when weighted_bipred_idc field is set to 1, while Default mode is indicated with weighted_bipred_idc field being set to 0.

The weighting factors and offsets used in a particular slice are included in the slice header when Explicit mode is indicated. The allowable range of parameter values is constrained to permit 16-bit arithmetic operations in the inter prediction process. The dynamic range and precision of weighting factors can be adjusted using the "luma_log_weight_denom" and "chroma_log_weight_denom" fields, which are the binary logarithm of the denominator of the Luma and Chroma weighting factors, respectively. The multiplicative weighting factors are coded as "luma_weight_l0," "luma_weight_l1," "chroma_weight_l0" and "chroma_weight_l1." The additive offsets are coded as "luma_offset_l0," "luma_offset_l1," "chroma_offset_l0" and "chroma_offset_l1." The use of such parameters is as follows:

$$c^{luma} = Clip1(((c_0 \times luma\_weight\_l0 + 2^{luma\_log\_weight\_denom-1}) >> luma\_log\_weight\_denom) + luma\_offset\_l0) \quad (6\text{-}34)$$

for single directional Luma prediction from List0.

$$c^{luma} = Clip1(((c_1 \times luma\_weight\_l1 + 2^{luma\_log\_weight\_denom-1}) >> luma\_log\_weight\_denom) + luma\_offset\_l1) \quad (6\text{-}35)$$

for single directional luma prediction from List1.

$$c^{luma} = Clip1(((c_0 \times luma\_weight\_l0 + c_1 \times luma\_weight\_l1 + 2^{luma\_log\_weight\_denom}) >> (luma\_log\_weight\_denom + 1)) + (luma\_offset\_l0 + luma\_offset\_l1 + 1) >> 1) \quad (6\text{-}36)$$

for bi-directional luma prediction from List0 and List1. Here, $c_0^{luma}$, $c_1^{luma}$ are the list0 and list1 initial predictors, and $c^{luma}$ is the weighted predictor. The Equation (6-34)~Equation (6-36) can be extended to chroma weighted predictions with corresponding parameters exchanged.

The number of MVs for a MB is counted by the variable MvCnt after the completion of the intra or inter prediction process for the MB. For B Direct modes, the rule is given slightly differently as follows:

- If subMbPartIdx is equal to 0, subMvCnt is set equal to 2.

- Otherwise (subMbPartIdx is not equal to 0), subMvCnt is set equal to 0.

### Spatial Direct Mode

In general, representation elements for a MB include MVs and residual data. For skipped mode, no data are required for MVs and residual data. For not coded mode, only MVs data are used. For direct mode, only residual data are inscribed. These are the three most efficient representations for MBs in modern video coding. The idea of Spatial direct mode in H.264 is to provide an efficient representation for scene changes as shown in Figure 6-36 in conjunction with Direct representation.

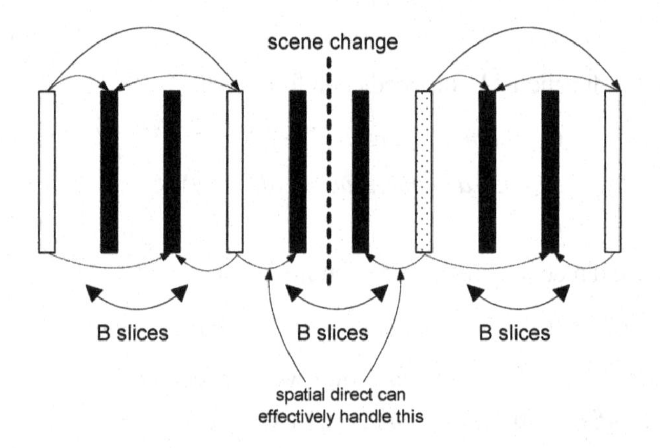


**Figure 6-36 Scene Change and Spatial Direct**

A big scene change typically decreases temporal co-relation between pictures. When a scene change happens as shown in Figure 6-36, the conventional way is to use either forward or backward MV modes adaptively. However, such representation contains both MVs and residual data, which is not as efficient as Direct mode representation.

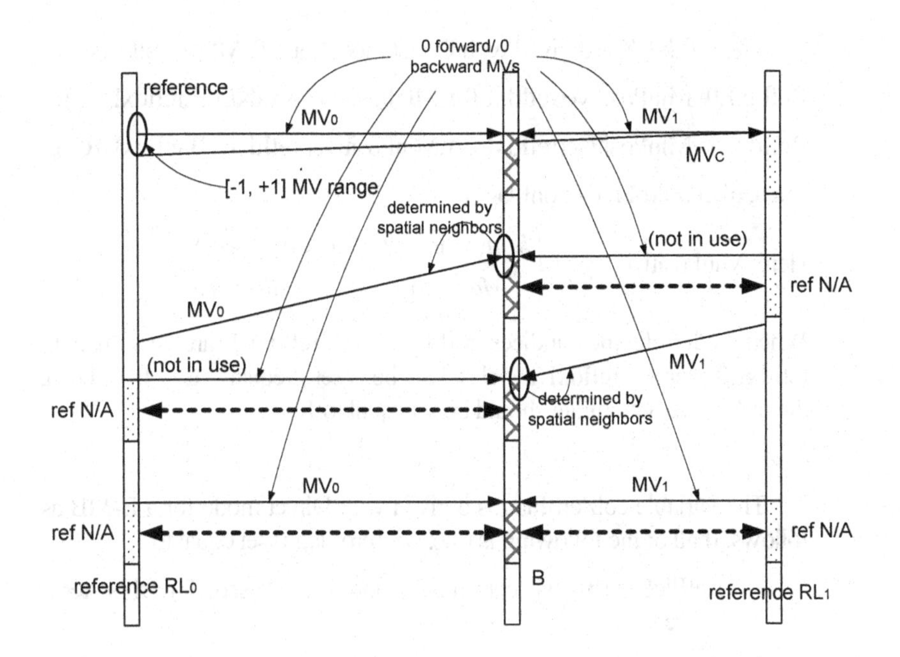


**Figure 6-37 Cases for Spatial Direct in H.264**

Spatial direct mode defines four cases as shown in Figure 6-37. The first scenario is to use 0 forward/ 0 backward MVs when its co-located MB has MV in the range of [-1,+1]. In other words, co-located MBs are copied from reference 0 and reference 1 to interpolate when the area is almost stationary in the period of time. The second and third scenarios are copying a MB based on MVs of spatial neighbors either from forward or backward directions depending on where the scene change occurs. Note that the decision of availability for reference pictures is made in the encoder side. The fourth scenario is that both co-located MBs are used for interpolation when no correlation can be found from reference picture 0 and reference picture 1. No availability of reference index can imply "scene change" between co-located MBs. The encoder uses the syntax element refIdxLX to describe a scene change scenario. The encoder can write a negative number for refIdxLX to tell a decoder which direction contains a scene change as shown in Figure 6-37. Spatial direct mode is interpreted for Direct mode when "direct_spatial_mv_pred_flag" is equal to 1 at a slice header. This tool is turned on with slice-based signals.

The refIdxLX is derived with Direct mode for a B-MB as follows:

RefIdxL0=MinPositive(refIdxL0A,MinPositive( refIdxL0B,refIdxL0C)),

RefIdxL1=MinPositive(refIdxL1A,MinPositive( refIdxL1B,refIdxL1C)).

Then, directZeroPredictionFlag=0.

Here, MinPositive(x,y)= $\begin{cases} Min(x,y) & if \quad (x \geq 0 \,\|\, y \geq 0) \\ Max(x,y) & otherwise. \end{cases}$

When both reference indices refIdxL0 and refIdxL1 are less than 0, refIdxL0 and refIdxL1 should be set equal to 0. Then, directZeroPredictionFlag should be set equal to 1.

The variable colZeroFlag is derived with Direct mode for a B-MB as follows. If all of the following is true, colZeroFlag is set equal to 1:

- RefPicList1[0] is currently marked as "used for short-term reference."
- RefIdxCol is equal to 0.
- Both MV components mvCol[0] and mvCol[1] lie in the range of –1 to 1 in units of ¼-pel luma frame resolution for frame MB. If field MB is used, the unit is ¼-pel luma field.

Both components of MV (i.e., mvLX) are set equal to 0 with Direct mode for a B-MB based on the above two flags, directZeroPredictionFlag and colZeroFlag.

- if directZeroPredictionFlag is equal to 1.
- if refIdxLX is equal to 0 and colZeroFlag is equal to 1.
- if refIdxLX is less than 0.

Otherwise, MVs are obtained through interpretation of spatial neighborhood MVs, as was in non-median or median MVP derivation for Skip bits or MVDs. The prediction utilization flags predFlagL0 and predFlagL1 shall be derived as specified using Table 6-2.

**Table 6-2 Assignment of Prediction Utilization Flags**

| refIdxL0 | refIdxL1 | predFlagL0 | predFlagL1 |
|---|---|---|---|
| $\geq 0$ | $\geq 0$ | 1 | 1 |
| $\geq 0$ | $< 0$ | 1 | 0 |
| $< 0$ | $\geq 0$ | 0 | 1 |

The variable subMvCnt for B Spatial direct mode is derived as follows:

- If subMbPartIdx is not equal to 0 or direct_8x8_inference_flag is equal to 0, subMvCnt is set equal to 0.
- Otherwise (subMbPartIdx is equal to 0 and direct_8x8_inference_flag is equal to 1), subMvCnt is set equal to predFlagL0+predFlagL1.

# 7. In-Loop and Out-Loop Filters

## 7.1 Deblocking Process

### Blocky Effect and Compression Efficiency

The Blocky Effect is the most irritating artifact created in compression/ decompression processes. The main cause of the effect is the quantization on a signal spanned over a fixed and limited length of Transform basis set. It causes not only visual artifacts, but also compression inefficiency. For example, such distorted block edges can influence the residual signal when a MB takes the best match over the blocky area as shown in Figure 7-1. In other words, compression efficiency can be hurt due to the blocky effect. MPEG-2 does not provide any deblocking tools to remove patterns of distorted block edges. However, later video standards such as H.263, VC-1 and H.264 tried to devise tools that enhance quality over blocky edges. Note that H.263 adopted Overlapped Block Motion Compensation (OBMC) to resolve this issue [ITU:H.263, rijkse:H.263], while VC-1 and H.264 provide the direct pixel domain process called "deblocking" process.

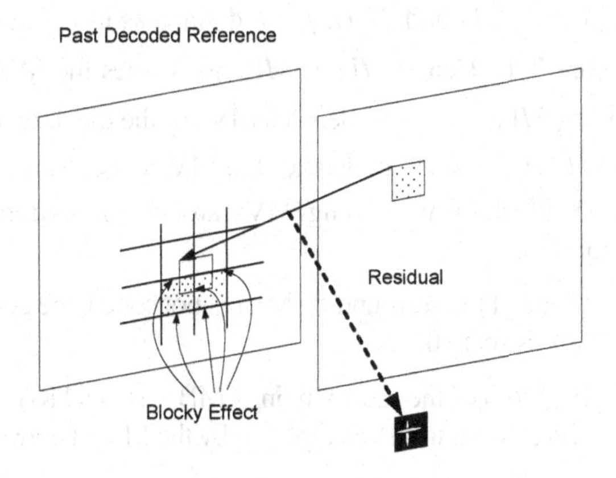


**Figure 7-1 Propagation of Blocky Effect in Compression**

These filtering processes can be called "in-loop" since the processing actually influences on the reference data, so that manipulated reference data could propagate the effect for upcoming picture coding.

### Overlapped Block Motion Compensation (OBMC)

One way to prevent decoded images from exhibiting the blocky effect is to use the OBMC technique. For example, H.263 utilizes OBMC as one of its Advanced Prediction modes defined in the standard. The technique is to obtain the predictor of each pixel within the 8x8 block as a weighted sum of neighborhood values around the best match given by:

$$\hat{p}(x,y) = (q(x,y)H_0(i,j) + r(x,y)H_1(i,j) + s(x,y)H_2(i,j) + 4)/8 . \quad (7\text{-}1)$$

The $q(x,y)$, $r(x,y)$, $s(x,y)$ are pixels from the reference picture obtained with neighborhood MVs as follows:

$$q(x,y) = p(x + MV_0.x, y + MV_0.y), \quad (7\text{-}2)$$

$$r(x,y) = p(x + MV_1.x, y + MV_1.y) \quad (7\text{-}3)$$

and $s(x,y) = p(x + MV_2.x, y + MV_2.y)$. (7-4)

$H_0(i,j)$, $H_1(i,j)$ and $H_2(i,j)$ are defined as in Figure 7-2, Figure 7-3 and Figure 7-4. Here, $(MV_0.x, MV_0.y)$ denotes the MV set for the current block. $(MV_1.x, MV_1.y)$ denotes MVs of the block either above or below, while $(MV_2.x, MV_2.y)$ denotes the MVs either to the left or right of the current block. Any missing MVs around the current block are derived as follows:

- If one of the surrounding MBs is not coded, the corresponding MV is set to 0.
- If one of the surrounding MBs is INTRA coded, the corresponding MV is replaced by the MV of current block.
- If the current block is at the border of the picture and therefore a surrounding block is not present, the corresponding MV is replaced by the current MV.

- In all cases, if the current block is at the bottom of the MB, the corresponding MVs in the MB below the current MB are replaced by the MV of current block.

| | | | | | | | |
|---|---|---|---|---|---|---|---|
| 4 | 5 | 5 | 5 | 5 | 5 | 5 | 4 |
| 5 | 5 | 5 | 5 | 5 | 5 | 5 | 5 |
| 5 | 5 | 6 | 6 | 6 | 6 | 5 | 5 |
| 5 | 5 | 6 | 6 | 6 | 6 | 5 | 5 |
| 5 | 5 | 6 | 6 | 6 | 6 | 5 | 5 |
| 5 | 5 | 6 | 6 | 6 | 6 | 5 | 5 |
| 5 | 5 | 5 | 5 | 5 | 5 | 5 | 5 |
| 4 | 5 | 5 | 5 | 5 | 5 | 5 | 4 |

**Figure 7-2 OBMC Weighting Values for $H_0$ H.263 Example**

| | | | | | | | |
|---|---|---|---|---|---|---|---|
| 2 | 2 | 2 | 2 | 2 | 2 | 2 | 2 |
| 1 | 1 | 2 | 2 | 2 | 2 | 1 | 1 |
| 1 | 1 | 1 | 1 | 1 | 1 | 1 | 1 |
| 1 | 1 | 1 | 1 | 1 | 1 | 1 | 1 |
| 1 | 1 | 1 | 1 | 1 | 1 | 1 | 1 |
| 1 | 1 | 1 | 1 | 1 | 1 | 1 | 1 |
| 1 | 1 | 2 | 2 | 2 | 2 | 1 | 1 |
| 2 | 2 | 2 | 2 | 2 | 2 | 2 | 2 |

**Figure 7-3 OBMC Weighting Values for $H_1$ H.263 Example**

| | | | | | | | |
|---|---|---|---|---|---|---|---|
| 2 | 1 | 1 | 1 | 1 | 1 | 1 | 2 |
| 2 | 2 | 1 | 1 | 1 | 1 | 2 | 2 |
| 2 | 2 | 1 | 1 | 1 | 1 | 2 | 2 |
| 2 | 2 | 1 | 1 | 1 | 1 | 2 | 2 |
| 2 | 2 | 1 | 1 | 1 | 1 | 2 | 2 |
| 2 | 2 | 1 | 1 | 1 | 1 | 2 | 2 |
| 2 | 2 | 1 | 1 | 1 | 1 | 2 | 2 |
| 2 | 1 | 1 | 1 | 1 | 1 | 1 | 2 |

**Figure 7-4 OBMC Weighting Values for H2 H.263 Example**

To perform OBMC, five MVs around the current block are considered as shown in Figure 7-5.

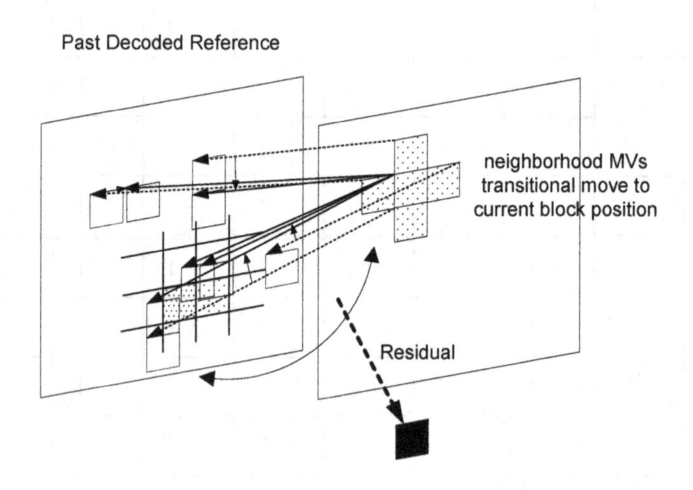


**Figure 7-5 Neighborhood Block Data used for OBMC**

The current block is pixel-by-pixel multiplied with $H_0(i,j)$. The MVs of the blocks above and below are used to extract each block to be pixel-by-pixel multiplied with $H_1(i,j)$. The MVs of the blocks to the left and right are used to extract each block to be pixel-by-pixel multiplied with $H_2(i,j)$. Note that $H_1(i,j)$ and $H_2(i,j)$ in Figure 7-2 and Figure 7-3 are divided with double lines to show which half is used for which position block. Notice that all the weighting factors for a fixed point are summed to be 8. For example, (0,0) position values are 4, 2 and 2 from $H_0(i,j)$, $H_1(i,j)$ and $H_2(i,j)$, respectively. When those values are added, the resultant value becomes 8. This is why 8 is used to divide in Equation (7-1). In other words, all weighting values are designed to be normalized with 8 to obtain an average valued predictor.

OBMC typically produces pleasing results on block boundaries. Research shows that OBMC with four MVs for a MB can improve coding gain by 1 dB.

## In-loop Deblocking Filter

OBMC is burdensome in terms of computation and bandwidth. To obtain one predictor, five blocks of data should be fetched out and pixel-by-pixel multiplication should be performed. Most seriously, the encoder has to take over a great deal of computation in ME process, considering five neighborhood blocks to calculate a 5-blocks weighted predictor at each and every pixel point at the reference picture for the search of best combination. In contrast, modern video coding standards such as VC-1 try to define direct pixel domain processing in terms of OverLapped Transform smoothing (OLT) and In-Loop deblocking Filtering (ILF). H.264 defines only ILF, though. Direct pixel domain processing provides benefits in terms of computation and bandwidth since it is not connected with ME/ MC as opposed to OBMC. On top of that, the processing is operated in the loop on the reference picture so that quality improvement of later coming pictures can be expected.

OLT in VC-1 is more or less an analytic way to reduce the blocky effect based on an accurately defined pre-/post-processing pair. The idea is that forward and inverse operations are defined in such a way that original data is recovered perfectly when forward and backward operations are serially applied. The forward Transform exchanges information across boundary edges in adjacent blocks before the main encoding stage. In the decoder side, the exact inverse operation is required to exchange the edge data back again to obtain original data after the main decoding stage.

On the other hand, ILFs in VC-1 and H.264 are more or less heuristic ways to reduce blocky effects. Blocky patterns are considered to be high frequency since abrupt value changes are happening around block edges. While original data patterns are preserved, relatively simple non-linear low pass filtering is applied around block edges in ILF.

## 7.2 VC-1 In-Loop Filtering

### OverLapped Transform (OLT) Smoothing Filter

There are two techniques used in VC-1 to eliminate blocky effects – OLT smoothing and ILF. To clearly tell the difference between the two operations in VC-1, the term "DBF" is defined to refer to OLT and ILF together as a one operation. There are three scenarios to tackle in VC-1 deblocking operation. High-low quality discontinuity as shown in Figure 7-6 (a) is enhanced with the OLT tool, while low-low quality discontinuity and high-high quality discontinuity as shown in Figure 7-6 (b) and (c) are enhanced with the ILF tool. Note that this kind of discontinuity comes from different quantization effects at neighboring blocks in different frequency components. For example, DC offset between two adjacent blocks can be rectified with ILF as in the case of Figure 7-6 (b).

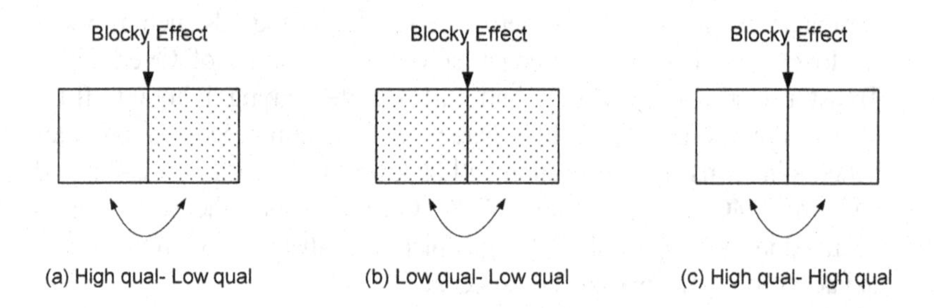


**Figure 7-6 Blocky Effect Due to Quality Discontinuity**

The basic idea of OLT is to switch over the edge data of two adjacent blocks, where both of them are of original quality as shown in Figure 7-7. When two such adjacent blocks undergo Transform/Quantization and inverse Quantization/ inverse Transform, quantization error and/or blocky effects can be introduced in one block more severely than the other in certain cases. At the decoder, two edge data should be switched over again due to recovering original data topology. Then, good quality block contains bad quality edges, while bad quality block contains good quality edges. In other words, good quality and bad quality blocks diffuse each other.

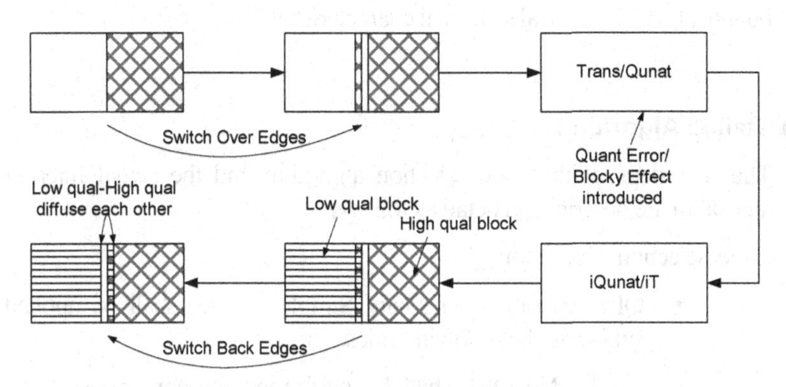


**Figure 7-7 Concept of OLT**

The idea of OLT is further extended from Figure 7-7. When the idea of Figure 7-7 is directly applied, high frequency components can be introduced due to edge exchange. Therefore, a filtering operation is defined as OLT instead of simple data switch-over. The filtering operation required at the decoder is defined as Equation (7-5). Apparently, the inverse matrix of Equation (7-5) should be applied at the encoder side. Note that the matrix is implementing a kind of low pass filter distributing original data around the edges of two adjacent blocks as was explained earlier. If some texture is lost in one block in the quantization, missing texture can be gained back due to inverse distribution operation at a decoder.

When data is exchanged, Intra block and Inter block should not be exchanged (i.e., Inter is about residual data). Therefore, OLT is applied only to Intra coded blocks, which is always 8x8 size. In addition, when the data of two blocks is almost saturated at 255, filtering might introduce overflow due to linear property of the operation. To avoid overflow, 128 level shift is defined for OLT in the standard. Note that OLT applies when PQUANT is greater than or equal to 9. This is because OLT would take effect only with blocks of distorted edges. For Advance Profile, OLT can be applied even with PQUANT less than 9 based on CONDOVER and OVERFLAGMB information. This provides controllability over OLT even for high quality images with some local areas distorted in quantization due to local statistical characteristics.

### Detailed Algorithm

The two steps of the edge selection algorithm and the actual filtering algorithm are described in detail as follows:

Edge Selection Algorithm:

- OLT smoothing in Main/ Simple Profiles shall be applied subject to the following rules:
    1. No OLT shall be performed for any frame if the OVERLAP=0; The remainder of these rules shall apply only when OVERLAP =1; OLT shall be applied only if the frame level quantization step size PQUANT is 9 or higher:
        1) All 8x8 block boundaries between adjacent 8x8 blocks shall be smoothed for I frames and BI frames.
        2) Only block boundaries separating two intra blocks shall be smoothed for P frames.
        3) No OLT shall be performed for predicted B frames., i.e., B frames that are not coded as BI.
        4) There shall be no dependence on DQUANT across MBs.

- OLT smoothing in Advanced Profile shall be applied subject to the following rules:
    1. No OLT shall be performed for any frame if OVERLAP=0; the remainder of these rules shall apply only when OVERLAP=1.
    2. No OLT shall be performed for predicted B frames. (i.e., non BI frames)
    3. Only block boundaries separating two intra blocks shall be smoothed for P frames such that:
        1) Picture level PQUANT is 9 or higher, regardless of HALFQP.
    4. For I frames and BI frames, 8x8 block boundaries (which are defined as boundaries between adjacent 8x8 blocks) shall be smoothed as per the following rules:
        1) When picture level PQUANT is 9 or higher (regardless of HALFQP), all 8x8 block boundaries shall be smoothed.
        2) When picture level PQUANT is 8 or lower (regardless of HALFQP), no 8x8 block boundaries shall be smoothed if the conditional overlap flag CONDOVER is 0 binary.
        3) When picture level PQUANT is 8 or lower (regardless of HALFQP), all 8x8 block boundaries shall be smoothed if the conditional overlap flag CONDOVER is 10 binary.
        4) When picture level PQUANT is 8 or lower (regardless of HALFQP), some 8x8 block boundaries shall be smoothed if the conditional overlap flag CONDOVER is 11 binary, and the following additional rules apply:

i. Internal 8x8 block boundaries within the Luma plane of a MB shall be smoothed when the decoded binary symbol from OVERFLAGS bitplane, or OVERFLAGMB when the raw mode is used to code OVERFLAGS bitplane, for the MB is 1.

ii. 8x8 block boundaries between adjacent MBs (both Luma and chroma) shall be smoothed only when the decoded binary symbols from OVERFLAGS bitplane, or OVERFLAGMB when the raw mode is used to code OVERFLAGS bitplane, for both adjacent MBs are 1.

5. There shall be no dependence on DQUANT across MBs.

6. There shall be no overlap across a block boundary, if the adjacent MBs belong to different slices.

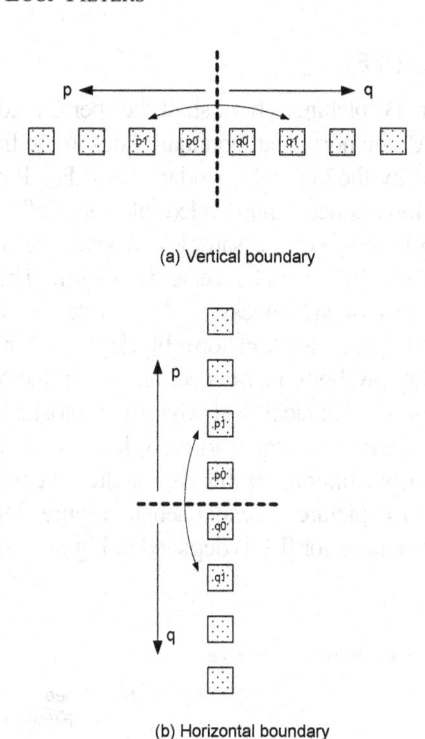


**Figure 7-8 Definition of Edge Pixels for OLT in VC-1**

Actual Filtering Algorithm:

A 4x4 matrix is applied with inputs *p1, p0, q0, q1* producing smoothed outputs $p1'$, $p0'$, $q0'$ and $q1'$. This matrix is the inverse matrix of a matrix applied in the encoder side. The composite operation of the two – forward and inverse – matrices is to be the identity matrix. The inverse computation at a decoder is as follows:

$$\begin{bmatrix} p1' \\ p0' \\ q0' \\ q1' \end{bmatrix} = \left[ \begin{bmatrix} 7 & 0 & 0 & 1 \\ -1 & 7 & 1 & 1 \\ 1 & 1 & 7 & -1 \\ 1 & 0 & 0 & 7 \end{bmatrix} \begin{bmatrix} p1 \\ p0 \\ q0 \\ q1 \end{bmatrix} + \begin{bmatrix} r_0 \\ r_1 \\ r_0 \\ r_1 \end{bmatrix} \right] >> 3 \qquad (7\text{-}5)$$

The rounding parameters r0 and r1 take alternate values (3, 4) or (4, 3) depending on the indexed columns/rows being, respectively, even or odd.

## In-Loop Filter (ILF)

For I or B pictures, ILF shall be performed at all 8x8 block boundaries. All the horizontal boundary lines in the frame shall be filtered first, followed by the vertical boundary lines. For P pictures, blocks may be Intra or Inter-coded. Intra-coded blocks shall use an 8x8 inverse Transform to reconstruct the samples, whereas Inter-coded blocks shall use an 8x8, 8x4, 4x8 or 4x4 inverse Transform. The boundary between transform blocks or sub-blocks shall be filtered, unless the following exception holds: when the transform blocks (or sub-blocks) on either side of the boundary are both inter-coded, and when the MVs of these blocks (or sub-blocks) are identical, and when both blocks (or sub-blocks) have all transform coefficients equal to zero, filtering shall not be performed. The reason for not filtering in this case is due to copying over an already filtered reference picture as explained in Figure 7-9. Typical boundary selection in P pictures for ILF is depicted in Figure 7-10.

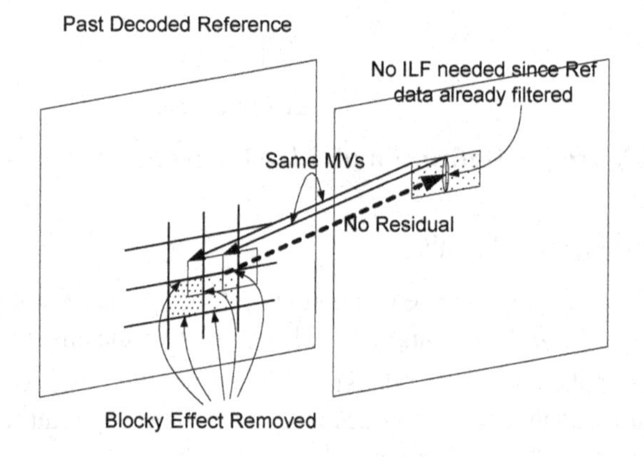


**Figure 7-9 Filtering Exception for ILF in VC-1**

Filtering degree is adaptively chosen based on certain criteria. In the first path, the 3rd pixel out four pixels in the block undergoes filtering. In the second path, all pixels are filtered when some criteria is further met. Otherwise, no more pixels are filtered. The idea of ILF is applied to 8 pixels dissected with an edge in the middle as shown in Figure 7-11.

First, the pixels are categorized into three groups, each of which is composed of four pixels as shown in Figure 7-11. Second, the criteria a0=(2 × (p3-p6)-5 × (p4-p5)+4)>>3 is applied in the center area. Note that the criterion is a measure to extract high frequency activity due to its alternative signs. When this criteria is bigger than PQUANT, it is determined not to contain the blocky effect, but to contain originally high frequency signal in the center area. This is because the quality of image is high with small PQUANT. Note that coefficients of the measure computation are designed in such a way that it can be directly compared with PQUANT. Third, if the high frequency activity measure in the center area is not such a big value, it is compared with the activity measure of the left side and the activity measure of the right side to see if the activity measure of the center area is the smallest one. Note that the left side and the right side measures are defined by a1=(2 × (p1-p4)-5 × (p2-p3)+4)>>3 and a2=(2 × (p5-p8)-5 × (p6-p7)+4)>>3, respectively. If the activity measure of the center area is the smallest one, filtering is not necessary since the center area, where the edge is located, is less active compared with the left side or the right side.

Once the filtering decision is made for a specific edge, edge value correction is performed in each side. The value correction method is suggested to be a function of the minimum activity measure and the center activity measure.

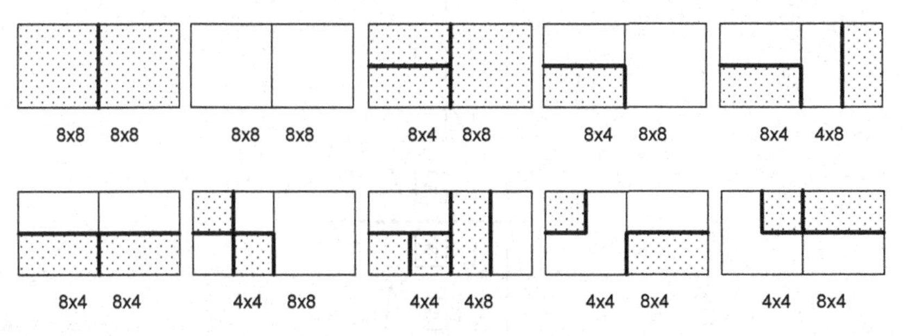


**Figure 7-10 Typical Block Boundaries in P for ILF in VC-1**

### Detailed Algorithm

A filtering algorithm is applied with inputs *p1, p0, q0, q1,* producing smoothed outputs $p0'$ and $q0'$ . Only two pixels are adjusted for progressive video. Two levels of filtering are taken – in the 1st path, every 3rd pixel out of four pixels undergoes filtering; in the 2nd path, all four pixels undergo filtering based on an additional operating condition – the result of the 3rd pixel filtering operation shall determine whether other three pixels in the segment are also filtered. The filtering algorithm is composed of three sub-steps – 1st sub-step of testing whether the video input is originally blocky looking, 2nd sub-step of comparing activities to see if the blocky effect on the edge is not serious compared with other two sides, and 3rd sub-step of actual adjustment at *p0* and *q0* when the determination was made.

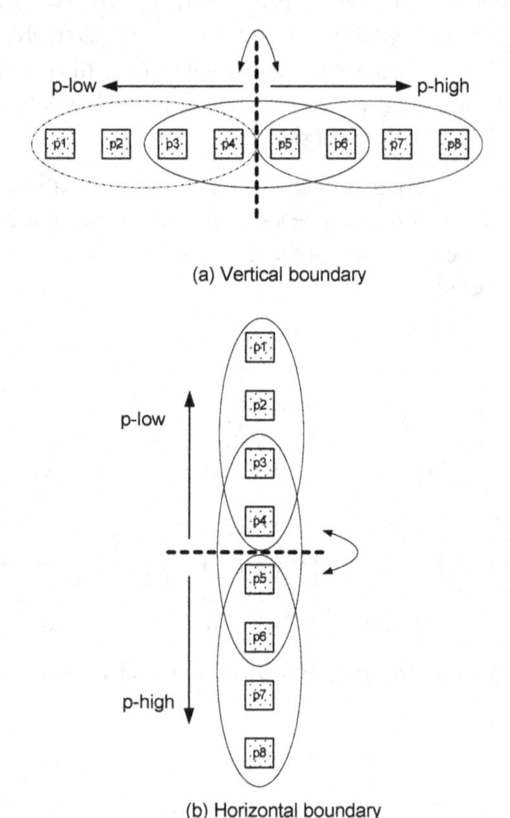


**Figure 7-11 Definition of Edge Pixels for ILF in VC-1**

The two steps of edge selection algorithm and actual filtering algorithm are described in detail as follows:

Edge Selection Algorithm:

- ILF edge selection shall be performed with the following rules:
  1. No ILF shall be performed for any frame if LOOPFILTER=0; the remainder of these rules shall apply only when LOOPFILTER=1.
  2. For I pictures, ILF is performed at all 8x8 block boundaries. The order of filtering is as follows: all the horizontal boundary lines in the frame are filtered first. Then, all the vertical boundary lines are filtered.
  3. For P pictures, the boundaries between Transform blocks shall be filtered unless the following exception holds: the filtering is not performed when the Transform blocks on either side of boundary are both Inter coded and when the MVs of these blocks are identical and when both blocks have zero residual error.
  4. For B pictures, the same ILF as that of I pictures is applied. Only 8x8 block boundaries shall be filtered, and MVs or 4x8/8x4/4x4 are not considered.

Actual Filtering Algorithm:

- ILF shall be applied subject to the following rules:
  1. Compute a0 first, compare, and exit or continue: (1st sub-step)
     1) $a0=(2\times(p3-p6)-5\times(p4-p5)+4)>>3$.
     2) If $(|a0| \geq PQUANT)$, exit.
     3) Otherwise, compute $a3 = \min(|a1|, |a2|)$.
  2. Compare, exit or continue: (2nd sub-step)

1) If (a3 ≥ |a0|), exit.
2) Otherwise, compute d=5 × ((sign(a0) × a3)-a0)/8.
3) Compute clip = (p4-p5)/2.

3. Compare, exit or continue: (3rd sub-step)
   1) If (clip==0), exit.
   2) If (clip > 0), then do the following:

      if (d<0) d=0; if (d>clip) d=clip;
   3) If (clip < 0), then do the following:

      if (d>0) d=0; if (d<clip) d=clip;
4. Correction: p4=p4-d, p5=p5+d. (3rd sub-step continued)

The ILF used for P pictures in the Main Profile is slightly different from the algorithm discussed above. The differences are described in Section 8.6.4.1 of the VC-1 standard. This difference exists primarily for compatibility with WMV-9.

## 7.3 H.264 In-Loop Deblocking Filtering

### In-Loop Deblocking Filter

After coding each block, a blocky effect is inevitable. The control of deblocking process is based on a flag "disable_deblocking_filter_idc." If the flag is equal to "0," all edges are filtered. If the flag is "1," the filtering should be disabled. If the flag is "2," the filtering is expected in all areas except slice boundaries. H.264 handles ILF as follows: the cases of blocky effects are prioritized in the order of strength in theory; and it then is computed in actuality with coded pixels. If the computed blocky effect is severe, the edge is filtered and the filter length is adaptively controlled based on strength measure the theory suggests. Otherwise, it is not filtered.

The filtering applied is adaptive linear filtering that looks at four pixels in each block to rectify up to three pixels on each side of the edge and block context. Basically, all transform edges are filtered except for minor cases. Filtering degree is adaptively chosen based on certain criteria. For little blocky effect, a 4-tap filter is mainly used for the filtering process. For large blocky effect, 3~5 tap filters are mainly used for the filtering process.

The two steps of the algorithm are edge selection and actual filtering– the filtering decision about whether a filtering should be applied, and implementation about how the actual filter works on determined edges. The filtering decision is based on two measures – boundary strength and absolute pixel differences that are compared with trained thresholds of $\alpha$ and $\beta$ given in the standard.

A key assumption is that significant change around transform block boundaries is unlikely due to blocky effect since QP has limited ability to destroy edge texture. Therefore, the proposed algorithm tries to preserve significant change around block boundaries and smooth out relatively small jumps in the area. The thresholds to determine for a significant change around the area are $\alpha$ and $\beta$ that are defined in the standard. Note that $\alpha$ and $\beta$ values increase with the average quantizer parameter QP of two adjacent blocks since blocking distortion is proportional to QP. When QP is small, the blocky effect across the boundaries is likely small due to high image quality. Thus, the thresholds $\alpha$ and $\beta$ are taken low. When QP is larger, blocky effect is likely to be more significant. Therefore, $\alpha$ and $\beta$ are taken high.

Let p and q be representations for (p0, p1, p2, p3) and (q0, q1, q2, q3), respectively, as shown in Figure 7-12.

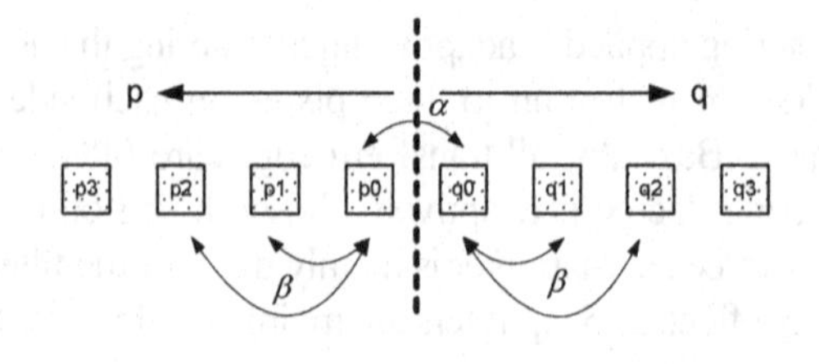


(a) Vertical boundary

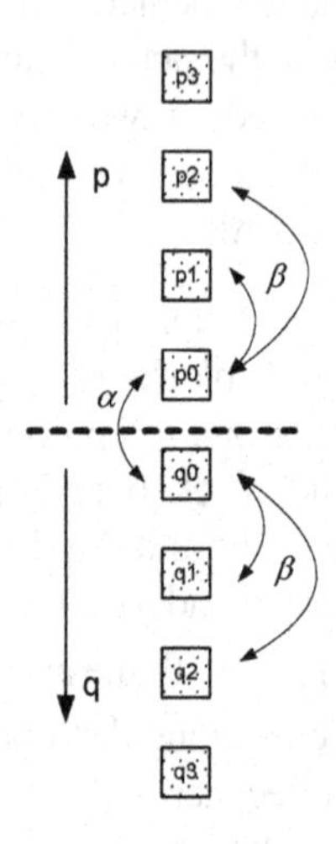


(b) Horizontal boundary

**Figure 7-12 Definition of Edge Pixels for ILF in H.264**

In theory, blocky effect is strong around boundaries in the following order:

- p and/or q is intra-coded and the boundary is a MB boundary. For this case, let's define bS=4.
- p and q are intra-coded, but the boundary is not on a MB boundary. For this case, let's define bS=3.
- neither p or q is intra-coded; p and q contain coded coefficients. For this case, let's define bS=2.

- neither p or q is intra-coded; neither p or q contain coded coefficients; p and q use different reference pictures or a different number of reference pictures or have motion vector values that differ by one Luma sample or more. For this case, let's define bS=1.
- Otherwise, let's define bS=0 for no filtering.

Typically the blocky effect is quite visible at Intra-Intra block boundaries. In particular, boundaries between Intra-Intra MBs look most serious. When residual is not coded, blocky effect between two blocks is not serious. Particularly, no filtering is needed when MVs of two blocks are the same – this implies that reference picture is just copied over, which is already block boundary filtered, as shown in Figure 7-9.

In actuality, the blocky effect is computed as follows:

$$|p0-q0| < \alpha \text{ (QP)}, \qquad (7\text{-}6)$$

$$|p1-p0| < \beta \text{ (QP)}, \qquad (7\text{-}7)$$

$$|q1-q0| < \beta \text{ (QP)}. \qquad (7\text{-}8)$$

The filtering is performed if (bS>0 && (7-6) && (7-7) && (7-8)). The filter length is chosen based on the value bS. Note that the value of $\alpha$ is much bigger than $\beta$ typically.

**Detailed Algorithm**

The variables fieldModeMbFlag, filterInternalEdgesFlag, filterLEftMbEdgeFlag and filterTopMbEdgeFlag are derived as follows:

- The variable fieldModeMbFlag (a.k.a., field MB indicator) is derived as follows:
    1. If any of the following conditions is true, fieldModeMbFlag is set equal to 1:
        1) field_pic_flag is equal to 1.

2) MbafFrameFlag is equal to 1 and the MB currMbAddr is field MB.

2. Otherwise, fieldModeMbFlag is set equal to 0.

- The variable filterInternalEdgesFlag is derived as follows:

  1. If disable_deblocking_filter_idc for the slice that contains the MB CurrMbAddr is set equal to 0, the variable filterInternalEdgesFlag is set equal to 0:
  2. Otherwise, filterInternalEdgesFlag is set equal to 1.

- The variable filterLeftMbEdgeFlag (a.k.a., no left edge filtering) is derived as follows:

  1. If any of following conditions is true, the variable filterLeftMbEdgeFlag is set equal to 0.

     1) MbaffFrameFlag is equal to 0 and CurrMbAddr%PicWidthInMbs is equal to 0.

     2) MbaffFrameFlag is equal to 1 and (CurrMbAddr>>1)%PicWidthInMbs is equal to 0.

     3) disable_deblocking_filter_idc for the slice that contains the MB CurrMbAddr is equal to 1.

     4) disable_deblocking_filter_idc for the slice that contains the MB CurrMbAddr is equal to 2 and the MB mbAddrA is not available.

  2. Otherwise, filterLeftMbEdgeFlag is set equal to 1.

- The variable filterTopMbEdgeFlag (a.k.a., no top edge filtering) is derived as follows:

  1. If any of following conditions is true, the variable filterTopMbEdgeFlag is set equal to 0.

1) MbaffFrameFlag is equal to 0 and CurrMbAddr is less than PicWidthInMbs.
2) MbaffFrameFlag is equal to 1 and (CurrMbAddr>>1) is less than PicWidthInMbs, and the MB CurrMbAddr is a Field MB.
3) MbaffFrameFlag is equal to 1 and (CurrMbAddr>>1) is less than PicWidthInMbs, the MB CurrMbAddr is a Frame MB, and CurrMbAddr%2 is equal to 0.
4) disable_deblocking_filter_idc for the slice that contains the MB CurrMbAddr is equal to 1.
5) disable_deblocking_filter_idc for the slice that contains the MB CurrMbAddr is equal to 2 and the MB mbAddrA is not available.

2. Otherwise, filterTopMbEdgeFlag is set equal to 1.

The two steps of edge selection algorithm and actual filtering algorithm are described in detail as follows:

Edge Selection Algorithm:

- When filterLeftMbEdgeFlag is equal to 1, the filtering of the left vertical Luma edge is selected as follows:
  1. The actual filtering process is invoked with chromaEdgeFlag = 0, verticalEdgeFlag = 1, fieldModeFilteringFlag = fieldModeMbFlag, and $(xE_k, yE_k) = (0,k)$ with k=0..15 as selected edge input.

- When filterInternalEdgesFlag is equal to 1, the edges for filtering of the internal vertical Luma data is selected as follows:

    1. When transform_size_8x8_flag is equal to 0, the filtering process is invoked with chromaEdgeFlag = 0, verticalEdgeFlag = 1, fieldModeFilteringFlag = fieldModeMbFlag, and ($xE_k$, $yE_k$) = (4,k) with k=0..15 as selected edge input.
    2. The filtering process is invoked with chromaEdgeFlag = 0, verticalEdgeFlag = 1, fieldModeFilteringFlag = fieldModeMbFlag, and ($xE_k$, $yE_k$) = (8,k) with k=0..15 as selected edge input.
    3. When transform_size_8x8_flag is equal to 0, the filtering process is invoked with chromaEdgeFlag = 0, verticalEdgeFlag = 1, fieldModeFilteringFlag = fieldModeMbFlag, and ($xE_k$, $yE_k$) = (12,k) with k=0..15 as selected edge input.

- When filterToplEdgesFlag is equal to 1, the edges for filtering of the top horizontal Luma data is selected as follows:

    1. If MbaffFrameFlag is equal to 1, (CurrMbAddr%2) is equal to 0, CurrMbAddr is greater than or equal to $2\times$ PicWidthInMbs, the MB CurrMbAddr is a frame MB, and the MB (CurrMbAddr-$2\times$ PicWidthInMbs+1) is a field MB, the following applies:

        1) The filtering process is invoked with chromaEdgeFlag = 0, verticalEdgeFlag = 0, fieldModeFilteringFlag = 1 and ($xE_k$, $yE_k$) = (k,0) with k=0..15 as selected edge input.

        2) The filtering process is invoked with chromaEdgeFlag = 0, verticalEdgeFlag = 0, fieldModeFilteringFlag = 1 and ($xE_k$, $yE_k$) = (k,1) with k=0..15 as selected edge input.

2. Otherwise, the filtering process is invoked with chromaEdgeFlag = 0, verticalEdgeFlag = 0, fieldModeFilteringFlag = fieldModeMbFlag and $(xE_k, yE_k) = (k,0)$ with k=0..15 as selected edge input.

- When filterInternalEdgesFlag is equal to 1, the edges for filtering of the internal horizontal Luma data is selected as follows:
  1. When transform_size_8x8_flag is equal to 0, the filtering process is invoked with chromaEdgeFlag = 0, verticalEdgeFlag = 0, fieldModeFilteringFlag = fieldModeMbFlag, and $(xE_k, yE_k) = (k,4)$ with k=0..15 as selected edge input.
  2. The filtering process is invoked with chromaEdgeFlag = 0, verticalEdgeFlag = 0, fieldModeFilteringFlag = fieldModeMbFlag, and $(xE_k, yE_k) = (k,8)$ with k=0..15 as selected edge input.
  3. When transform_size_8x8_flag is equal to 0, the filtering process is invoked with chromaEdgeFlag = 0, verticalEdgeFlag = 0, fieldModeFilteringFlag = fieldModeMbFlag, and $(xE_k, yE_k) = (k,12)$ with k=0..15 as selected edge input.

The Chroma filtering operations are performed in the same way defined above with the only exception that processing would need to be adjusted according to Chroma format information in the syntax "chroma_format_idc" as described in the section 8.7 of the standard. Note that Chroma filtering process is invoked when chromaEdgeFlag is set equal to 1 in above algorithm.

Actual Filtering Algorithm:

The two sub-steps described above are explained in detail as follows:

- Compute Boundary Strength (bS) (1st sub-step)

1. bS=4 if the block edge is on a MB edge and any of the following conditions are true:

    1) The samples p0 and q0 are both in Frame MBs and either (or both) of the samples p0 and q0 is in a MB coded using an Intra MB prediction mode.

    2) The samples p0 and q0 are both in Frame MBs and either (or both) of the samples p0 and q0 is in a MB that is in a slice with slice_type equal to SP or SI.

    3) MbaffFrameFlag is equal to 1 or field_pic_flag is equal to 1, and verticalEdgeFlag is equal to 1, and either (or both) of the samples p0 and q0 is in a MB coded using an Intra MB prediction mode.

    4) MbaffFrameFlag is equal to 1 or field_pic_flag is equal to 1, and verticalEdgeFlag is equal to 1, and either (or both) of the samples p0 and q0 is in a MB that is in a slice with slice_type equal to SP or SI.

2. Otherwise, bS=3 if any of the following conditions are true:

    1) MixedModeEdgeFlag is equal to 0 and either (or both) of the samples p0 and q0 is in a MB coded using an Intra MB prediction mode.

    2) MixedModeEdgeFlag is equal to 0 and either (or both) of the samples p0 and q0 is in a MB that is in a slice with slice_type equal to SP or SI.

    3) MixedModeEdgeFlag is equal to 1, verticalEdgeFlag is equal to 0, and either (or both) of the samples p0 and q0 is in a MB coded using an Intra MB prediction mode.

4) MixedModeEdgeFlag is equal to 1, verticalEdgeFlag is equal to 0, and either (or both) of the samples p0 and q0 is in a MB that is in a slice with slice_type equal to SP or SI.

3. Otherwise, bS=2 if the following condition is true:

   1) The Luma block containing sample p0 or the Luma block containing sample q0 contains non-zero transform coefficient levels.

4. Otherwise, bS=1 if any of the following conditions are true:

   1) MixedModeEdgeFlag is equal to 1.

   2) MixedModeEdgeFlag is equal to 0 and for the prediction of the MB/sub-MB partition containing the sample p0 different reference pictures or a different number of MVs are used than for the prediction of the MB/sub-MB partition containing the sample q0.

   3) MixedModeEdgeFlag is equal to 0 and one MV is used to predict the MB/sub-MB partition containing the sample p0 and 1 MV is used to predict the MB/sub-MB partition containing the sample q0, and the absolute difference between the horizontal or vertical component of the MVs used is greater than or equal to 4 in units of quarter Luma frame samples.

4) MixedModeEdgeFlag is equal to 0 and two MVs and two different reference pictures are used to predict the MB/sub-MB partition containing the sample p0. And, two MVs for the same two reference pictures are used to predict the MB/sub-MB partition containing the sample q0, and the absolute difference between the horizontal or vertical component of the two MVs used in the prediction of the two MB/sub-MB partitions for the same reference picture is greater than or equal to four in units of quarter Luma frame samples.

5) MixedModeEdgeFlag is equal to 0 and 2 MVs for the same reference picture are used to predict the MB/sub-MB partition containing the sample p0 and two MVs for the same reference picture are used to predict the MB/sub-MB partition containing the sample q0, and both of the following conditions are true:

    i. The absolute difference between the horizontal or vertical component of List 0 MVs used in the prediction of the two MB/sub-MB partitions is greater than or equal to 4 in quarter Luma frame samples OR the absolute difference between the horizontal or vertical component of List 1 MVs used in the prediction of the two MB/sub-MB partitions is greater than or equal to 4 in quarter Luma frame samples.

ii. The absolute difference between the horizontal or vertical component of the List 0 MV used in the prediction of the MB/sub-MB partition containing the sample p0 and the List 1 MV used in the prediction of MB/sub-MB partition containing the sample q0 is greater than or equal to 4 in quarter Luma frame samples OR the absolute difference between the horizontal or vertical component of List 1 MV used in the prediction of the MB/sub-MB partition containing the sample p0 and the List 0 MV used in the prediction of MB/sub-MB partition containing the sample q0 is greater than or equal to 4 in quarter Luma frame samples.

5. bS=0, otherwise.

Here, the variable mixedModeEdgeFlag is derived as follows:

i. If MbaffFrameFlag is equal to 1 and the samples p0 and q0 are in different MB pairs, one of which is a field MB pair and the other is a frame MB pair, mixedModeEdgeFlag is set equal to 1.

ii. Otherwise, mixedModeEdgeFlag is set equal to 0.

- Compare measures with $\alpha$ and $\beta$

1. The filtering process turned on when BS>0 with $|p0-q0|<\alpha$ && $|p1-p0|<\beta$ && $|q1-q0|<\beta$.

- The values of $\alpha$ and $\beta$ are dependent on pre-computed $\alpha'$ and $\beta'$ in the Table 7-1 (IS: Table 8-16) as a function of indexA or indexB as follows: $\alpha = \alpha' \times$ (1<<(BitDepthY-8)) and $\beta = \beta' \times$ (1<<(BitDepthY-8)) at a Luma block, while $\alpha = \alpha' \times$ (1<<(BitDepthC-8)) and $\beta = \beta' \times$ (1<<(BitDepthC-8)) at a chroma block. Note that indexA or indexB can be computed with: indexA = Clip3(0, 51, qPav + FilterOffsetA) and indexB = Clip3(0, 51, qPav + FilterOffsetB), where qPav=(qPp+qPq +1)>>1, FilterOffsetA = slice_alpha_c0_offset_div2<<1 and FilterOffsetB = slice_beta_offset_div2 << 1 . Such data are available in Slice header. Note that Clip3( ) operation is defined in Chapter 6.

2. No filtering, otherwise (including bS=0).

- Deblocking filtering processing (2nd sub-step)
  1. If( bS==1||2||3) → reasonable processing

  $p0'$ and $q0'$ are produced by p1, p0, q0 and q1

  a) $\Delta$ = Clip3(-tC, tC, ((((q0-p0)<<2)+(p1-q1)+4)>>3))
  b) $p0'$ = Clip1(p0+ $\Delta$ )
  c) $q0'$ = Clip1(q0- $\Delta$ )

  where the threshold tC is defined as

  tC= $tC0'$ $\times$ (1<<(BitDepthY-8))+((|p2-p0|< $\beta$ )?1:0) + (( |q2-q0|< $\beta$ )?1:0) at a Luma block or tC= $tC0'$ $\times$ (1<<(BitDepthC-8))+1 at a chroma block. Note that $tC0'$ is given in Table 7-2 (IS: Table 8-17) as a function of indexA and bS.

- If |p2-p0|< $\beta$ && this is a Luma block:

a) $p1' = \text{p1} + \text{Clip3(-tC0, tC0, (p2} + \text{((p0+q0+1)>>1)-(p1<<1))>>1)}$

- Else (a.k.a., $|\text{p2-p0}| \geq \beta$ || this is a chroma block):

  a) No processing

Plus,

- If $|\text{q2-q0}| < \beta$ && this is a Luma block:

  a) $q1' = \text{q1} + \text{Clip3(-tC0, tC0, (q2} + \text{((p0+q0+1)>>1)-(q1<<1))>>1)}$

- Else (a.k.a., $|\text{q2-q0}| \geq \beta$ || this is a chroma block):

  a) No processing

2. If( bS==4) → strong processing

- If $|\text{p2-p0}| < \beta$ && $|\text{p0-q0}| < ((\alpha >> 2) + 2)$ && this is a Luma block:

  a) $p0' = (\text{p2}+2\times\text{p1}+2\times\text{p0}+2\times\text{q0}+\text{q1}+4)>>3$

  b) $p1' = (\text{p2+p1+p0+q0}+2)>>2$

  c) $p2' = (2\times\text{p3}+3\times\text{p2+P1+p0+q0}+4)>>3$

- Else (a.k.a., $|\text{p2-p0}| \geq \beta$ || $|\text{p0-q0}| \geq ((\alpha >> 2) + 2)$ || this is a chroma block):

  a) $p0' = (2\times\text{p1+p0+q1}+2)>>2$

Plus,

- If $|\text{q2-q0}| < \beta$ && $|\text{p0-q0}| < ((\alpha >> 2) + 2)$ && this is a Luma block:

  a) $q0' = (\text{p1}+2\times\text{p0}+2\times\text{q0}+2\times\text{q1}+\text{q2}+4)>>3$

  b) $q1' = (\text{p0+q0+q1+q2}+2)>>2$

  c) $q2' = (2\times\text{q3}+3\times\text{q2+q1+q0+p0}+4)>>3$

- Else (a.k.a., $|q2\text{-}q0| \geq \beta$ || $|p0\text{-}q0| \geq ((\alpha >> 2) + 2)$ || this is a chroma block):

  a) $q0' = (2 \times q1 + q0 + p1 + 2) >> 2$

**Table 7-1 Derivation of $\alpha'$ and $\beta'$ from indexA and indexB (IS: Table 8-16)**

| | indexA (for $\alpha'$) or indexB (for $\beta'$) | | | | | | | | | | | |
|---|---|---|---|---|---|---|---|---|---|---|---|---|
| | 0 | 1 | 2 | ... | 22 | 23 | 24 | ... | 48 | 49 | 50 | 51 |
| $\alpha'$ | 0 | 0 | 0 | ... | 9 | 10 | 12 | ... | 203 | 226 | 255 | 255 |
| $\beta'$ | 0 | 0 | 0 | ... | 3 | 4 | 4 | ... | 17 | 17 | 18 | 18 |

**Table 7-2 Derivation of $tC0'$ from indexA and bS (IS: Table 8-17)**

| | indexA | | | | | | | | | | | |
|---|---|---|---|---|---|---|---|---|---|---|---|---|
| | 0 | 1 | 2 | ... | 22 | 23 | 24 | ... | 48 | 49 | 50 | 51 |
| bS=1 | 0 | 0 | 0 | ... | 0 | 1 | 1 | ... | 9 | 10 | 11 | 13 |
| bS=2 | 0 | 0 | 0 | ... | 1 | 1 | 1 | ... | 12 | 13 | 15 | 17 |
| bS=3 | 0 | 0 | 0 | ... | 1 | 1 | 1 | ... | 18 | 20 | 23 | 25 |

Since the filtering process changes values on the fly in an applied MB neighborhood, it is important to define processing order. In the standard, the filtering is applied to vertical or horizontal edges of 4x4 blocks in a MB. The choice of operation on slice boundaries is given as options as already discussed. The definition of the processing order is shown in Figure 7-13.

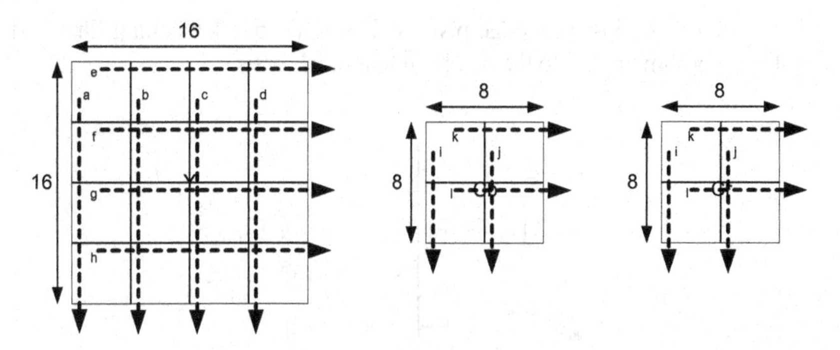


**Figure 7-13 The Order of Edge Filtering for ILF in H.264**

Prior to the operation of the deblocking filter process for each MB, the deblocked samples of the MB or MB pair above (if any) and the MB or MB pair to the left (if any) of the current MB are always available because the deblocking filter process is performed after the completion of the picture construction process.

In addition, the deblocking filter process is invoked for the Luma and chroma components separately. Chroma filtering position is dependent on the value of chroma_format_idc since the transform boundaries change based on the chroma format such as 4:2:0/ 4:2:2/ 4:4:4.

## 7.4 Out-Loop Filtering

### Deblocking Filter

Generally, video codecs introduce blocking and ringing artifacts. Some standards recommend deblocking and/or deringing processes as post filters out of the loop. These out-loop filters mean to work on display pictures after decoding, not on reference pictures. Typically, smaller or more highly compressed video needs post filters to enhance visual quality. VC-1 defines special deblocking and deringing filters as post processing filters other than OLT and ILF. The application of either one or both of them may be signaled in the bitstream with the POSTPROC field of the picture layer. This section explains about those two filters in VC-1. However, the same explanation can be applied to similar filters defined in other standards such as MPEG-4.

Figure 7-14 depicts edge pixel definition for the deblocking filter and the algorithm applies to the pixels on 8x8 block basis.

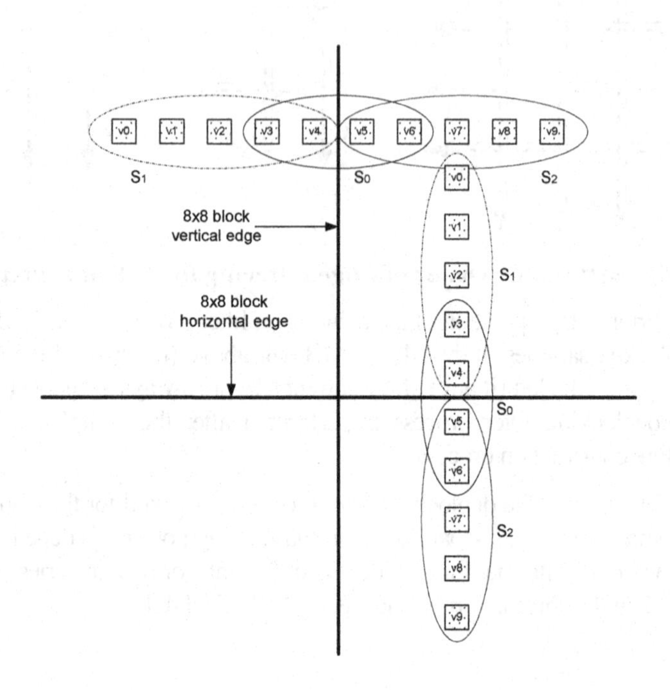


**Figure 7-14 Typical Block Boundaries for Deblock Post-processing**

In the filter operations, two modes are used separately depending on the pixel conditions around boundaries – DC offset and Default modes.

$$eq\_cnt = \phi(v0 - v1) + \phi(v1 - v2) + \phi(v2 - v3) + \phi(v3 - v4) + \phi(v4 - v5) + \phi(v5 - v6) + \phi(v6 - v7) + \phi(v7 - v8) + \phi(v8 - v9) \quad (7\text{-}9)$$

where $\phi(\gamma) = 1$ if $|\gamma| \leq \text{thr1}$. Otherwise, $\phi(\gamma) = 0$.

The $\phi(\gamma)$ is incremented when the difference of adjacent values is pretty small (i.e., smooth region). If $eq\_cnt$ is greater than or equal to thr2, DC offset mode is applied. If not, Default mode is applied. In other words, DC offset mode is selected when the region is smooth. The threshold values are given as 2 or 6 for thr1 or thr2 in the standard.

In the Default mode, a signal adaptive smoothing scheme is applied by differentiating image details at the block discontinuities using the frequency information of neighbor pixel arrays, $S_0$, $S_1$ and $S_2$. The algorithm for Default mode is similar to that of ILF in VC-1. The filtering scheme in Default mode is executed by replacing the boundary pixel values $v_4$ and $v_5$ with $v_4'$ and $v_5'$ as follows:

$$v4' = v4 - d, \tag{7-10}$$

$$v5' = v5 + d \tag{7-11}$$

and

$$d = Clip3(0, (v4 - v5)/2, 5 \times (a_{3,0}' - a_{3,0}) // 8) \times \delta(|a_{3,0}| < QP) \tag{7-12}$$

$$\text{where } a_{3,0}' = sign(a_{3,0}) \times \min(|a_{3,0}|, |a_{3,1}|, |a_{3,2}|). \tag{7-13}$$

Here, activity measures $a_{3,0}$, $a_{3,1}$ and $a_{3,2}$ can be evaluated from the simple inner product of the approximated high-frequency DCT kernel [2 –5 5 –2] with the pixel vectors:

$$a_{3,0} = ([2 \ \ -5 \ \ 5 \ \ -2] \cdot [v3 \ \ v4 \ \ v5 \ \ v6]^t)//8, \tag{7-14}$$

$$a_{3,1} = ([2 \ \ -5 \ \ 5 \ \ -2] \cdot [v1 \ \ v2 \ \ v3 \ \ v4]^t)//8, \tag{7-15}$$

$$a_{3,2} = ([2 \ \ -5 \ \ 5 \ \ -2] \cdot [v5 \ \ v6 \ \ v7 \ \ v8]^t)//8. \tag{7-16}$$

Note that $a_{3,0}$, $a_{3,1}$ and $a_{3,2}$ are the center, the left-side, the right-side activity measures, respectively. In this context, QP means PQUANT, where PQUANT is passed as a meta data parameter to post processing. And $\delta(condition) = 1$ if the condition is true. Otherwise, $\delta(condition) = 0$.

In a very smooth region, the filtering in the Default mode is not good enough to reduce the blocking artifact due to DC offset. This case is specially treated with DC offset mode to apply a stronger filter as follows:

Max=max (*v1, v2, v3, v4, v5, v6, v7, v8*), (7-17)

Min=min (*v1, v2, v3, v4, v5, v6, v7, v8*), (7-18)

If(|Max-Min|)<2 $\times$ *QP*){

$$vn' = \sum_{k=-4}^{4} b_k \cdot p_{n+k}, \quad 1 \leq n \leq 8,$$

} else  No change will be done.

Here, input data and the filter coefficients are defined as follows:

$$p_m = \begin{cases} (|v1 - v0| < QP)\,?\,v0 : v1, & if \quad m < 1 \\ vm & if \quad 1 \leq m \leq 8 \\ (|v8 - v9| < QP)\,?\,v9 : v8, & if \quad m > 8 \end{cases} \quad (7\text{-}19)$$

and $\{b_k : -4 \leq k \leq 4\}$ ={1, 1, 2, 2, 4, 2, 2, 1, 1}//16. (7-20)

Note that two far distant values (a.k.a, (*v0* or *v1*) and (*v8* or *v9*)) from the center are adaptively selected based on quantization parameter QP. If values of *v0* and *v1* are close enough, the most distant one, *v0*, is selected as input data. The same logic applies to *v8* and *v9*, either. The order of processing is horizontal and then vertical directions. Any pixel changed on the fly is used for future processing.

### Deringing Filter

The ringing effect is most evident along high contrast edges within smooth texture in reconstructed image under high compression. It appears as a shimmering and rippling outwards from the edge up to the encompassing block boundary as shown in Figure 7-15. The idea of the deringing filter is to leave abrupt edge alone, but to work on nearest ripples as shown in Figure 7-15. The deringing filtering comprises three sub-processes: threshold determination, index acquisition and adaptive smoothing. The algorithm applies to the pixels on an 8x8 block basis, where neighborhood of 10x10 pixels at each block are considered as shown in Figure 7-16.

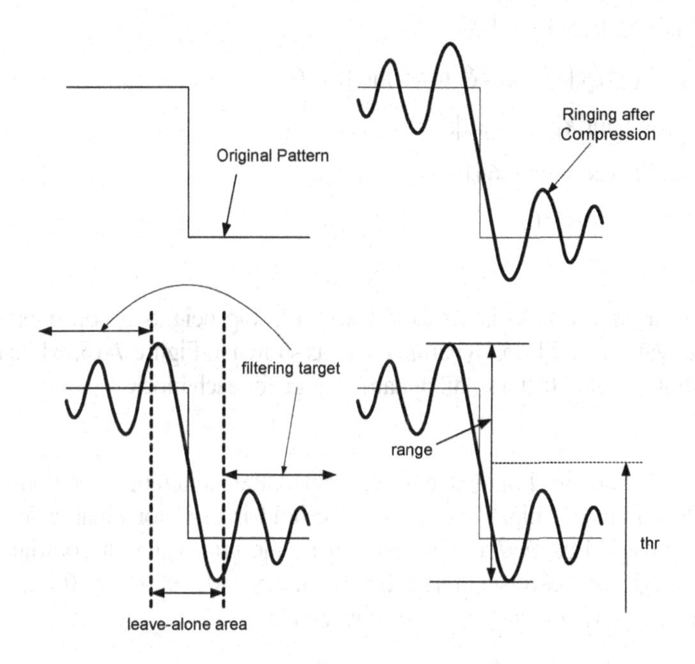


**Figure 7-15 Ringing Effect and Deringing Filter**

First, computation for maximum and minimum gray value within a Luma block is performed in the decoded image. Then, the threshold denoted by $thr[k]$ and the dynamic range of gray scale denoted by $range[k]$ are set:

$$thr[k] = (\text{maximum}[k] + \text{minimum}[k] + 1)/2, \qquad (7\text{-}21)$$

$$range[k] = \text{maximum}[k] - \text{minimum}[k]. \qquad (7\text{-}22)$$

Here *thr* is taken as the average of the maximum and minimum values, while range value is the maximum difference among any two values (i.e., dynamic range) in a block as shown in Figure 7-15. The value $k$ ranges from 1 to 4. In addition, *max_range* is defined to be the maximum value of the dynamic range among four Luma blocks.

$$max\_range = \text{range}[k_{max}]. \qquad (7\text{-}23)$$

Then, an arrangement process follows:

```
for(k=1; k<5; k++){
    if(range[k]<32 && max_range≥ 64)
        thr[k]=thr[kmax];
    if(max_range<64)
        thr[k]=0;
}
```

Here, a threshold is set to thr[$k_{max}$] at a big neighborhood dynamic range with a small local dynamic range as shown in Figure 7-15, while the threshold is set to 0 at a small dynamic range for each block.

Once the threshold value is determined, the remaining operations are purely on an 8x8 block basis. The threshold is used for binarization of given image. Let *rec(h,v)* and *bin(h,v)* be the gray value at coordinates *(h,v)* and the corresponding binary index where $h$, $v=0,1,2,\dots,7$, respectively. Then, *bin(h,v)* can be obtained by:

$$bin(h,v) = \begin{cases} 1 & if \quad rec(h,v) \geq thr \\ 0 & \quad otherwise. \end{cases} \qquad (7\text{-}24)$$

In other words, pixel value above that threshold is set to 1, while pixel value under the threshold is set to 0 in binarization. An example of binarization is given in Figure 7-16. Note that *(h, v)* is used to address a pixel in a block, while *(i,j)* is for accessing a pixel in a 3x3 window.

The filter is applied only if the binary indices in a 3x3 window are all the same value – all "0" indices or all "1" indices. In other words, the filtering is not performed in magnitude transitional areas, but applied in similar magnitude areas as explained in Figure 7-15. The adaptive filter is depicted in the right side of Figure 7-16. Note that 10x10 binary indices are obtained with a single threshold which corresponds to the 8x8 block shown in Figure 7-16 and that the shaded area represents the pixels to be filtered. The filter indices are defined as *coef(i,j)* where *ij*=-1, 0, 1.

Here the coefficient at the center pixel (i.e., *coef(0,0)*) corresponds to the pixel to be filtered. The output is obtained by:

$$flt'(h,v) = \{8 + \sum_{i=-1}^{1}\sum_{j=-1}^{1} coef(i,j) \cdot rec(h+i, v+j)\} // 16. \quad (7\text{-}25)$$

| 0 | 0 | 0 | 0 | 0 | 0 | 0 | 0 | 0 | 0 |
|---|---|---|---|---|---|---|---|---|---|
| 0 | 0 | 0 | 0 | 0 | 0 | 0 | 0 | 0 | 0 |
| 1 | 0 | 0 | 0 | 0 | 0 | 0 | 0 | 1 | 1 |
| 1 | 1 | 0 | 0 | 0 | 0 | 0 | 1 | 1 | 1 |
| 1 | 1 | 1 | 0 | 0 | 0 | 1 | 1 | 1 | 1 |
| 1 | 1 | 1 | 1 | 1 | 1 | 1 | 1 | 1 | 1 |
| 1 | 1 | 1 | 1 | 1 | 1 | 1 | 1 | 1 | 1 |
| 1 | 1 | 1 | 1 | 1 | 1 | 1 | 1 | 1 | 1 |
| 1 | 1 | 1 | 1 | 1 | 1 | 1 | 1 | 1 | 0 |
| 1 | 1 | 1 | 1 | 1 | 1 | 1 | 1 | 0 | 0 |

| 1 | 2 | 1 |
|---|---|---|
| 2 | 4 | 2 |
| 1 | 2 | 1 |

**Figure 7-16 Example of Adaptive Filtering and Filter Mask**

The maximum gray level change between the reconstructed pixel and the filtered one is clipped according to the quantization parameter *QP*. The $flt(h,v)$ and $flt'(h,v)$ are the filtered pixel value and the pixel value before limitation, respectively.

If( $flt'(h,v) - rec(h,v)$ >max_diff)

$$flt(h,v) = rec(h,v) + \text{max\_diff}$$

Else if( $flt'(h,v) - rec(h,v) < -\text{max\_diff}$)

$$flt(h,v) = rec(h,v) - \text{max\_diff}$$

Else $flt(h,v) = flt'(h,v)$

where max_diff=$QP/2$ for both Intra and Inter MBs.

# 8. Interlace Handling

## 8.1 MPEG-2 Interlace Handling

### Progressive, Interlace, Frame- and Field-Picture

The main focus of MPEG-2 as opposed to MPEG-1 was to code interlaced CCIR 601 video at a bitrate for broadcast and consumer applications since video archives for several decades are captured/ stored in interlaced format [schafer:digital]. A coded video sequence consists of a sequence of coded pictures, where a coded picture can represent either an entire frame or a single field in MPEG-2. If the two fields of a frame are captured at different time points, the frame is referred to as an interlaced or interlace frame. Otherwise it is called a progressive frame.

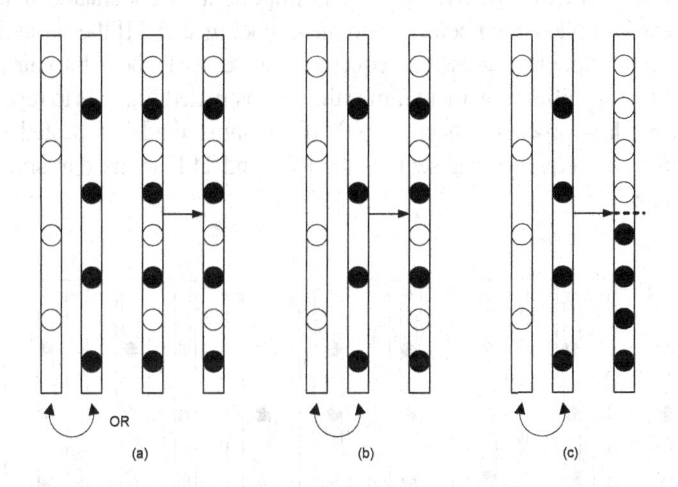


**Figure 8-1 Progressive and Non-Progressive Sequences**

Progressive sequences can only contain "Progressive Frame-Pictures." This structure is "mandatory" in MPEG-1, but "optional" in MPEG-2. As shown in Figure 8-1 (a), interlaced fields or simple frame can constitute a "Progressive Frame-Picture" in MPEG-2. Two fields can be combined for an "Interlaced Frame-Picture" as shown in Figure 8-1 (b). As depicted in

Figure 8-1 (a) and Figure 8-1 (b), there is no pictorial difference between Progressive and Interlaced Frame-Pictures. However, a different set of compression tools are allowed to be applied for these pictures. Two fields can represent two "Interlaced Field-Pictures" as shown in Figure 8-1 (c). When a frame is coded with two Field-Pictures, the "temporal reference" associated with each coded picture shall be the same. Note that the same picture constructions are represented with the Frame Coding Mode (FCM) in VC-1 as "Progressive" frame, "Frame-Interlace" frame and "Field-Interlace" frame, respectively.

### Repeat First Field (RFF) and Top Field First (TFF)

If the same frame or field data is repeated, an encoder doesn't need to code it, thus saving bandwidth and computation. To compactly represent such a situation, RFF and TFF flags are used in the Picture Coding Extension structure of MPEG-2. It is important to understand that this process is applied even before the main encoding stage. If the same frame is repeated under progressive_sequence flag set to 1, the 2-bit pair (RFF and TFF) specifies how many times the reconstructed frame is to repeat by the decoding process – from 1 to 3. If the same field is repeated under progressive_sequence flag set to 0, the RFF and TFF are used as shown in Figure 8-2.

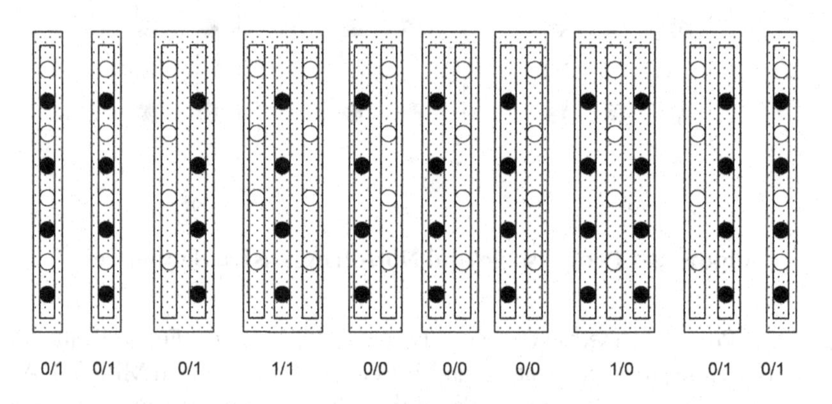


**Figure 8-2 RFF and TFF**

In any frame grouping shown in Figure 8-2, TFF is set to "1" when the first field is "top" field (i.e., white circles). Especially when two fields in the same polarity contain the same data (i.e., top field first-top field repeat or bottom field first-bottom field repeat), RFF is set to "1." Since the 3rd field is the same as the 1st one, it is dropped from bitstream representation (thus from transmission) only leaving the RFF/TFF combination. To have the represented picture be self-contained, picture_structure data is included in the Picture Coding Extension structure. The data describes whether current picture is Frame-Picture or Field-Picture. If the current picture is Field-Picture, it further tells whether a top field is the first field or a bottom field is the first one.

The interpretation of RFF/TFF is performed after decoding and before displaying decoded data. These flags can be used to convert frame rate if such approximation is acceptable. For example, film content (24 fps) can be converted to broadcasting content (29.97 fps) with certain intentional field repeat – this is sometimes called "3:2 Pulldown." This is beneficial since it doesn't really increase bandwidth to get a higher frame rate. This only requires careful use of the 2-bit pair flags RFF/TFF.

### Prediction Modes for Frame-Pictures

There are two main ways to inter-predict in Frame-Pictures as shown in Figure 8-3 and Figure 8-4 – Frame Prediction and Field Prediction. The most common concept for Frame Prediction is to get a 16x16 best match in the reference frame-picture, where 1 MV represents both top and bottom fields simultaneously as shown in Figure 8-3 and Figure 8-4.

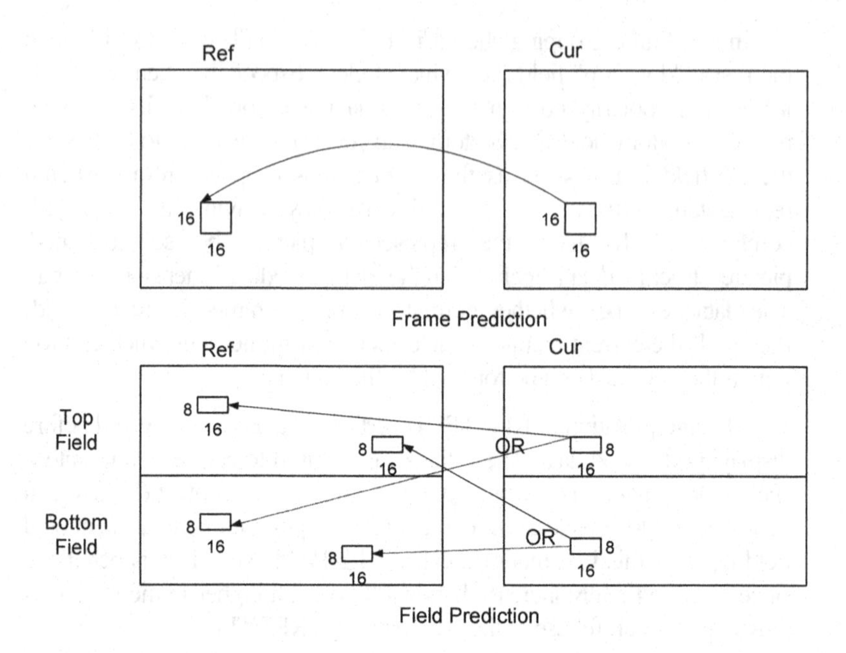


**Figure 8-3 Frame Prediction and Field Prediction in Frame Pictures in Space Representation**

The most common concept for Field Prediction is to get two 16x8 best matches in the reference field-pictures for decomposed current top field and current bottom field's 16x8 blocks, where 1 MV is used for the top field and a separate 1 MV is used for the bottom field independently as shown in Figure 8-3 and Figure 8-4.

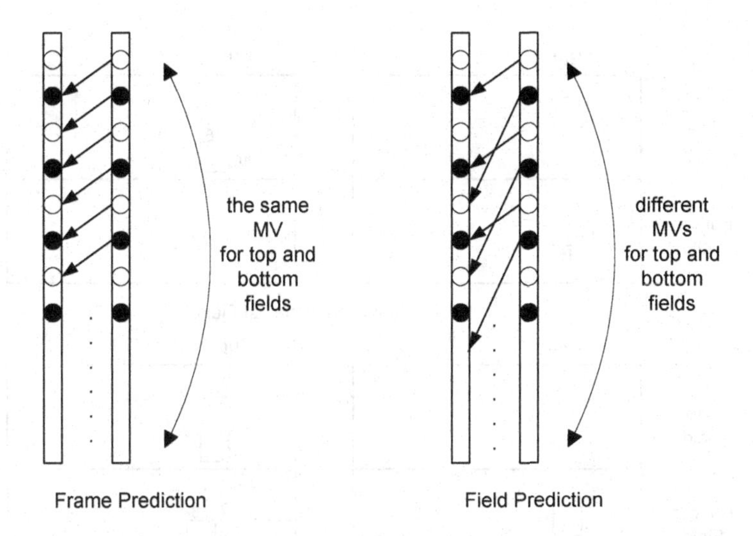


**Figure 8-4 Frame Prediction and Field Prediction in Frame Pictures in Time Representation**

Frame Prediction and Field Prediction can be adaptively chosen in each MB, depending on which provides a better prediction performance. However, there are more options for Frame-Pictures and Field-Pictures developed in MPEG-2 as explained in the next sections. Among all possible options, an encoder can choose the best one in each MB adaptively.

## Prediction Modes for Field-Pictures

The way to inter-P predict in Field-Pictures is shown in Figure 8-5 and Figure 8-6. Fundamentally the prediction gets a 16x16 best match in a reference field-picture, where 1 MV is used for both top and bottom fields simultaneously with the option of reference field-pictures as shown in Figure 8-5 and Figure 8-6. For P Prediction, the most recent two previous reference field-pictures are used both for 1$^{st}$ field and 2$^{nd}$ field as shown in Figure 8-5 and Figure 8-6. Note that the 2$^{nd}$ field is allowed to reference the 1$^{st}$ field for P Prediction in MPEG-2. The predictor for the current MB at any field can easily come from either the 1$^{st}$ field or the 2$^{nd}$ field without much bias since sometimes top and bottom field inversion can happen due to RFF/TFF use.

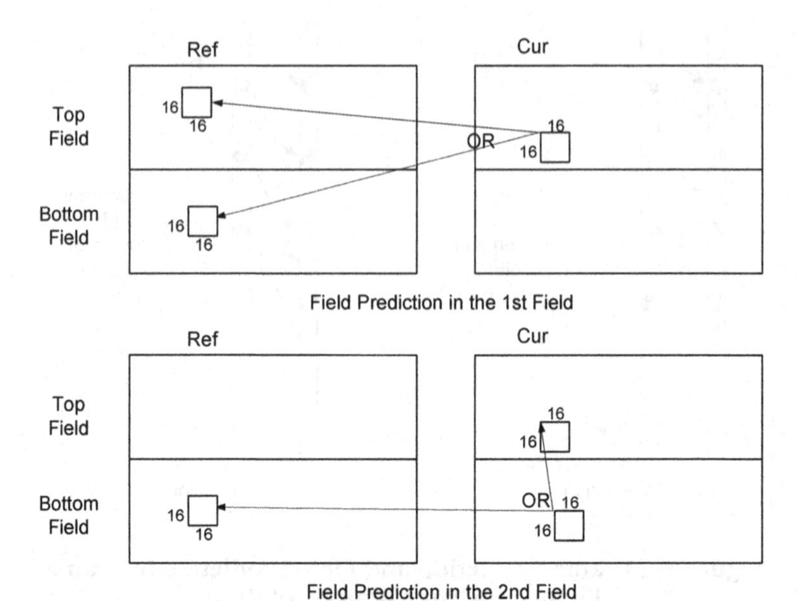


**Figure 8-5 Field Prediction in Field Pictures in Space Representation**

The way to inter-B predict in Field-Pictures is shown in Figure 8-6. The prediction gets a 16x16 best match from two reference field-pictures in a combined way, where 1 MV is used for the forward direction to a reference field and 1 MV is used for the backward direction to the other reference field simultaneously with option of reference field-pictures as shown in Figure 8-6. For B Prediction, different reference field-pictures are used for $1^{st}$ field and $2^{nd}$ field. Note that the $2^{nd}$ field is not allowed to reference the $1^{st}$ B field for B Prediction in MPEG-2.

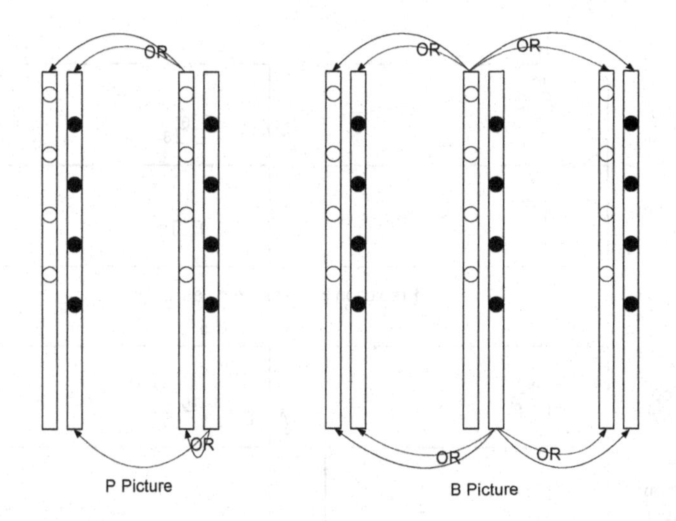


**Figure 8-6 Field Prediction in Field Pictures in Time Representation**

## Dual Prime Prediction

Dual Prime Prediction is defined for P Pictures only as shown in Figure 8-7 and Figure 8-8. This technique can be used for both Field-Pictures and Frame-Pictures as shown in Figure 8-7. The idea here is to obtain a predictor that is the average of two candidates from the top and bottom fields. To take advantage of compact representation, one important assumption is taken to be that the space displacement between top and bottom field candidates ranges only from –0.5 to 0.5 vertically. In such a case, Dual Prime Prediction can only utilize one MV per MB and one or two differential MVs (dmvector) that work on vertical MV component correction. For a decoder, the MV per MB is used to derive 2 MVs to get candidates from both top and bottom fields for Field-Pictures. The MV in the bitstream determines the displacement to the same polarity field as shown in Figure 8-8. The differential MVs are then used to obtain vertically adjusted MVs to opposite polarity field reference pictures as shown in Figure 8-8. Note that the adjustment is performed on half-pel resolution since such interpolation is required for MPEG-2 MC processing.

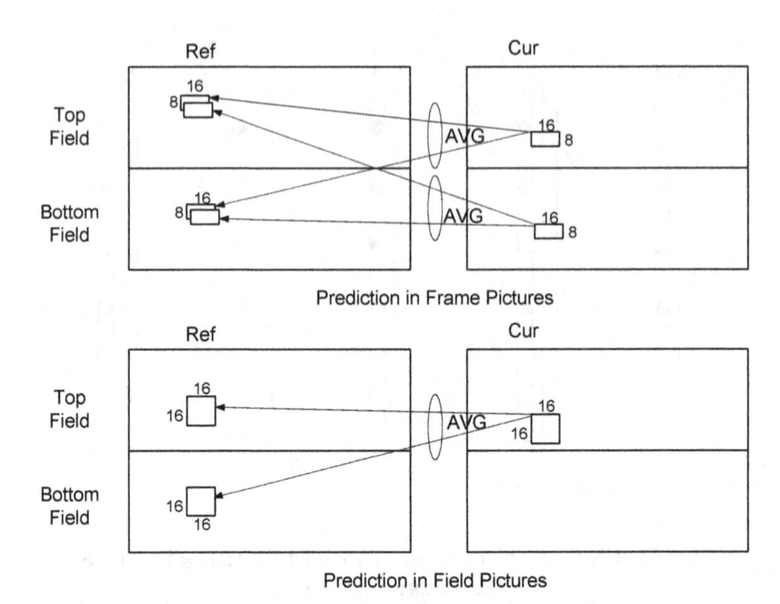


**Figure 8-7 Dual Prime Prediction in Field Pictures in Space Representation**

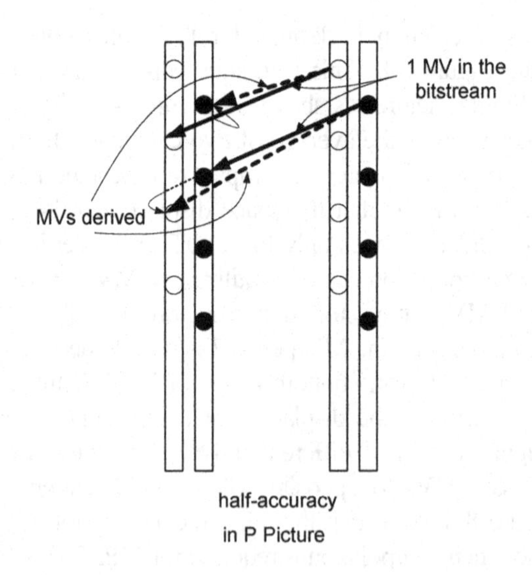


**Figure 8-8 Dual Prime Prediction in Field Pictures in Time Representation**

### 16x8 Motion Compensation (16x8 MC)

When Field-Pictures are handled with Field Prediction, the normal size of prediction (i.e., 16x16) is too large. The field data is gathered from one out of every other line. Therefore, the prediction size, 16x16, in Field-Pictures can cover the size of 16x32 in Progressive video. Such a size is too large to assume to be uniform in terms of motion characteristics, which in turn, looses the benefit of Inter-Prediction for signal de-correlation. To overcome this difficulty, 16x8 MC is introduced in MPEG-2 for interlaced video. A MB in the current field is broken into two 16x8 size blocks for MC and 16x8 MC search is performed independently for both top 16x8 portion and bottom 16x8 portion as shown in Figure 8-9. Two MVs are used for P-MBs, while two or four MVs are used for B-MBs.

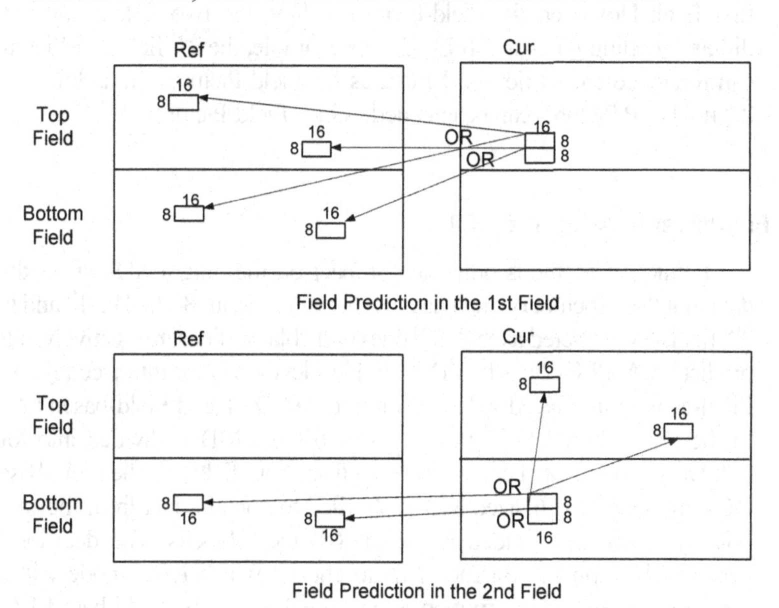


**Figure 8-9 16x8 MC Prediction in Field Pictures**

### Prediction Defined in MPEG-2

The aforementioned prediction types are allowed in specific picture constructions. The classification can be shown according to Field-Pictures and Frame-Pictures as follows:

**Table 8-1 Interlace Prediction Allowed in MPEG-2**

| Field-Pictures | Frame-Pictures |
|---|---|
| Field Prediction | Frame Prediction |
| Dual Prime | Field Prediction |
| 16x8 MC Prediction | Dual Prime |

Typically, the 2nd field of a frame is to be the same coding type as the first field. However, for Field-Picture coding, the two fields may have different coding types in MPEG-2. For example, the 2nd field in I Pictures can reference the 1st field in I Pictures for Field-Pictures. In addition, the 2nd field of P Pictures can be encoded as I for Field-Pictures.

### Field/Frame Adaptive DCT

Prediction methods only have impact on the magnitudes of residual data, not the structure of such data as shown in Figure 8-10. The 1st and the 2nd fields are depicted as "white" lines and "black" lines, respectively, after prediction. MPEG-2 has two DCT modes developed for Interlaced Frame-Pictures with interlaced video – Frame-based DCT and Field-based DCT. In the Frame-based DCT mode, a 16x16 sized MB is divided into four 8x8 DCT blocks, each containing data from both fields. In the Field-based DCT mode, a 16x16 sized MB is shuffled to contain data from the same polarity fields and divided into four 8x8 DCT blocks. The decision is performed adaptively on each MB to choose which DCT mode will be used. Generally, if high motion is contained in a MB, Field-based DCT provides better rate-distortion characteristics than that of Frame-based DCT. When high motion happens, the combined Frame MB would contain high spatial-frequency patterns of edges. Therefore, separation of fields through Field-based DCT would help a lot to reduce high spatial-frequency in this case, thus providing higher compression than that of Frame-based DCT. If the video contains little motion, Frame-based DCT seems to provide better rate-distortion characteristics than that of Field-based DCT. When little or no motion happens, the combined Frame MB would contain low spatial-frequency patterns of edges. Note that the combined Frame MB would contain lower spatial-frequency patterns of edges than that of the Field MB since field-down sampling in the Field MB would increase the spectral effect of aliasing. Therefore, Frame MB through Frame-based DCT would help to reduce high spatial-frequency in this case, thus making higher compression than that of Field-based DCT.

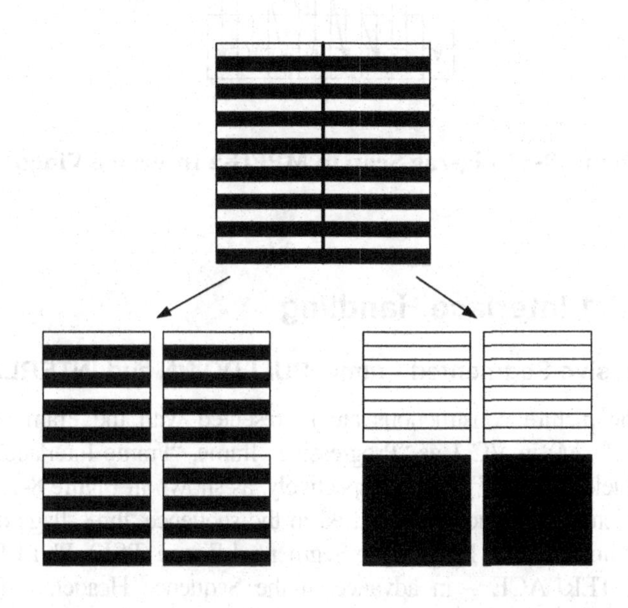

**Figure 8-10 Frame DCT and Field DCT**

When progressive_sequence=1, frame_pred_and_ frame_dct flag in Picture Coding Extension structure of MPEG-2 is set to 1. This indicates that only Frame-based DCT and Frame Prediction are used for the picture. The flag should be set to 0 for Field-Pictures.

## Zig-zag Scan Pattern for Interlace Video in MPEG-2

Zig-zag scan in MPEG-2 maps 2-D Transform coefficients to a series of pairs (zero-run, level), while inverse zig-zag scan interprets pairs into 2-D Transform coefficients. Figure 8-11 illustrates the MPEG-2 scanning pattern for interlace video input.

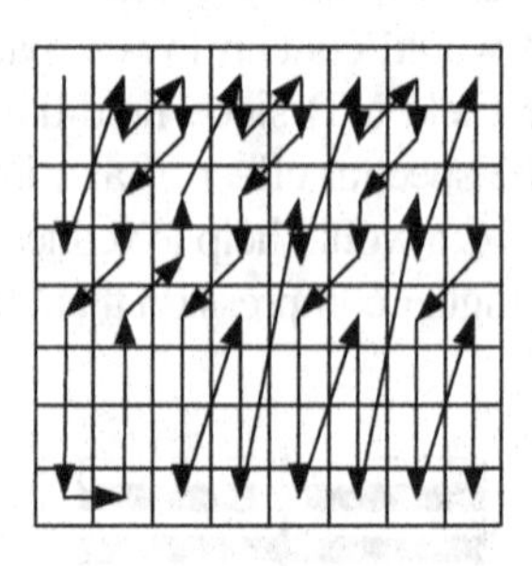

**Figure 8-11 Zig-zag Scan in MPEG-2 Interlace Video**

## 8.2 VC-1 Interlace Handling

### Progressive Segmented Frame, PULLDOWN and INTERLACE

The picture constructions are represented with the Frame Coding Mode (FCM) in VC-1 as "Progressive" frame, "Frame Interlace" frame and "Field Interlace" frame, respectively, as shown in Figure 8-1. Before such picture constructions are used in the sequence, three flags describe syntax limitations – Progressive Segmented Frame (PSF), PULLDOWN and INTERLACE – in advance in the Sequence Header. PSF tells whether the original video captured is progressive video or not. PULLDOWN "1" means that RFF/TFF or RPTFRM syntax is used. Here, RFF/TFF flags are the same information used in MPEG-2, while the 2-bit RPTFRM syntax element replacing the 2-bit pair RFF/TFF tells how many times the current frame data is consecutively repeated. INTERLACE "0" means an individual frame is handled based on Progressive Tools/ Syntax, while "1" means an individual frame is based on Progressive or Interlace Tools/ Syntax. The use of RFF/TFF or RPTFRM is the same as that of MPEG-2 as shown in Figure 8-2.

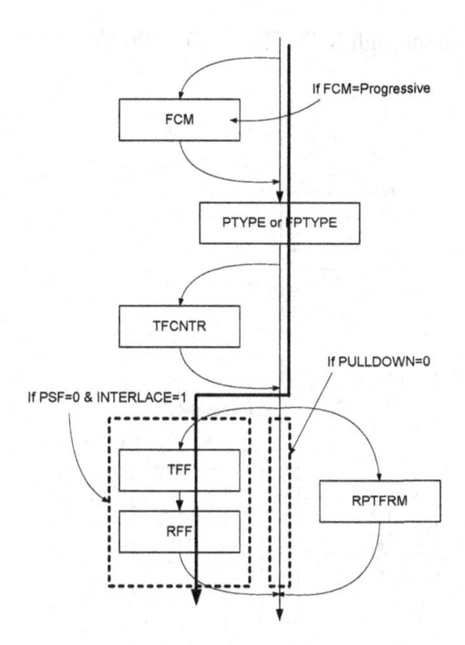


**Figure 8-12 Progressive Syntax on Picture Header for VC-1**

Figure 8-12 shows a scenario where original input video is captured in the Interlaced manner, but frame construction is done in the Progressive manner with the Interlace Tools on. If PULLDOWN=0, neither frame Pulldown through RPTFRM or field Pulldown through RFF/TFF is used. Here, Temporal Reference Frame Counter (TFCNTR) is an 8-bit fixed length value in display order. This increases in each frame with modulo 256 operation. Picture Type (PTYPE) or Field Picture Type (FPTYPE) indicates the construction of the current frame: I, P, B, BI, I/I, I/P, P/I, P/P, B/B, B/BI, BI/B or BI/BI.

Figure 8-13 shows a scenario where original input video is captured in the Progressive manner, but frame construction is done in the Interlace Frame manner with the Interlace Tools on. If PULLDOWN=0, neither frame Pulldown through RPTFRM or field Pulldown through RFF/TFF is used.

Figure 8-14 shows a scenario where original input video is captured in the Progressive manner, but frame construction is done in the Interlace Field manner with the Interlace Tools on. If PULLDOWN=0, neither frame Pulldown through RPTFRM or field Pulldown through RFF/TFF is used.

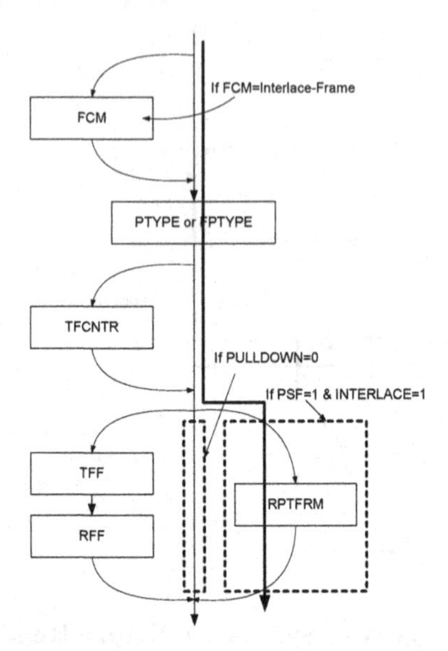


**Figure 8-13 Interlace-Frame Syntax on Picture Header for VC-1**

### BFRACTION and REFDIST

BFRACTION signals the relative temporal position of the B frame within the interval formed by its anchors and it is used to scale for Direct mode MVs. In VC-1, the relative temporal position is inscribed as BFRACTION only for Progressive and Interlace Frame-Pictures. This is not considered in VC-1 for Interlace Field-Pictures since hidden factors such as RFF/TFF or RPTFRM can compactly represent abnormal displacement in terms of temporal position. RFF/TFF or RPTFRM can also impact Interlace Frame-Pictures, but an encoder then might need to carefully select picture structures between Interlace Frame-Pictures or Interlace Field-Pictures.

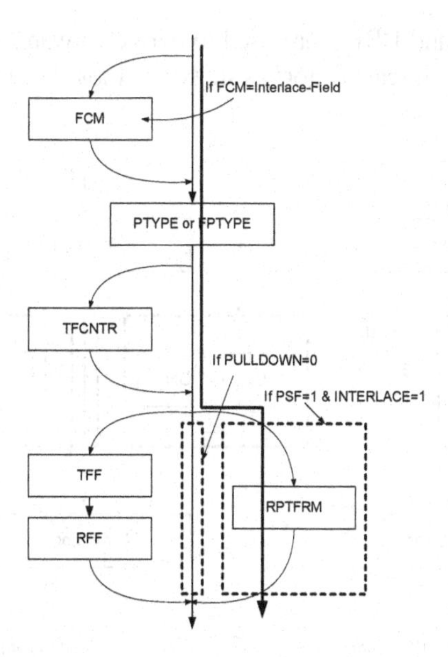


**Figure 8-14 Interlace-Field Syntax on Picture Header for VC-1**

For Interlace Field-Pictures, the reference selection or position of reference pictures is described clearly with REFDIST or BFRACTION. REFDIST defines the number of frames between the current frame and the reference frame. REFDIST mainly helps determine pre-meditated scale factors to obtain MVPs among neighborhood MV candidates, where difficulty of geometric division is caused by field(s) time difference between dominant and non-dominant referencing neighborhood candidates. The REFDIST syntax element is used for I/I, I/P, P/I and P/P, while BFRACTION syntax element is used for B/B, B/BI, BI/B and BI/BI. If the entry point level flag REFDIST_FLAG=0, REFDIST shall be set to the default value of 0. BFRACTION in Interlace Frame-Pictures and Interlace Field-Pictures shall be the same. The Forward Reference Frame Distance (FRFD) and Backward Reference Frame Distance (BRFD) are derived based on BFRACTION to help compute forward/backward MVPs as shown in Equation (8-2) and Equation (8-3). In the case of two reference P Field-Pictures of P Interlace Field-Pictures, REFDIST is used to derive MVP through pre-defined scale factors as shown in Table 113 and Table 114 in the standard. In B Interlace Field-

Pictures, FRFD and BRFD are used to derive forward/ backward MVPs through pre-defined scale factors as shown in Table 115 in the standard.

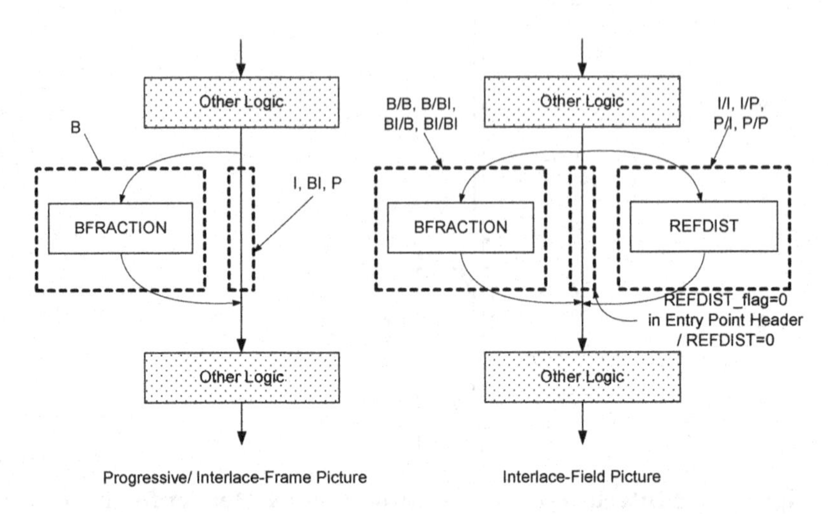


**Figure 8-15 Interpretation of BFRACTION/REFDIST for VC-1**

## Prediction Modes for P Frame-Pictures

There are four main ways to inter-predict in P Frame-Pictures as shown in Figure 8-16 and Figure 8-17 – 16x16 or 8x8 blocks for Frame Prediction or Field Prediction. The most common concept for P Frame Prediction is to get the best 16x16 match in the reference frame-picture, where one MV is used for both top and bottom fields simultaneously as shown in Figure 8-16 and Figure 8-18. A corresponding color-difference MV shall be derived to represent the displacement of the one 8x8 CbCr MV. A finer prediction is introduced to get the best 8x8 match for a MB — a MB is broken down to four individual MVs with 8x8 blocks for prediction as shown in Figure 8-16. Similarly, each color-difference block shall be motion compensated using four derived CbCr MVs that describe the displacements of the four "4x4" sub-blocks. This allows compression systems to capture more detail about motion behavior. In P Frame-Pictures, the reference selection is straightforward– the Previous I or P Picture is used as reference.

The most common concept for P Field Prediction is to get two best 16x8 matches in the reference field-pictures for decomposed current top and current bottom 16x8 blocks, where one MV is used for the top field and a separate MV is used for the bottom field as shown in Figure 8-17 and Figure 8-18. Using the two field MVs, a corresponding top field color-difference MV shall be derived that describes the displacement of the even lines of the CbCr blocks. Similarly, a bottom field color-difference MV shall be derived from the bottom field MV that describes the displacements of the odd lines of the CbCr blocks.

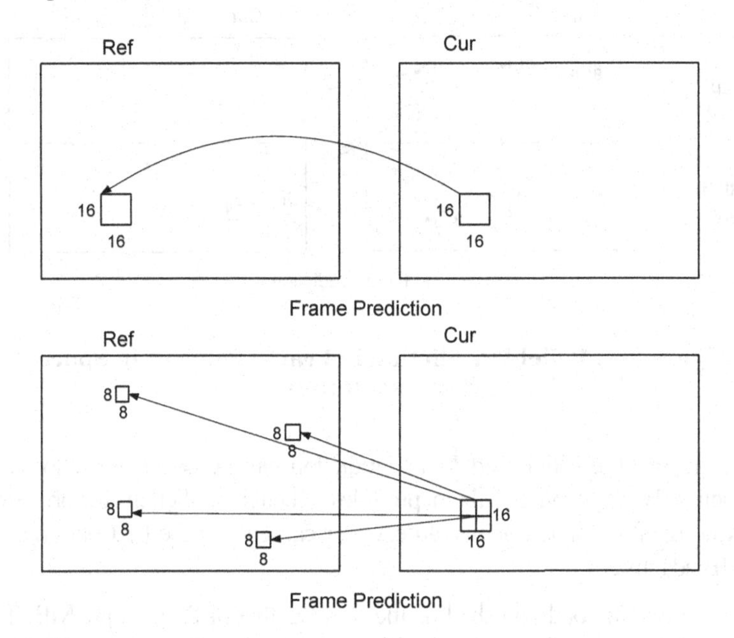


**Figure 8-16 P Frame Prediction in Frame Pictures in Space Representation**

A finer prediction is introduced to get the best 8x8 matches for a MB — the field 16x8 size is divided into right/ left 8x8 blocks for four Field MV Prediction shown in Figure 8-17. This allows compression systems to capture more detail about field motion behavior. Field selection is implicit based on the value of (MV.y & 4). Each color-difference block shall be partitioned into four regions in the same way as the Luma blocks and each 4x4 region shall be motion compensated using a derived Field CbCr MV.

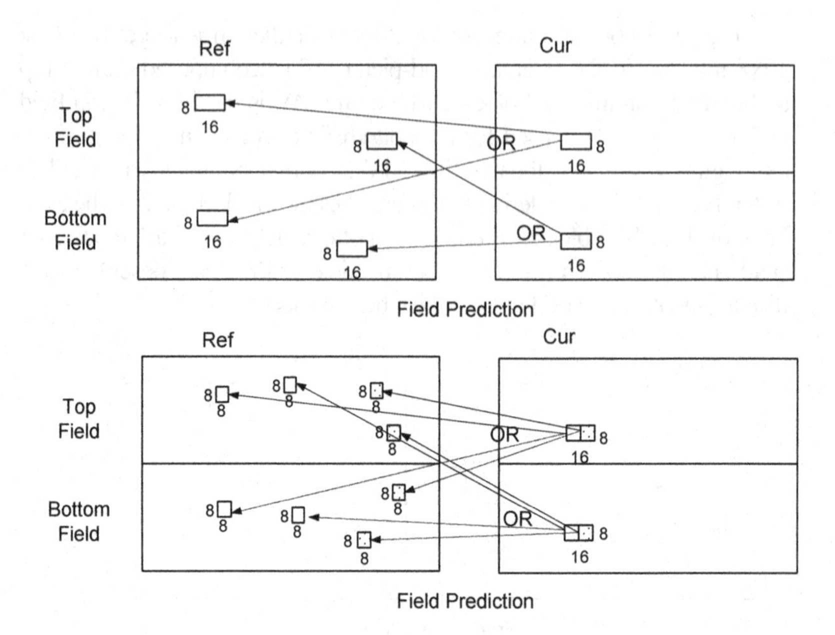


**Figure 8-17 P Field Prediction in Frame Pictures in Space Representation**

Frame Prediction and Field Prediction can be adaptively chosen in each MB when one of them provides a better prediction performance. Among all possible options, an encoder can choose the best one in each MB adaptively.

The Skip condition shall be the same as that of Progressive MB. The SKIPMB syntax element shall indicate the skip condition for a MB. If the SKIPMB syntax element is 1, then the current MB shall be skipped and there shall be no other information sent after the SKIPMB field. The Skip condition implies that the current MB is 1-MV with zero MVD and there are no coded blocks (CBP=0). In addition, HYBRIDMVP is not used for P Interlace Frame-Pictures.

The MB mode signaling is performed with the MBMODE syntax element that specifies:

- types of MB: 1MV, 4MV, 2 Field MV, 4 Field MV or Intra

- types of Transform for Inter-coded MB: 1 of field or frame or zero coded blocks (i.e., CBP=0), and
- whether there is differential MV for 1MV

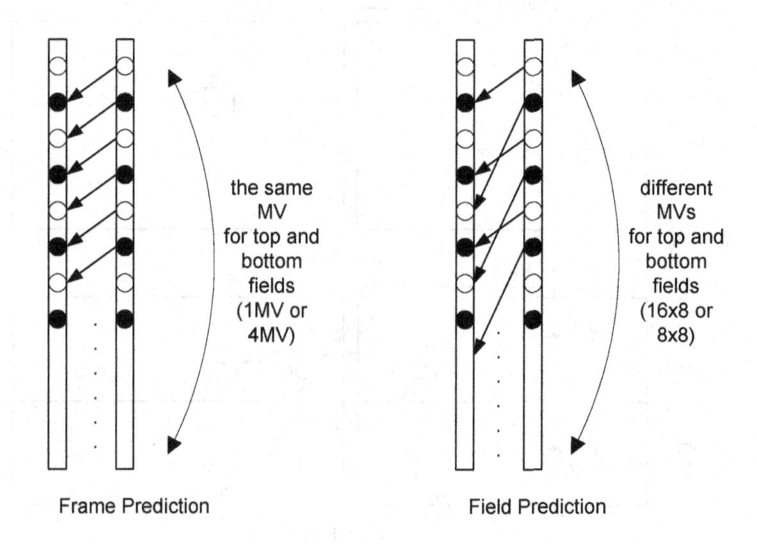


**Figure 8-18 P Frame Prediction and Field Prediction in Frame Pictures in Time Representation**

## Prediction Modes for B Frame-Pictures

There are two main ways to inter-predict in B Frame-Pictures as shown in Figure 8-19 – 16x16 or 16x8 blocks for Frame Prediction or Field Prediction, respectively. The most common concept for B Frame Prediction is to get one or two best 16x16 match(es) in the reference Frame-Picture(s), where one MV is used for both top and bottom fields simultaneously as shown in Figure 8-18 and Figure 8-19. In B Frame-Pictures, the two previous I or P pictures are used as reference.

The most common concept for B Field Prediction is to get two or four best 16x8 matches in the reference Field-Pictures for decomposed current top and current bottom 16x8 blocks, where one or two MVs are used for top field and a separate one or two MVs are used for bottom fields independently as shown in Figure 8-19. Field selection is implicit based on the value of (MV.y & 4). Forward/ backward MVs are populated for prediction. In B Field Prediction, the two previous I or P Field-Pictures are used as references.

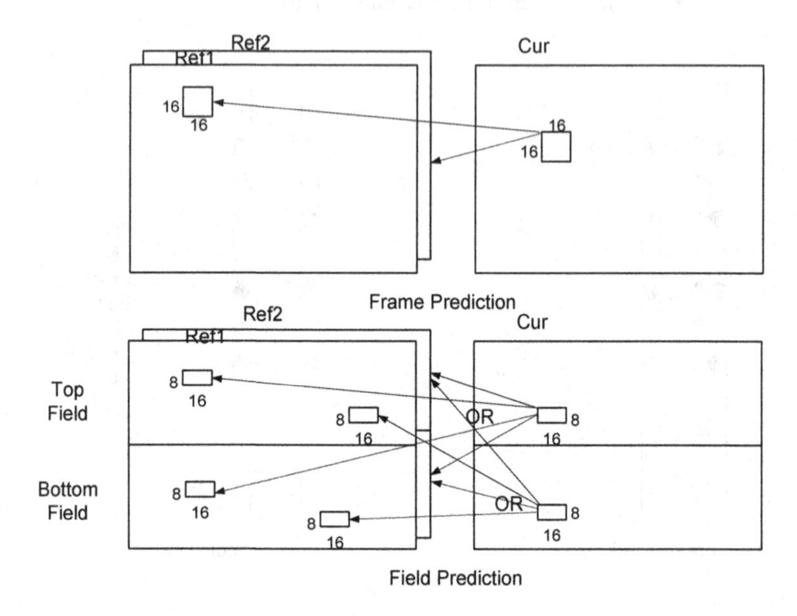


**Figure 8-19 B Frame Prediction and Field Prediction in Frame Pictures in Space Representation**

The Skip condition shall be similar to that of P Interlace Frame-Pictures. The SKIPMB syntax element shall indicate the skip condition for a MB. If a MB is skipped, then only the BMVTYPE and DIRECTBBIT information shall be sent for that MB, so that the MVs are correctly predicted as forward, backward, interpolative or direct modes. In addition, HYBRIDMVP is not used for B Interlace Frame-Pictures.

The MB mode signaling is performed with the MBMODE syntax element that specifies:

- types of MB: 1MV, 2 Field MV or Intra
- types of Transform for Inter-coded MB: 1 of field or frame or zero coded blocks (i.e., CBP=0), and
- whether there is differential MV for 1MV

The MB prediction mode signaling is performed with the BMVTYPE and the DIRECTMB syntax elements that jointly specify:

- types of MB inter-prediction: forward, backward, interpolative or direct modes
- the prediction type may be different for the top and bottom fields with field prediction

If a MB is of type 1MV and either forward or backward, it uses a single MV. If it is of type 1MV and either direct or interpolative, it uses two MVs. If it is of type 2 Field MV and either forward or backward, it uses two MVs. If it is of type 2 Field MV and direct or interpolative, it uses four MVs.

The calculation of direct mode MVs is similar to the methods used with Progressive B Pictures. The MVs from the previously decoded co-located MB shall be used to compute the direct mode MVs for the current B Frame-Pictures. The procedure to obtain direct mode MVs is as follows: first, MVTs and MVBs are buffered for I or P interlace Frame-Pictures. Second, the scaling of direct mode MVs is given in Figure 150 in the standard to provide high precision computation. The ScaleFactor shall be computed as was given in Equation (6-11) and Equation (6-12) for Progress B Frames. Nominally, one MV (denoted by MVT) shall be buffered for the top field of the MB, and one MV (denoted by MVB) shall be buffered for the bottom field of the MB that is used as the co-located MB. The rules for buffering are defined to embrace 4MVs (regardless of field or frame coding mode) in the standard.

### Prediction Modes for P Field-Pictures

Reference selection in P Field-Pictures is determined based on two syntax elements – NUMREF and REFFIELD. Figure 8-20 depicts examples of reference selection for P Field-Pictures. NUMREF "0" indicates use of only one reference picture, while NUMREF "1" implies the use of two reference pictures as shown in Figure 8-20. REFFIELD "0" means that the temporally "closest" I or P field is used as a reference picture, while REFFIELD "1" implies that the temporally "second-closest" I or P field is used as a reference picture as shown in Figure 8-20.

The way to inter-predict in P Field-Pictures is as shown in Figure 8-20, Figure 8-21 and Figure 8-22. Fundamentally the prediction is to get the best 16x16 match in a reference field-picture, where one MV is used for both top and bottom fields exclusively with an option of reference Field-Pictures as shown in Figure 8-20. A finer prediction is introduced to get the best 8x8 matches for a MB — a MB is broken down to four individual MVs with 8x8 blocks for prediction as shown in Figure 8-22.

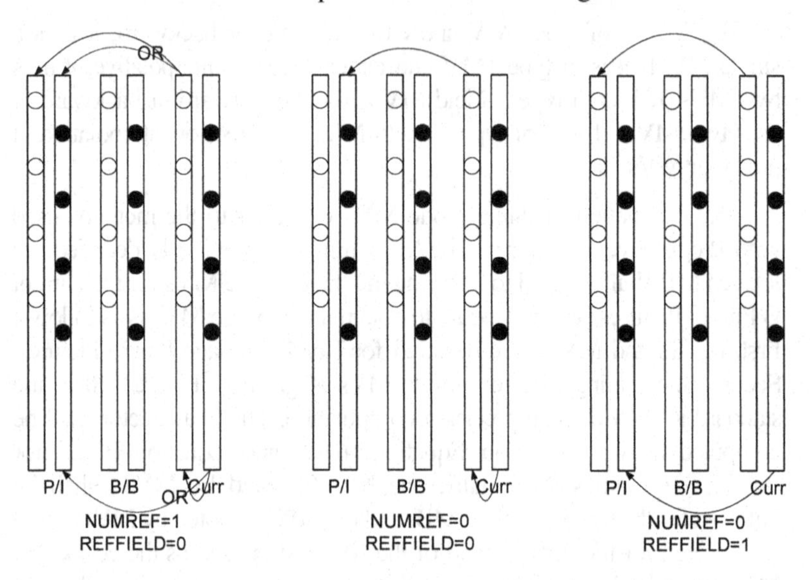


**Figure 8-20 P Field Prediction in Field Pictures in Time Representation**

For P Field Prediction, two previous reference Field-Pictures can be used for both 1st field and 2nd field as shown in Figure 8-20. In addition, only one reference Field-Picture can be used as shown in Figure 8-20. Note that the 2nd field is allowed to reference the 1st field for P Field Prediction in VC-1. In the two reference P Field-Pictures, the predictor for the current MB at any field can easily come from either the 1st field or the 2nd field, without much bias, since sometimes the top and the bottom field inversion can happen due to RFF/TFF use. However, VC-1 provides priority to the "opposite" polarity at the beginning. The same polarity is called "non-dominant," while the opposite polarity is called "dominant."

Since the dominant field shall be the field containing the majority of the MVP candidates, it can vary in the course of coding/ decoding process. In the case of a tie, the MV derived from the opposite field shall be the dominant predictor. When two reference P Field-Pictures are used, whether the dominant or non-dominant predictor is used is indicated in each MB level syntax "dmv_y." Note that REFDIST is considered to obtain pre-determined scale factors for MVP computation since two reference P Field-Pictures can most likely generate irregular geometric distances for dominant and non-dominant references. It is important to understand that such irregularity does not occur for one reference P Field-Picture case.

Figure 8-21 and Figure 8-22 show examples for 1MV and 4MV cases when NUMREF=1 and REFFIELD=0 are configured. Note that two previous pictures are used as references.

Picture layer syntax elements MVMODE and MVMODE2 indicate about use of prediction type information and use of intensity prediction, respectively. These signal for either 1-MV P Picture or Mixed-MV P Picture. In 1-MV P Pictures, a single MV shall be used to indicate the displacement of the predicted blocks for all six blocks in the inter-coded MB. In Mixed-MV P Pictures, each inter-coded MB shall be coded as a 1MV or 4MV MBs. In 4MV MBs, each of the four luma blocks shall have an associated MV.

In the algorithms that define the luma MV, all MVs are expressed in ½ or ¼ pel units (depending on the value of MVMODE), while all MVs are expressed in ¼ pel units for color-differential MVs.

The MB mode signaling is performed with the MBMODE syntax element that specifies:

- types of MB: 1MV, 4MV or Intra
- whether the CBPCY syntax element is present, and
- whether there is MV differential.

For 1MV MBs, the MVDATA syntax element is present in the MB layer to represent MV differential. For 4MV MBs, BLKMVDATA is present in the 8x8 block layer to represent MV differentials. HYBRIDMVP is used for P Interlace Field-Pictures. The detailed algorithm to compute HYBRID MV Predictor is described in Figure 120 in the standard. Also, the MV differential describes the reference selection implicitly in each MB through dmv_y. The predictor_flag is a binary flag indicating whether the dominant or the non-dominant MVP is used. Here, the flag being equal to 0 means that the dominant predictor is used. Otherwise, the non-dominant predictor is used. To be specific, predictor_flag is typically provided as follows:

$$\text{predictor\_flag} = \text{dmv\_y} \& 1. \tag{8-1}$$

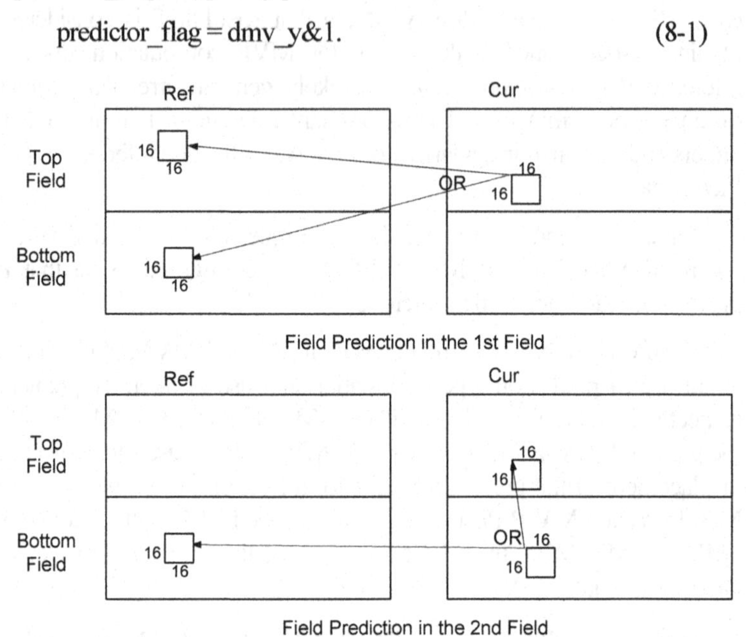


**Figure 8-21 1MV P Field Prediction in Field Pictures in Space Representation (NUMREF=1, REFFIELD=0)**

Intensity Compensation (IC) can be also applied to P Interlace Field-Pictures. If the MVMODE syntax element indicates that IC information is present in the bitstream then the INTCOMPFIELD syntax element shall be present and shall indicate which reference fields undergo IC. If INTCOMPFIELD indicates that both reference fields undergo IC, the LUMSCALE1, LUMSHIFT1, LUMSCALE2 and LUMSHIFT2 syntax elements shall be present in the bitstream. The first two elements shall be used to control the IC of the top reference field, while the last two elements shall be used to control the IC of the bottom reference field.

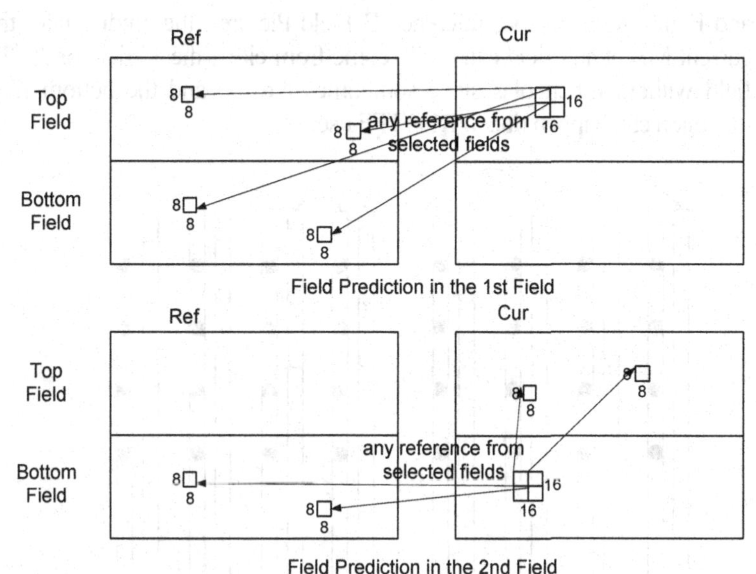


**Figure 8-22 4MV P Field Prediction in Field Pictures in Space Representation (NUMREF=1, REFFIELD=0)**

## Prediction Modes for B Field-Pictures

Indication of reference selection in B Field-Pictures is depicted in Figure 8-23, where B fields use four reference fields in all (top and bottom forward, top and bottom backward) to predict the current MB. Note that the 1st B field may be used as reference for the 2nd B field being decoded.

The way to inter-predict in B Field-Pictures is shown in Figure 8-24 and Figure 8-25. Fundamentally the prediction gets the best 16x16 match in a reference Field-Picture, where one MV is used for both top and bottom fields exclusively and two MVs are used for interpolative/direct modes for top and/or bottom fields with reference Field-Pictures as shown in Figure 8-24. A finer prediction is introduced to get the best 8x8 match for a MB — a MB is broken down into four individual MVs with 8x8 blocks for either forward or backward prediction as shown in Figure 8-25.

For B Field Prediction, four previous reference field pictures can be used for both 1st field and 2nd field as shown in Figure 8-23, Figure 8-24 and Figure 8-25. In four reference B Field-Pictures, the predictor for the current MB at any field can easily come from either the 1st field or the 2nd field without much bias since sometimes the top and the bottom field inversion can happen due to RFF/TFF use.

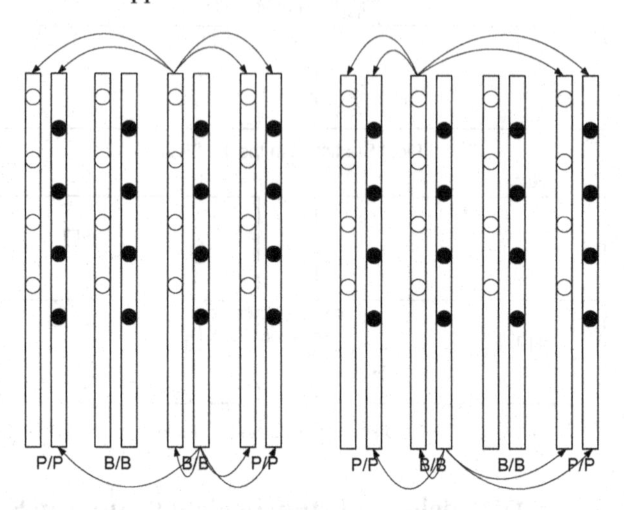


**Figure 8-23 B Field Prediction in Field Pictures in Time Representation**

Note that REFDIST is considered to obtain pre-determined scale factors for MVP computation since two reference Field-Pictures are used to generate irregular geometric distances for dominant and non-dominant references. It is important to understand that such irregularity always occurs for B Field-Picture coding.

Picture layer syntax MVMODE element indicates the use of prediction type information. This signals for either a 1-MV B Picture or a Mixed-MV B Picture. In 1-MV B Pictures, a single MV shall be used to indicate the displacement of the predicted blocks for all six blocks in the inter-coded MB. In Mixed-MV B Pictures, each inter-coded MB shall be coded as a 1MV or 4MV MBs. In 4MV MBs, each of the four Luma 8x8 blocks shall have an associated MV.

1MV can take forward, backward, interpolative or direct mode, while 4MV can take only either forward or backward mode. The MB mode signaling is performed with the MBMODE syntax element that specifies:

- types of MB: 1MV, 4MV or Intra
- whether the CBPCY syntax element is present, and
- whether there is MV differential.

For 1MV MBs, the BMV1 syntax element is present in the MB layer to represent MV differential. If the MB type is 1MV and the prediction type of the MB is decoded as "interpolative," the INTERPMVP (e.g., non-zero interpolative MV) syntax element shall be used to signal whether or not the 2nd MV differential BMV2 is present. For 4MV MBs, BLKMVDATA is present in the 8x8 block layer to represent MV differentials. The decoding of predictor_flag is performed in the same manner in Equation (8-1) of P Field-Pictures.

The MB prediction mode signaling is performed with BMVTYPE and FORWARDMB syntax elements that jointly specify:

- types of MB inter-prediction: forward, backward, interpolative or direct mode.

The calculation of direct mode MVs is similar to the methods used with Progressive B Pictures. The MVs from the previously decoded co-located MB shall be used to compute the direct mode MVs for the current B field. The MVs corresponding to the top field shall be used to compute the direct mode MVs in the top B field, while those corresponding to the bottom field shall be used to compute MVs of the bottom B field. While the direct mode MV utilizes the MV of the co-located MB for 1MV mode, the direct mode MV utilizes the MV of the co-located MB for 4MV mode based on reference field polarities of the four MVs and favors the dominant polarity. If the number of MVs, out of four, that point to field of the same polarity as the current field outnumbers those that point to field of the opposite polarity, the median4, median3 and arithmetic mean of 2 shall be used as described in Figure 125 in the standard. The scaling of direct mode MVs is given in Figure 126 in the standard to adjust field(s) time difference between dominant and non-dominant referencing neighborhood candidates. However, there shall be no computation of direct mode MVs for MBs that are not using the direct mode prediction (i.e., forward or backward) unlike Progressive B Pictures.

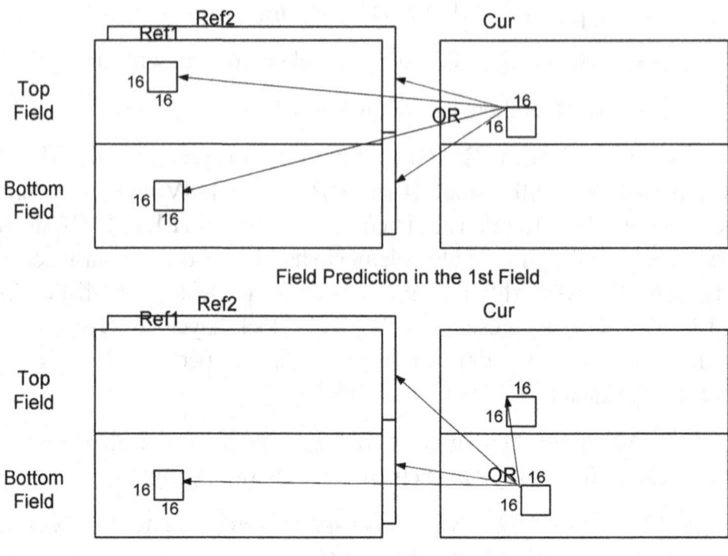


**Figure 8-24 1MV B Field Prediction in Field Pictures in Space Representation**

Forward MV Prediction for both B fields shall be identical to that for P field MV Prediction except for the method of deriving the reference frame distance. The forward reference frame distance shall be computed from the BFRACTION syntax element in the B interlace field picture header and from the REFDIST syntax element in the backward reference frame picture header. Backward MV Prediction for the second B fields shall be identical to that for P field MV Prediction. Forward/ backward MVs are populated for prediction.

The Forward Reference Frame Distance (FRFD) and the Backward Reference Frame Distance (BRFD) shall be computed as:

$$FRFD=((Numerator \times FrameReciprocal \times REFDIST)>>8), \quad (8\text{-}2)$$

$$BRFD=REFDIST-FRFD-1. \quad (8\text{-}3)$$

Here, Numerator and FrameReciprocal values shall be derived from BFRACTION defined in Equation (6-11) and Equation (6-12). And if BRFD<0, BRFD=0.

HYBRIDMVP is not used for B Interlace Field-Pictures.

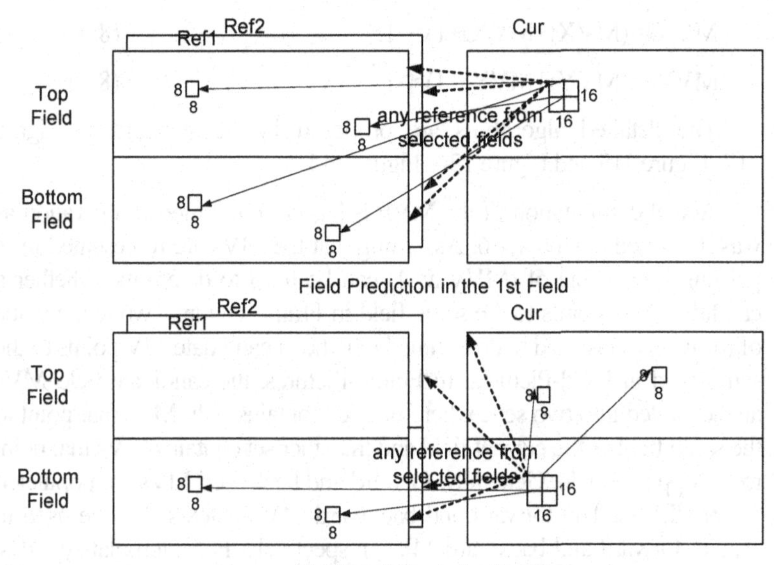


**Figure 8-25 4MV B Field Prediction in Field Pictures in Space Representation**

### Motion Vector Predictor

The process of computing the MVPs for the current MB in Interlace video input consists of two steps:

- First, three candidate MVs from the current MB are gathered from its neighboring MBs. Here, neighborhood MBs are defined as shown in Figure 6-21 that was depicted for progressive video.
- Second, the MVP for the current MB shall be computed through median operations from the set of valid candidate MVs. Scale factors may be applied if needed.

These two processes are similar to those for Progressive Pictures except that in Frame-Pictures MVs from top and bottom fields for the same region are proposed to be averaged as follows:

$$MVX_A=(MVX_1+MVX_2+1)>>1, \quad (8\text{-}4)$$

$$MVY_A=(MVY_1+MVY_2+1)>>1. \quad (8\text{-}5)$$

The detailed algorithms are proposed in the standard in Figure 117~Figure 119 and Figure 134~Figure 144.

Actual computation of the MVP is based on the median operation as was described in Chapter 6. Assuming that the MVs are represented in ¼ pel units, the value of (MV.y & 4) can be used to determine whether a candidate MV points to the same field in Frame-Pictures, while the value of (dmv_y&1) is used to determined whether a candidate MV points to the same field in Field-Pictures. In Frame-Pictures, the candidate field MVs are separated into two sets, where one set contains only MVs that point to the same field as the current field and the other set contains MVs that point to the opposite field. Generally, forward and backward MVs are populated for prediction. The forward and backward MV contexts shall be used to predict forward and backward MVs, respectively. For interpolative MBs, the forward and backward prediction buffer shall be used to predict the forward MV in BMV1 and the backward MV in BMV2. When the MB is of type direct or interpolative, the forward MV component shall be buffered in the forward buffer and the backward buffer.

### Prediction Defined in VC-1

Aforementioned prediction types are summarized in Table 8-2.

**Table 8-2 Interlace Prediction Allowed in VC-1**

| Field-Pictures | Frame-Pictures |
|---|---|
| 16x16 Field Prediction | 16x16 Frame Prediction |
| 8x8 Field Prediction | 16x8 Field Prediction |
| | 8x8 Field Prediction |

Typically, the 2nd field of a frame is to be the same coding type as the first field. VC-1 allows the 2nd field to reference the 1st field in I pictures. In addition, the 2nd field of P pictures can be encoded as I.

### Field/Frame Adaptive Transform

The Field/Frame Adaptive Transform tool is adopted for Interlace Frame-Pictures in VC-1 as it was in MPEG-2. The only difference is that the size of the Transform in VC-1 can be 8x8, 8x4, 4x8 or 4x4. Figure 8-26 depicts that an interlace 16x16 data can be decomposed in Frame Transform or Field Transform, respectively with various size Transform applied. The data shuffling decision and action happens before the Transform application.

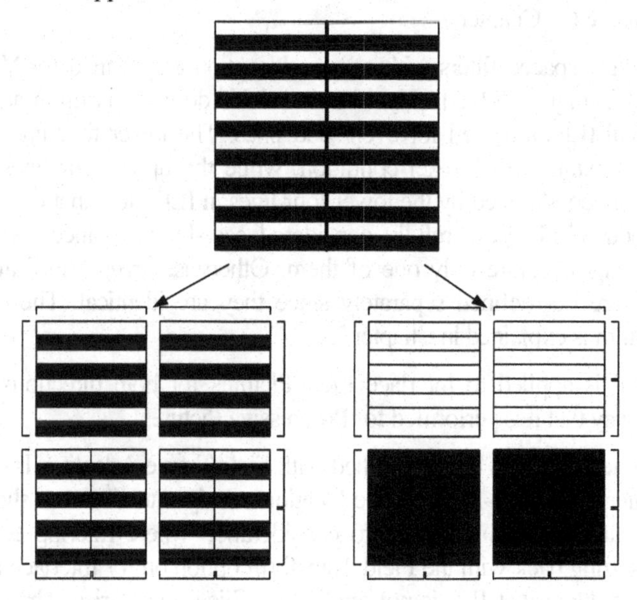

**Figure 8-26 Frame Transform and Field Transform**

### OLT with Interlace Video

OLT is not applied to Interlace Frame-Pictures for the horizontal direction. Otherwise, top and bottom field data might be mixed up to blur the edges. OLT is applied to Interlace Frame-Pictures for vertical direction in the same way that it is performed for Progressive Pictures. OLT is applied to Interlace Field-Pictures for both directions in the same way that it is performed for Progressive Pictures.

### ILF with Interlace Video

New ILF definitions are applied to Interlace Frame-Pictures as shown in Figure 8-27. HFbF means Horizontal Field-based Filter for ILF, while VFbF means Vertical Field-based Filter for ILF. Top field ILF and bottom field ILF are defined independently as shown in Figure 8-27. Each ILF requires four pixels on each side, so both top and bottom field ILFs require eight pixels on each side as shown in Figure 8-27. The detailed algorithm is explained in Chapter 7.

Single spaced filters with even/ odd option are defined for VFbF as shown in Figure 8-27. Top and bottom fields do not mix up in any case. Identical ILFs are used for even/ odd lines. The lower four lines of the block contain the 3rd pixel condition, while the upper four lines in the block are conditioned by the lower four lines in ILF. Note that even VFbF and odd VFbF are carefully considered case-by-case since sometimes some edges require only one of them. Otherwise, even VFbF and odd VFbF are not defined separately since they are identical. The detailed algorithm is explained in Chapter 7.

ILF is applied to Interlace Field-Pictures for both directions in the same way that it is performed for Progressive Pictures.

When Interlace ILF is handled with Field/Frame Adaptive Transform for Interlace Frame-Pictures, the location of edges to filter out should be done cautiously. This takes into consideration where filtering edges go after sorting back with the Field Transform option on for Interlace Frame-Pictures. Note that ILF is not applied on Slice boundaries. The detailed algorithm is presented in Figure 153~ Figure 156 in the standard for both horizontal and vertical edges in addition to I/P/B Picture types. Note that the size of Transform also impacts the algorithm since VC-1 adopted various sizes of Transforms.

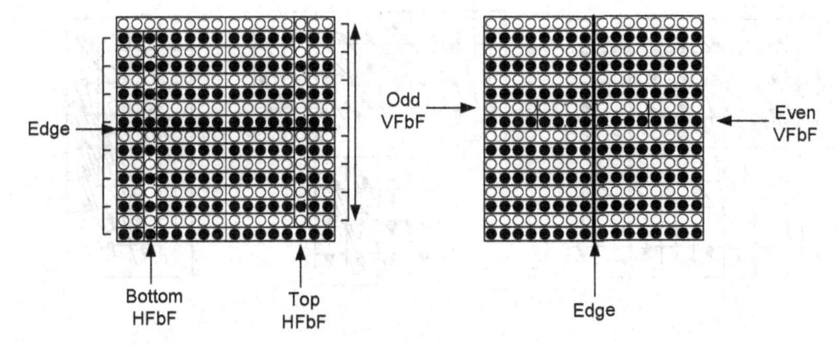


**Figure 8-27 Frame Transform and Field Transform**

## Zig-zag Scan Pattern for Interlace Video in VC-1

Zig-zag scan maps 2-D Transform coefficients to a series of triplets (zero-run, level, last), while inverse zig-zag scan interprets triplets into 2-D Transform coefficients. Figure 8-28 illustrates the VC-1 scanning pattern for interlace video input.

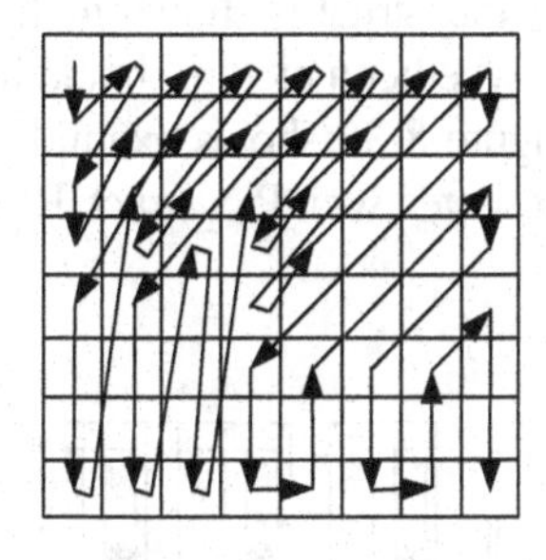

**Figure 8-28 Zig-zag Scan for 8x8 Intra Block in VC-1 Interlace Video**

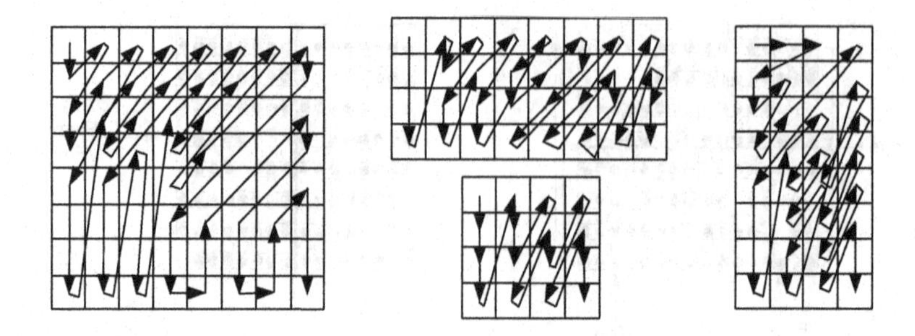

**Figure 8-29 Zig-zag Scan for Inter Blocks in VC-1 Interlace Video**

## 8.3 H.264 Interlace Handling

### Pic_struct and RFF/ TFF

The concept of RFF/TFF is replaced by Pic_struct in H.264. The predefined parameters for Pic_struct is shown in Table 2-8. Pic_struct in the SEI in H.264 indicates whether a picture should be displayed as a frame or one or more fields. Figure 8-30 shows examples where redundant fields can be compactly represented with Pic_struct. RFF/TFF pair is not defined in H.264.

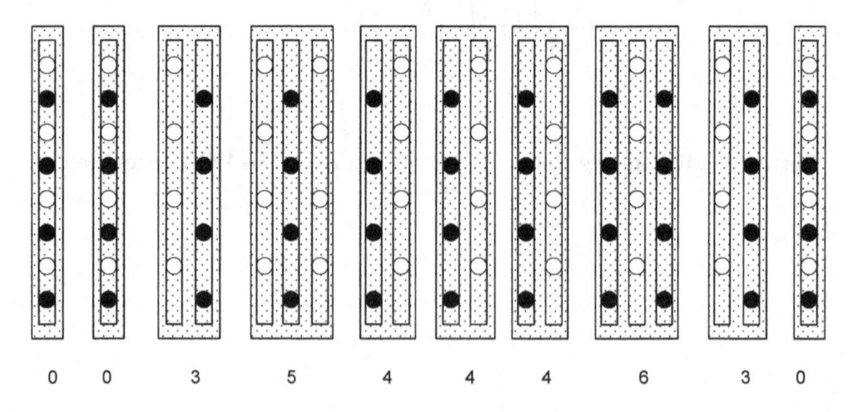


**Figure 8-30 Pic_struct**

### Adaptive Frame/ Field Coding

Generally important issues in Motion Compensation (MC) of Interlaced video include the size of coverage and uniformity about motion details. Since H.264 breaks down the size of MC down to 4x4, the granularity of motion capture is more than enough. In addition, H.264 defines a pair of MBs as a coding unit for Interlaced video to switch between Frame Transform and Field Transform. Therefore, there might be almost no need to develop dedicated Interlace tools such as MPEG-2 and VC-1. Once frame or field repeats are captured with Pic_struct, no special compression tools are applied with minor exceptions. Such exceptions are Picture Adaptive Frame Field (PAFF) coding and Macro-Block Adaptive Frame Field (MBAFF) coding. PAFF is shown in Figure 8-31, while MBAFF is depicted in Figure 8-32.

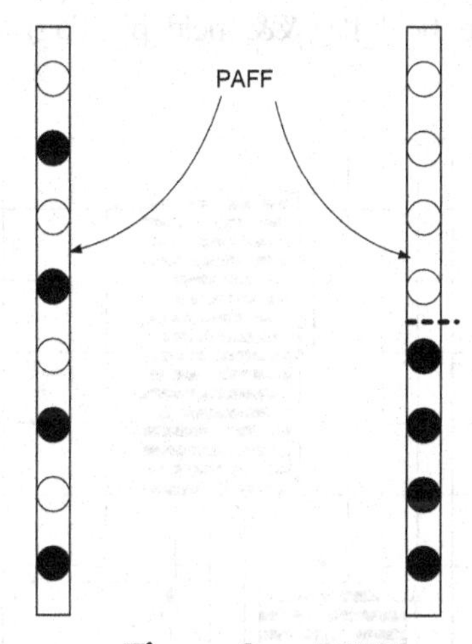


**Figure 8-31 PAFF**

As was explained in aforementioned Frame/ Field DCT, two adjacent rows of Interlaced Pictures tend to show smaller statistical dependency compared with that of Progressive Pictures. In this case, it may be more efficient to compress each field separately. To this end, H.264 proposes picture coding modes as follows:

- Combination of two fields together to code them as a single coded frame

- Separate two fields to code them as separate coded fields
- Combination of two fields together to code them as a single coded frame, but two vertically paired MBs are defined as a unit for coding as either a pair of two Field MBs or a pair of two Frame MBs.

The picture coding mode decision is made adaptively for each frame in a sequence. Coding with a choice between single coded frame and separate coded fields is called PAFF coding. Coding with vertically paired two MB units is called MBAFF. Note that Frame/ Field decision is made at the MB pair level rather than MB level as shown in Figure 8-32. The "mb_field_decoding_flag" in Slice level indicates if the current MB pair is a Frame MB pair. In addition, the reading pattern inside a paired MBs is shown in Figure 8-32. Figure 8-32 depicts a typical Interlaced Frame-Picture that undergoes MBAFF mode. MbaffFrameFlag is given as (mb_adaptive_frame_field_flag && !field_pic_flag).

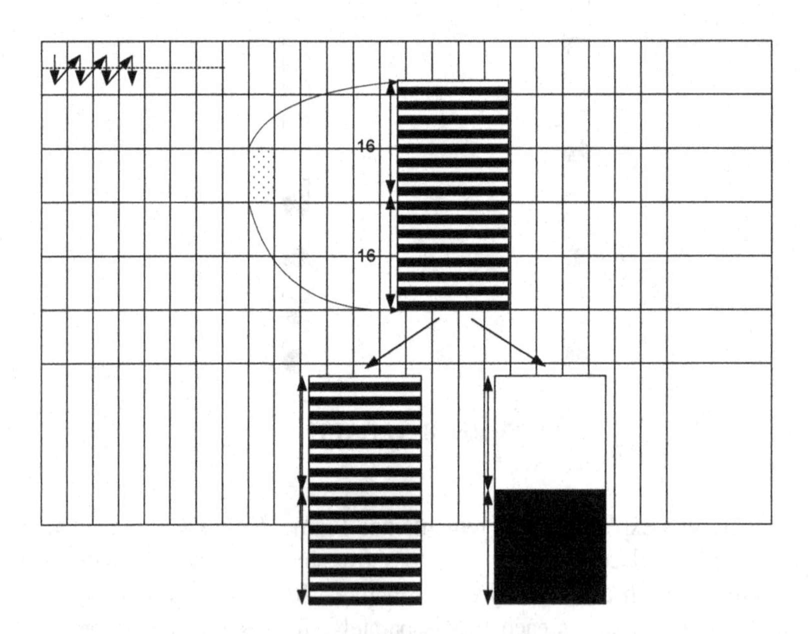


**Figure 8-32 MBAFF**

In summary, there is no difference in terms of applying compression tools for Interlaced video – when a frame is coded as two fields, each field is partitioned into MBs and is coded in a manner very similar to a frame. The only exceptions are as follows:

- Fields instead of reference frames for MC
- Zig-zag scanning pattern
- Field-based ILF for horizontal edges filtering.

## ILF with Interlaced Video

To redefine ILF for field-based processing, the edge pixel definitions shown in Figure 7-12 are modified and redefined in the standard for Interlaced video.

First, the variable dy is derived as follows:

- If fieldModeFilteringFlag is equal to 1 and MbaffFrameFlag is equal to 1, dy is set equal to 2.
- If fieldModeFilteringFlag is equal to 0 or MbaffFrameFlag is equal to 0, dy is set equal to 1.

For the sample location ($xE_k$, $yE_k$), k=0,1..,nE-1, the following applies. The filtering process is applied to a set of 8 samples across a 4x4 block horizontal or vertical edge denoted as p*i* and q*i* with i=0..3 as shown in Figure 7-12, where the edge lies between p0 and q0. The p*i* and q*i* with i=0..3 are specified as follows:

If verticalEdgeFlag is equal to 1,

$$q_i = s'[xP+xEk+i, yP+dy \times yE_k], \quad (8\text{-}6)$$

$$p_i = s'[xP+xEk-i-1, yP+dy \times yE_k] \quad (8\text{-}7)$$

where i=0..3. This falls back to the original definition in Figure 7-12 since these pixels are all aligned with horizontal direction.

If verticalEdgeFlag is equal to 0,

$$q_i = s'[xP+xEk, yP+dy \times (yE_k+i)-(yE_k\%2)], \quad (8\text{-}8)$$

$$p_i = s'[xP+xE_k, yP+dy \times (yE_k-i-1)-(yE_k\%2)] \quad (8\text{-}9)$$

where i=0..3. This implies pixels defined vertically are doubly spaced in the vertical direction compared with Figure 7-12. The same re-definition is extended to Chroma components.

For example, let's say that $yE_k$ is taken as 4. In addition, let's say that $(xE_k, yE_k)=(k,4)$. Then values can be computed through Equation (8-8) and Equation (8-9) as follows: q0= $s'$ [xP+k,8], q1= $s'$ [xP+k,10], q2= $s'$ [xP+k,12], q3= $s'$ [xP+k,14], p0= $s'$ [xP+k,6], p1= $s'$ [xP+k,4], p2= $s'$ [xP+k,2] and p3= $s'$ [xP+k,0]. Samples doubly spaced in the vertical direction are observed. Note that the ILF filtering process explained in Chapter 7 is still the same, but the filtering is performed with pixels newly defined for Interlaced video.

### Zig-zag Scan Pattern for Interlace Video in H.264

A zig-zag scan maps 2-D Transform coefficients to a series of separate data (level) and (run_before), while inverse zig-zag scan interprets them into 2-D Transform coefficients. Figure 8-33 illustrates the H.264 scanning pattern for interlace video input.

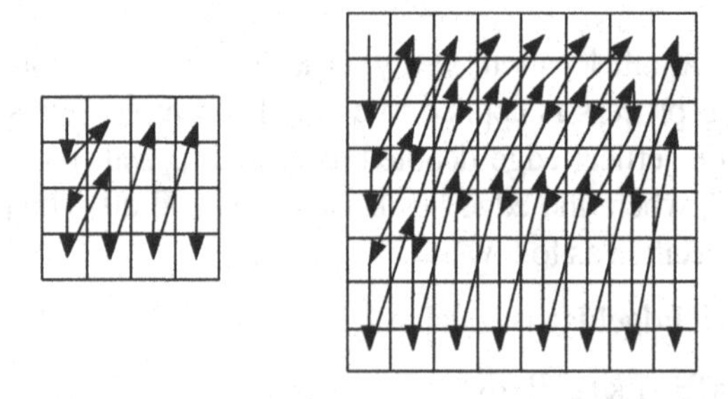

**Figure 8-33 Zig-zag Scan in H.264 Interlace Video**

## Reference Lists Development for Interlace Video in H.264

For P field slice prediction, RefPicList0 is used to capture recently decoded fields with short-term reference pictures and long-term reference pictures. Each field included in the reference picture list RefPicList0 has a separate index in the reference picture list RefPicList0. When a field is decoded, at least twice number of reference pictures is available in the list compared with the number of reference pictures for frame decoding.

Two ordered lists of reference frames, refFrameList0ShortTerm and refFrameList0LongTerm, are temporarily derived as follows:

- All frames having one or more fields marked "used for short-term reference" are included in the list of short-term reference frames refFrameList0ShortTerm. When the current field is the second field in the decoding order of a complementary reference field pair and the first field is marked as "used for short-term reference," the first field is included in the list of short-term reference frames refFrameList0ShortTerm.
- The refFrameList0ShortTerm for P field slice contains FrameNum in decreasing order, where FrameNum is a variable "wrapped around" with MaxFrameNum. And, PicNum is defined by either $PicNum = 2 \times FrameNumWarp + 1$ or $PicNum = 2 \times FrameNumWarp$ depending on whether the reference field has the same polarity or the opposite polarity of the current field, respectively.
- All frames having one or more fields marked "used for long-term reference" are included in the list of long-term reference frames refFrameList0LongTerm. When the current field is the second field in the decoding order of a complementary reference field pair and the first field is marked as "used for long-term reference," the first field is included in the list of long-term reference frames refFrameList0LongTerm.
- The refFrameList0LongTerm for P field slice contains FrameNum in increasing order.

For B field slice prediction, RefPicList0 and RecPicList1 are used to capture recently decoded fields with short-term reference pictures and long-term reference pictures. Each field included in the reference picture list RefPicList0 and RefPicList1 has a separate index in the reference picture list RefPicList0 and RefPicList1. When a field is decoded, at least twice number of reference pictures is available in the lists compared with the number of reference pictures for frame decoding.

Three ordered lists of reference frames, refFrameList0ShortTerm, refFrameList1ShortTerm and refFrameListLongTerm, are temporarily derived as follows:

- When the current field follows in decoding order, a coded field fldPrev that forms a complementary reference field pair is included into the list refFrameList0ShortTerm and the list refFrameList1ShortTerm.
- The refFrameList0ShortTerm for B field slice contains PicOrderCount in decreasing order for the first part, where PicOrderCount is POC earlier than current picture, and PicOrderCount in increasing order for the later part, where PicOrderCount is POC later than current picture. In contrast, refFrameList1ShortTerm for B field slice contains PicOrderCount in increasing order for the first part, where PicOrderCount is POC later than current picture, and PicOrderCount in decreasing order for the later part, where PicOrderCount is POC earlier than current picture.
- The refFrameListLongTerm for B field slice contains FrameNum in increasing order.

For P field slice prediction, refFrameList0ShortTerm and refFrameList0LongTerm are given as inputs in the following procedure and the output is assigned to RefPicList0. For B field slice prediction, refFrameList0ShortTerm and refFrameListLongTerm are given as inputs in the following procedure and the output is assigned to RefPicList0, while refFrameList1ShortTerm and refFrameListLongTerm are given as inputs in the following procedure and the output is assigned to RefPicList1.

The following procedure takes refFrameListXShortTerm (with X may be 0 or 1) and refFrameListLongTerm as inputs. The reference picture list

RefPicListX is a list ordered such that short-term reference fields have lower indices than long-term reference fields as follows:

- Short-term reference fields are ordered by selecting reference fields from the ordered list of frames refFrameListXShortTerm by alternating between fields of different polarity – starting with a field that has the same polarity as the current field. When one field of a reference frame was not decoded or is not marked as "used for short-term reference," the missing field is ignored and instead the next available stored reference field of the chosen polarity from the ordered list of frames refFrameListXShortTerm is inserted into RefPicListX. When there are no more short-term reference fields of the alternate polarity in the ordered list of frames refFrameListXShortTerm, the next not yet indexed fields of the available polarity are inserted into RefPicListX in the order in which they occur in the ordered list of frames refFrameListXShortTerm.
- Long-term reference fields are ordered by selecting reference fields from the ordered list of frames refFrameListLongTerm by alternating between fields of different polarity – starting with a field that has the same polarity as the current field. When one field of a reference frame was not decoded or is not marked as "used for long-term reference," the missing field is ignored and instead the next available stored reference field of the chosen polarity from the ordered list of frames refFrameListLongTerm is inserted into RefPicListX. When there are no more long-term reference fields of the alternate polarity in the ordered list of frames refFrameListLongTerm, the next not yet indexed fields of the available polarity are inserted into RefPicListX in the order in which they occur in the ordered list of frames refFrameListLongTerm.

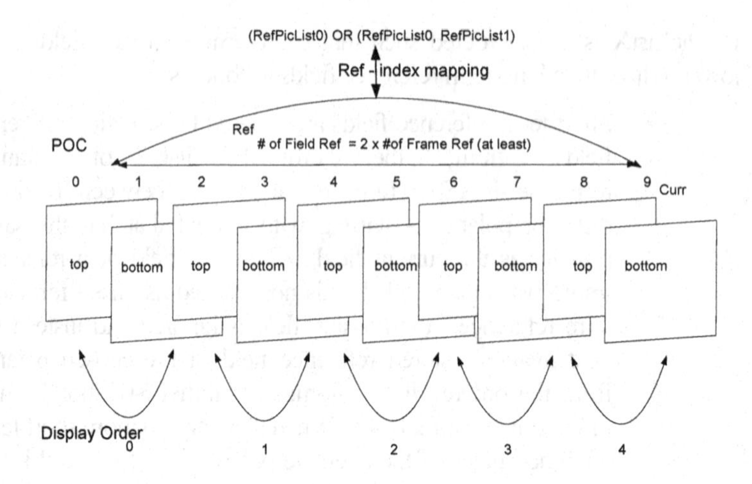


**Figure 8-34 An Example of POC/Lists for Interlace Video**

# 9. Syntax and Parsing

## 9.1 Table-based and Computation-based Codes

### Bitstream Parsing, Syntax Flow and Popular Codes

When a bitstream is received, the first task a decoder has to perform is to parse it into a series of Symbol data. Symbol here means certain number data that can be mapped into a practical meaning to continue decoding compressed video. For example, a certain Symbol can represent Picture type data, while some other Symbols can represent Transform coefficients. Parsing is the operation in the decoder that corresponds to the Coding operation in the encoder. The Coding operation in the encoder is a task to optimally (in terms of average length of codes) represent certain sequences of Symbols in binary codes. To continue parsing, the decoder has to be able to locate the boundaries of the binary codes, since typical codes are just concatenated continuously. Such codes that can be decoded without boundary delimiters are called "Prefix" codes. The decoder then has to be able to foresee what kind of Symbols would happen in the bitstream to parse the next Symbol. This interpretation is performed based on the current value parsed for the Symbol being decoded. In other words, the decoder has to fully understand the "Syntax Flow" defined in the standard. This is important since the order of Symbols is not always fixed. For example, while a MB generally contains Transform coefficients, specific MBs may not. In this case, the MB header will tell the decoder not to seek Transform coefficients in the parsing process.

There are two kinds of codes – table-based and computation-based codes. The well-known Huffman codes are table-based since the bitstream parsing process requires certain pre-defined tables in the memory to map a binary representation to a Symbol representation. Golomb and Arithmetic codes are computation-based since the bitstream parsing process requires certain calculation operations to map a binary representation to a Symbol representation. These do not require any kind of pre-defined tables in the memory. Video standards adopt variants of the three popular codes – Huffman, Golomb and Arithmetic codes. High Level headers are typically inscribed mainly with Huffman Variable Length Codes (VLC) or Fixed Length Codes (FLC). Since Chapter 2 discussed overall syntax hierarchies in VC-1 and H.264, this Chapter focuses on Syntax and Code analysis for MB level and below.

### Huffman Codes

If the shortest length binary representation is used for the highest probable Event for the Symbol, the average code length could be pretty small. The idea of Huffman coding is to compose/sort codes in such a way that the length of binary representation increases as the probability of corresponding Symbol decreases. For example, let's say, there are five possible values for a Symbol: A, B, C, D and E. And, let's assume that the probabilities for those values are 0.5, 0.15, 0.14, 0.12 and 0.09. The Huffman codes are suggested to be 1(A), 011(B), 010(C), 001(D) and 000(E) as shown in Figure 9-1. In this example, the probability is biased toward the value A event. This implies only 1 bit can be used to represent the event almost 50% of the time, thus making average code length very short. The average code length $l$=(1bit $\times$ Prob(A) + 3bits $\times$ Prob(B) + 3bits $\times$ Prob(C) + 3bits $\times$ Prob(D) + 3bits $\times$ Prob(E)) is 2. In fact, the average code length of Huffman codes was proven to be between Entropy (bit) and (Entropy+1) (bit) in Information Theory [cover:IT]. Since the performance of Huffman codes is extremely close to the theoretical bound (a.k.a., Entropy) of lossless compression, Huffman coding is sometimes called "Entropy" coding.

The Huffman code construction is composed of two steps as shown in Figure 9-1 – probability merge and re-ordering the tree to assign codes. Probability merge is continuously performed from the bottom two probabilities to construct a new probability re-arrangement, including a new summed probability element. Note that codes composed are Prefix codes.

| values of a Symbol | Prob | Merg. Prob | Merg. Prob | Merg. Prob | Merg. Prob |
|---|---|---|---|---|---|
| A | 0.5 | 0.5 | 0.5 | 0.5 | 1.0 |
| B | 0.15 | 0.21 | 0.29 | 0.5 | |
| C | 0.14 | 0.15 | 0.21 | | |
| D | 0.12 | 0.14 | | | |
| E | 0.09 | | | | |

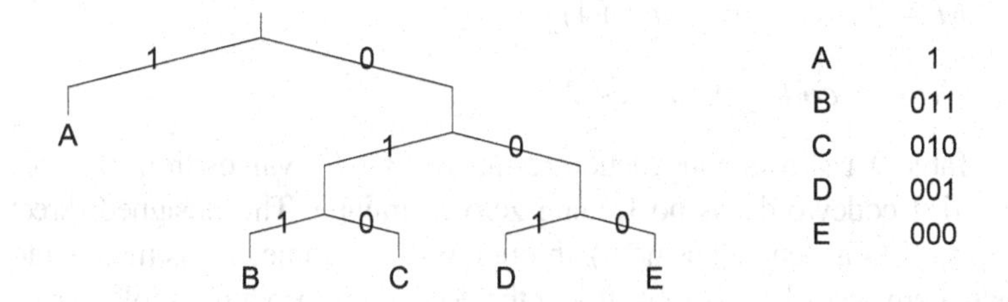


**Figure 9-1 Huffman Code Composition**

### Exponential Golomb Codes

Exponential Golomb (EG) code is an optimal Entropy code with sources of a geometric probability distribution:

$$p(n) = (1 - p_0)p_0^n, \; n \geq 0, \; 0 < p_0 < 1 \tag{9-1}$$

where $p_0$ is constant. Note that an EG code is not generally optimal for a source with any given probability distribution, unlike Huffman codes. It is only optimal when the source follows a geometric probability distribution. In H.264, EG codes are typically adopted for the "binarizer" as a pre-processor. This is adopted without identifying original signal probability distribution. This difference between geometric probability distribution and original signal probability distribution is rectified in a later stage of CA-BAC. Therefore, Bit Stream Parser (BSP) for EG codes will not be as effective as that for Huffman codes. This implies that the average length $\overline{L}$ of code words will be longer than that of Huffman. Since compactness of representation is not as efficient as that of Huffman, the target bit rate $R_{Golomb}$ will be higher than the target bit rate $R_{Huffman}$ to produce the same amount of symbols/s. In other words, BSP needs to parse code words at a higher speed to generate the same amount of symbols/s.

EG codes are variable length codes based on computation. The construction rule is as follows:

[M 0's][1][INFO]

where INFO is an M-bit field carrying information. The code can be constructed by the encoder based on its index "code_num."

$$M = \lfloor \log_2(code\_num + 1) \rfloor$$

$$INFO = code\_num + 1 - 2^M .$$

Table 9-1 shows construction results of EGs for values from 0 to 63. The first codeword has no leading zero or trailing. The unsigned direct mapped EG is denoted as ue(v), through which parameter v is mapped to code_num according to Table 9-1 or the decoding procedure as follows:

```
leadingZeroBits=-1;
for( b=0; !b; leading ZeroBits++)
        b= read_bits(1);
```

Then, the variable code_num is assigned as follows:

$$code\_num = 2^{leadingZeroBits} - 1 + read\_bits(leadingZeroBits)$$

where the value returned from *read_bits( leadingZeroBits)* is interpreted as a binary representation of an unsigned integer with most significant bit written first. In other words, it simply means INFO.

When computation takes more time than memory access, the pre-computed Table 9-1 can be used directly. This representation is used for MB type/ reference frame index/ etc.

**Table 9-1 ue(v)**

| | | | | | |
|---|---|---|---|---|---|
| 0 | 1 | 23 | 000011000 | 46 | 00000101111 |
| 1 | 010 | 24 | 000011001 | 47 | 00000110000 |
| 2 | 011 | 25 | 000011010 | 48 | 00000110001 |
| 3 | 00100 | 26 | 000011011 | 49 | 00000110010 |
| 4 | 00101 | 27 | 000011100 | 50 | 00000110011 |
| 5 | 00110 | 28 | 000011101 | 51 | 00000110100 |

| 6 | 00111 | 29 | 000011110 | 52 | 00000110101 |
|---|---|---|---|---|---|
| 7 | 0001000 | 30 | 000011111 | 53 | 00000110110 |
| 8 | 0001001 | 31 | 00000100000 | 54 | 00000110111 |
| 9 | 0001010 | 32 | 00000100001 | 55 | 00000111000 |
| 10 | 0001011 | 33 | 00000100010 | 56 | 00000111001 |
| 11 | 0001100 | 34 | 00000100011 | 57 | 00000111010 |
| 12 | 0001101 | 35 | 00000100100 | 58 | 00000111011 |
| 13 | 0001110 | 36 | 00000100101 | 59 | 00000111100 |
| 14 | 0001111 | 37 | 00000100110 | 60 | 00000111101 |
| 15 | 000010000 | 38 | 00000100111 | 61 | 00000111110 |
| 16 | 000010001 | 39 | 00000101000 | 62 | 00000111111 |
| 17 | 000010010 | 40 | 00000101001 | 63 | 0000001000000 |
| 18 | 000010011 | 41 | 00000101010 | | |
| 19 | 000010100 | 42 | 00000101011 | | |
| 20 | 000010101 | 43 | 00000101100 | | |
| 21 | 000010110 | 44 | 00000101101 | | |
| 22 | 000010111 | 45 | 00000101110 | | |

## Signed Exponential Golomb Codes

The signed, mapped EG is denoted as se(v), through which parameter v is mapped to code_num according to the following decoding procedure as follows:

*decoding as ue(v);*

Then, the variable code_num is assigned as follows:

$code\_num=(v<0)?(2|v|):(2|v|-1).$

When computation takes more time than memory access, the pre-computed Table 9-2 can be used directly. Signed EG codes that range from -31 to 32 under Slice data are mb_qp_delta.

**Table 9-2 se(v)**

| 0 | 1 | 12 | 000011000 | -23 | 00000101111 |
|---|---|---|---|---|---|
| 1 | 010 | -12 | 000011001 | 24 | 00000110000 |
| -1 | 011 | 13 | 000011010 | -24 | 00000110001 |
| 2 | 00100 | -13 | 000011011 | 25 | 00000110010 |
| -2 | 00101 | 14 | 000011100 | -25 | 00000110011 |
| 3 | 00110 | -14 | 000011101 | 26 | 00000110100 |
| -3 | 00111 | 15 | 000011110 | -26 | 00000110101 |
| 4 | 0001000 | -15 | 000011111 | 27 | 00000110110 |
| -4 | 0001001 | 16 | 00000100000 | -27 | 00000110111 |
| 5 | 0001010 | -16 | 00000100001 | 28 | 00000111000 |
| -5 | 0001011 | 17 | 00000100010 | -28 | 00000111001 |
| 6 | 0001100 | -17 | 00000100011 | 29 | 00000111010 |
| -6 | 0001101 | 18 | 00000100100 | -29 | 00000111011 |
| 7 | 0001110 | -18 | 00000100101 | 30 | 00000111100 |
| -7 | 0001111 | 19 | 00000100110 | -30 | 00000111101 |
| 8 | 000010000 | -19 | 00000100111 | 31 | 00000111110 |
| -8 | 000010001 | 20 | 00000101000 | -31 | 00000111111 |
| 9 | 000010010 | -20 | 00000101001 | 32 | 0000001000000 |
| -9 | 000010011 | 21 | 00000101010 | | |
| 10 | 000010100 | -21 | 00000101011 | | |
| -10 | 000010101 | 22 | 00000101100 | | |
| 11 | 000010110 | -22 | 00000101101 | | |
| -11 | 000010111 | 23 | 00000101110 | | |

### Mapped Exponential Golomb Codes

The mapped EG is denoted as me(v), through which parameter v is mapped to code_num according to the Table specified in the standard. For example, EG codes for Coded_block_pattern ranges from 0 to 63. In addition, EG codes that range from 0 to 63 under Slice data are Mb_type/ intra_chroma_pred_mode/ sub_mb_type, etc. The ue(v) decoding is performed at the first stage and a mapping process is followed later with defined Tables, such as Table 9-4, in the standard.

### Truncated Exponential Golomb Codes

**Table 9-3 te(v) with Binary Variable**

| 0 | 1 |
|---|---|
| 1 | 0 |

The truncated EG is denoted as te(v), through which parameter v is mapped to code_num. The range of this syntax element may be between 0 and x, with x being greater than or equal to 1. It is used in the derivation of the value of a syntax element as follows:

- *If x is greater than 1, code_num and the value of the syntax element shall be derived in the same way as for syntax elements coded as ue(v).*

- *Otherwise (x is equal to 1), the parsing process for code_num which is equal to the value of the syntax element is given by a process equivalent to:*

*b=read_bits(1).*

*code_num=!b.*

Truncated EG codes that may range from 0 to 1 under Slice data are ref_idx_l0/ ref_idx_l1. For this, Table 9-3 can be directly used. Truncated Exp-Golomb codes that may range from 0 to x (x>1) under Slice data are ref_idx_l0/ ref_idx_l1.

## Shannon-Fano-Elias Codes

Shannon-Fano-Elias (SFE) code is an Entropy code that uses the cumulative distribution function to allocate code words in a composition procedure. The idea of SFE coding is exactly the same as Huffman coding – composition/sorting of codes is performed in such a way that the length of binary representation increases as the probability of corresponding Symbol decreases. However, the probability cumulative function is used to make sure that code-bases are disjoint in terms of the intervals corresponding to code words to compose "Prefix" codes. For example, let's say, there are four possible values for a Symbol: A, B, C and D. And, let's assume that the probabilities of those values are 0.25, 0.5, 0.125, and 0.125. The SFE codes are suggested to be 001(A), 10(B), 1101(C) and 1111(D) as shown in Figure 9-2. In this example, the probability is biased to the value B event. This implies only two bits can be used to represent the event almost 50% of the time, thus making average code length very short. The average code length $l$=(3bits $\times$ Prob(A) + 2bits $\times$ Prob(B) + 4bits $\times$ Prob(C) + 4bits $\times$ Prob(D)) is 2.75. In fact, the average code length of SFE codes was proven to be between Entropy (bit) and (Entropy+2) (bit) in Information Theory [cover:IT]. Since the performance of SFE codes is pretty close to the theoretical bound (a.k.a., Entropy) of lossless compression, SFE coding is also sometimes called "Entropy" coding. Note that the performance of Huffman coding is better than that of SFE coding, since the upper bound of the average length is tighter than that of SFE coding.

The SFE code construction is composed of two steps as shown in Figure 9-2 – code-base computation and code capture to assign codes. Code-base is computed as follows:

$$\widetilde{F}(x) = F(x) - \tfrac{1}{2}\, p(x) \qquad (9\text{-}2)$$

where $p(x)$ and $F(x)$ are probability and its cumulative distribution function, respectively. Note that code-base is taken as the mid-points of probabilities of Symbol events as shown in Figure 9-2.

Once the cumulative distribution function is represented in binary, the length to capture is performed as follows:

$$\lceil L \rceil + 1 \qquad (9\text{-}3)$$

where $L = \log \frac{1}{p(x)}$. Note that codes composed are Prefix codes.

| values of a Symbol | Prob(x) | $F(x)$ | $\tilde{F}(x)$ | $\tilde{F}(x)$(binary) | length to cap |
|---|---|---|---|---|---|
| A | 0.25 | 0.25 | 0.125 | 0.001 | 3 |
| B | 0.5 | 0.75 | 0.5 | 0.10 | 2 |
| C | 0.125 | 0.875 | 0.8125 | 0.1101 | 4 |
| D | 0.125 | 1.0 | 0.9375 | 0.1111 | 4 |

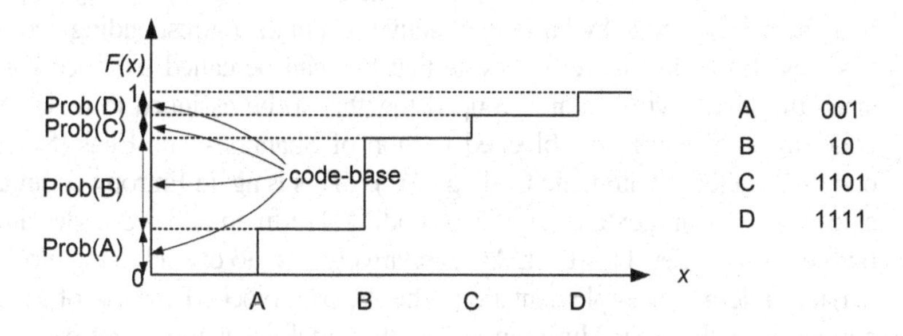


**Figure 9-2 SFE Code Composition**

The designed codes are continuously used for streamlined input Symbols. For example, if the input Symbols come as A, B, C, C, …, the binary output is sent as 0011011011101….

## Arithmetic Codes

Generally, it is called "block" coding when the input Symbols are grouped together to be assigned for a single code. Block Huffman coding can provide a much better compression performance than that of scalar Huffman coding, based on Information Theory. However, a difficulty lies with the number of combinations when Symbols are grouped together. For example, three 8-bit represented Symbols makes the number of combination $256^3$. To obtain statistical data and compose Huffman codes for this group is impractical. On top of that, if new codes are required to be generated due to a change of statistics, such an adaptation is even harder. Straightforward extension of block coding to SFE has the same problem as block Huffman coding. However, simple modification of the data handling in SFE can provide both a better compression performance and a better framework for statistics adaptation. For a new blocked version, mapping of a Symbol into a probability decomposition table, which was exactly used in SFE coding, is performed repeatedly as shown in Figure 9-3. The code-base will be either limited or left alone in the pre-defined resolution, based on necessity. The lower part of Figure 9-3 implies magnifying the resolution in practical implementation. This can be achieved by left-shifting resultant data appropriately. Note that a Symbol does not generate a binary representation, but a group of Symbols generates a final code by limiting effective bits in the corresponding code-base, as shown in Figure 9-3. Note that this can be called block coding since the input Symbols are grouped together to be assigned to a single code. In other words, the blocked version of Shannon-Fano-Elias (SFE) coding is called "Arithmetic Coding (AC)." This is again Entropy coding, since it is a natural extension of SFE coding. Huffman and AC codes are both entropy codes. However, AC is relatively easy to extend to the block version in terms of implementation. The AC is a blocked version of SFE compared with scalar Huffman codes, thus making it more compact in compression. This is why video communities started adopting AC in the standards.

Let's define R, L and H as the interval range R, the base (lower limit) L and upper limit H of the current code interval. In AC, the current interval range R and the base L of the current code interval are continuously computed/updated with the probability model as a new binary input is applied. The update rules are as follows:

$$R = Prev_{high} - Prev_{low} \tag{9-4}$$

$$L = Prev_{low} + R \times SubIntv_{low} \tag{9-5}$$

$$H = Prev_{low} + R \times SubIntv_{high}. \tag{9-6}$$

Each interval is split into several sub-intervals, each of which corresponds to a probable Event. Here, $SubIntv_{high}$ is the cumulative probability up to and including the corresponding Event, while $SubIntv_{low}$ is the cumulative probability up to but not including the Event.

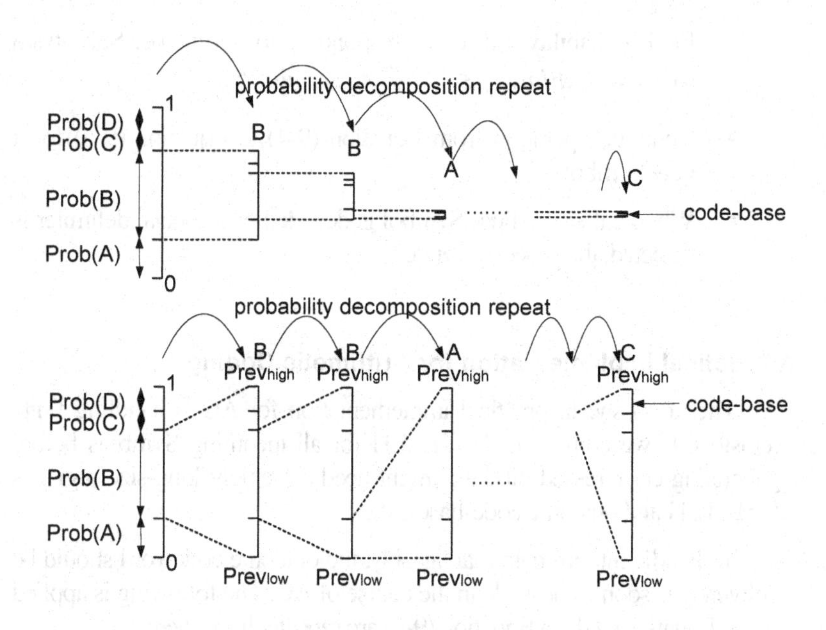


**Figure 9-3 Arithmetic Code Composition**

From above explanation, the encoding process can be summarized as follows:

- At the start of the process, R, $Prev_{low}$ and $Prev_{low}$ are set to 1, 0 and 1, respectively.
- Continue operation from Equation (9-4) to Equation (9-6) for a new Symbol.
- At the end of process, there is no need to transmit or save R, L and H for the last interval. Instead, a value within the final range can be sent or saved.

From above observations about the encoding process, the decoding process can be summarized as follows:

- At the start of the process, R, $Prev_{low}$ and $Prev_{low}$ are set to 1, 0 and 1, respectively.

- Find probability index $i$ corresponding to $SubIntv_{low}$, $SubIntv_{high}$ such that $SubIntv_{low} \le \frac{value-Prev_{low}}{R} \le SubIntv_{high}$.
- Continue operation from Equation (9-4) to Equation (9-6) for a new Symbol.
- When either the final Symbol is decoded or a special delimiter is detected, the process is ended.

## A Practical Implementation for Arithmetic Coding

The key issue in practical implementation for AC is handling finite registers. If we computed R, L and H for all incoming Symbols before generating compressed bits, we might need extremely long-size registers for R, L, H and outcome code-base data.

To handle interim data manageably, the outcome codeword should be delivered as soon as possible in the course of AC. The following is applied when Equation (9-4) to Equation (9-6) are repeatedly applied:

If L and H agree in the first m bits, those bits are the bits of the codeword. For example, if the content of L and H are "01100" and "01010" in binary, the first two bits are the same and will be removed from the registers. Registers should be able to shift data for this purpose.

When identical data is removed from L and H, the interval R should be scaled to be the whole range. If both L and H are both greater than ½, the top 1-more bit must be "1" in both L and H registers. If both L and H are all less than ½, the top 1 bit more must be "0" in both L and H registers. Therefore, 1 more bit can be removed from the L and H registers. A left shift by one means that L is replaced with $2 \times L$ and H is replaced with $2 \times H$.

When the interval R straddles ½, a codeword cannot be generated and no shift operation can be performed. However, if $1/4 \le L < 1/2 \le H < 3/4$, the first two bits are removed. And L then is replaced with $(L-0.25) \times 2$ and H is replaced with $(H-0.25) \times 2$.

After this data resolution management, either the encoding or the decoding process continues until the end. In many cases, a special End of String code is used to terminate the AC process.

When input Symbols are binary values, such coding scheme is called Binary Arithmetic Coding (BAC). Basically the same idea can be used to renormalize R and L for CA-BAC of H.264 [marpe:context], discussed in the next sections.

## 9.2 Codes in MPEG-2

### Codes above MB-Level

The video stream is organized as a hierarchy of headers and payload data that provide all the necessary information to parse syntax and decode them. Each header starts with its own distinctive Start Code (SC), e.g., sequence_header_code, picture_start_code, slice_start_code, etc. Each SC consists of the Psc (Prefix of Start Code) and a SC ID. The SC IDs are two digit hexadecimal numbers (e.g., B2, B3, AF) to identify the type of SC. Several Extensions are also defined for extra information over existing headers. An extension SC consists of an extention_start_code and an Extension ID that is a single digit hexadecimal number (e.g., 1, 2, A).

A Video Sequence is composed of Sequence_header( ), Sequence_extension( ), Extension_and_user_data_V( ), Group_of_picture(GOP)_header( ), Extension_and_user_data_G( ), Picture_header( ), Picture_coding_extension( ), Extension_and_user_data_P( ), Picture_data( ), etc. Each type of information can be found based on parsing the SC.

Sequenc_header( ) contains information about resolution, pel aspect ratio, bit rate, vbv buffer size, intra or non intra quantization matrix, etc. Sequence_extension( ) contains information about profile/level, progressive, chroma format, low delay, extra bits of information defined in Sequence_header( ), etc. Extension_and_user_data_V( ) contains information about Sequence_display_extension ( ) and User_data( ). Sequence_display_extension( ) contains information about display transfer characteristics such as color primaries. User_data( ) in the Video Sequence level may contain proprietary data for special applications, so decoders and simply discard the user data. GOP_header( ) contains information about time code, closed GOP and broken link. Picture_header( ) contains information about temporal reference, picture coding type, vbv_delay, full_pel_forward_ vector, forward_f_code, full_pel_backward_ vector, backward_f_code, etc. Picture_coding_extension( ) contains forward horizontal f_code, forward vertical f_code, backward_horizontal f_code, backward vertical f_code, intra_dc_precision, picture_structure, top_field_first, frame_pred_ frame_dct, concealment_motion_vectors, q_scale_ type, intra_vlc_ format, alternative_scan, repeate_first_ field, chorma_420_type, progressive_frame, etc. Extension_and_user_data_P( ) contains information about extra bits of information defined previously such as quant_matrix_extension( ), copyright_extenstion( ) and picture_display_extension( ). User_ data( ) in the Picture level may contain proprietary data for special applications, so decoders can simply discard the user data. The key data structure in Picture_data( ) is Slice( ). Slice( ) contains information about slice_vertical_position_extension, quanizer_scale_code, intra_slice_flag, intra_slice, etc.

## Codes below MB-Level

### *Parsing Process for Motion Vector*

Each component of data, $data[r][s][t]$, shall be interpreted as follows: Here, variable $r$ means 1[st] MV or 2[nd] MV, while variable $s$ means forward MV or backward MV. And variable $t$ means horizontal MV or vertical MV. MV part syntax is described in Figure 9-4, where two key data elements are motion_code and motion_residual. The motion_code is presented in VLC as shown in Table 9-4, while motion residual is a single binary number of "r_size" bits. Note that r_size=f_code-1, where f_code can be found in Picture header.

```
motion_vector(r,s) {
    motion_code[r][s][0]
    if((f_code[s][0]!=1) && (motion_code[r][s][0]!=0))
            motion_residual[r][s][0]
    if(dmv==1)
            dmvector[0]
    motion_code[r][s][1]
    if((f_code[s][1]!=1) && (motion_code[r][s][1]!=0))
            motion_residual[r][s][1]
    if(dmv==1)
            dmvector[1]
}
```

**Figure 9-4 MV Syntax Parsing**

**Table 9-4 motion_code (IS: Table B-10)**

| VLC | motion_code |
|---|---|
| 00000011001 | -16 |
| 00000011011 | -15 |
| …… | …… |
| 011 | -1 |
| 1 | 0 |
| 010 | 1 |
| …… | …… |
| 00000011010 | 15 |
| 00000011000 | 16 |

***Decoding Process for Motion Vector***

In MPEG-2, representation of MV differential (MVD) was specially designed to provide two powerful cases—first, the motion_code can directly imply full- or half-pel MVD when f_code=1. In other words, motion_residual=0 when f_code=1. Second, the motion_code can directly imply ($f$)-pel MVD when f_code!=1. Here, $f = 2^{r\_size}$ and ($f$)-pel MVD means that MVDs only take place in ($f$)-pel positions. In other words, MVD is given as (motion_code-1)$\times f$.

To develop a general representation to include the aforementioned cases, MVD is defined as follows:

$$|MVD|=(motion\_code-1) \times 2^{r\_size} + motion\_residual+1. \quad (9\text{-}7)$$

Note that the term $2^{r\_size}$ serves as a kind of quantization step size for MVD and motion_code ranges from –16 to +16 as shown in Table 9-4. Therefore, the range MVD can express is as follows:

$$-16 \times 2^{r\_size} \leq MVD < 16 \times 2^{r\_size}. \quad (9\text{-}8)$$

In other words, the highest value MVD can take is ($16\times f$)-1, while the lowest value MVD can take is ($-16\times f$). Therefore, the range must be ($32\times f$).

The standard defines the motion_residual as always being a positive value. In conjunction with such a definition, interpretation of the value should be performed as a negative value when motion_code is negative, as shown in Figure 9-5.

Each MV component, $vector'[r][s][t]$, shall be calculated by any process that is equivalent to the following one.

```
r_size=f_code[s][r]-1
f=1<<r_size
high=(16×f)-1
low=((-16)×f)
range=(32×f)

if((f==1)||(motion_code[r][s][t]==0))
   delta=motion_code[r][s][t]
else{
   delta=((abs(motion_code[r][s][t]-
1)×f)+motion_residual[r][s][t]+1
   if(motion_code[r][s][t]<0)
         delta= - delta
}
prediction=PMV[r][s][t]
if((mv_format==                                          "field")
&&(t==1)&&(picture_structure=="Frame picture"))
   prediction= PMV[r][s][t] DIV 2
vector'[r][s][t]=prediction + delta
if(vector'[r][s][t]< low)
   vector'[r][s][t]=vector'[r][s][t] + range
if(vector'[r][s][t]>high)
   vector'[r][s][t]=vector'[r][s][t] - range

if((mv_format==                                          "field")
&&(t==1)&&(picture_structure=="Frame picture"))
   PMV[r][s][t] = vector'[r][s][t]×2
Else
   PMV[r][s][t] = vector'[r][s][t]
```

**Figure 9-5 MV Decoding**

The subtraction operation requires that MVD contains one more bit in representation compared with those of MV and MVP.

### *Parsing and Decoding Process for Residual Data*

In MPEG-2, intra DC is treated differently than any other coefficients. Intra DC is coded jointly in two parts – dct_dc_size and dct_dc_differential. The dct_dc_size is the information about how many bits are assigned for the next coming DC differential value. Table 9-5 defines VLC codes for dct_dc_size in Luma data.

**Table 9-5 DCT Coefficients Option 0 (IS: Table B-14)**

| VLC | dct_dc_size_luminance |
|---|---|
| 100 | 0 |
| 00 | 1 |
| 01 | 2 |
| 101 | 3 |
| …… | …… |
| 11111110 | 9 |
| 111111110 | 10 |
| 111111111 | 11 |

Once size information is parsed, the bits indicated are read for DC differential data. Then, the interpretation shown in Figure 9-6 is performed to decode the DC value. Note that a dc_dct_differential is always interpreted as a positive value. The following decoding procedure as shown in Figure 9-6 can be considered as the way to map positive values to half positive and half negative values to rectify predictors. In other words, a dc_dct_differential value greater than or equal to half_range is still interpreted as the same positive value, while a value smaller than half_range is remapped to a negative value. Here, cc means 0, 1 or 2 for color components.

```
If(dc_dct_size==0)
     dct_diff=0
else {
    half_range=2^(dc_dct_size-1)
    if(dc_dct_differential≥half_range)
            dct_diff=dc_dct_differential
    else
            dct_diff=(dc_dct_differential+1)-(2×half_range)
}
QFS[0]=dc_dct_pred[cc]+dct_diff
dc_dct_pred[cc]=QFS[0]
```

**Figure 9-6 DC Decoding**

To encode other DCT coefficients, two Run-level combination VLC options are provided in the standard as shown in Table 9-7 and Table 9-8. Here, Run-level combination means the representation gives the number of zeros in a row (in zigzag order) before a Transform coefficient level and the Level of the coefficient itself. The encoder chooses which VLC is used for Intra blocks. Once selected, intra_vlc_format is to set by the encoder as shown in Table 9-6.

**Table 9-6 Selection of DCT Table (IS: Table 7-3)**

| intra_vlc_format | 0 | 1 |
|---|---|---|
| intra blocks (macroblock_intra=1) | Table 9-7 | Table 9-8 |
| non-intra blocks (macroblock_intra=0) | Table 9-7 | Table 9-7 |

**Table 9-7 DCT Coefficients Option 0 (IS: Table B-14)**

| VLC | Run | Level |
|---|---|---|
| 10 | End Of Block | |
| 1s (used for DC) | 0 | 1 |
| 11s (used for all others) | 0 | 1 |
| 011s | 1 | 1 |
| 0100s | 0 | 2 |

| | | |
|---|---|---|
| ...... | ...... | ...... |
| 0000000000011101s | 29 | 1 |
| 0000000000011100s | 30 | 1 |
| 0000000000011011s | 31 | 1 |

**Table 9-8 DCT Coefficients Option 1 (IS: Table B-15)**

| VLC | Run | Level |
|---|---|---|
| 0110 | End Of Block | |
| 10s (used for DC) | 0 | 1 |
| 010s (used for all others) | 1 | 1 |
| 110s | 0 | 2 |
| 00101s | 2 | 1 |
| ...... | ...... | ...... |
| 0000000000011101s | 29 | 1 |
| 0000000000011100s | 30 | 1 |
| 0000000000011011s | 31 | 1 |

If a run-level combination cannot be found from the Tables, ESCAPE code will be used to switch VLC mode to a FLC mode. ESCAPE will come with 6-bit Run data, followed by 12-bit signed Level data as follows:

ESCAPE ("000001") + Run (6 bit-FLC) + Signed_level (12 bit-FLC). (9-9)

***Examples***

Let's assume the following 8x8 block to encode (or decode if reverse procedure applies).

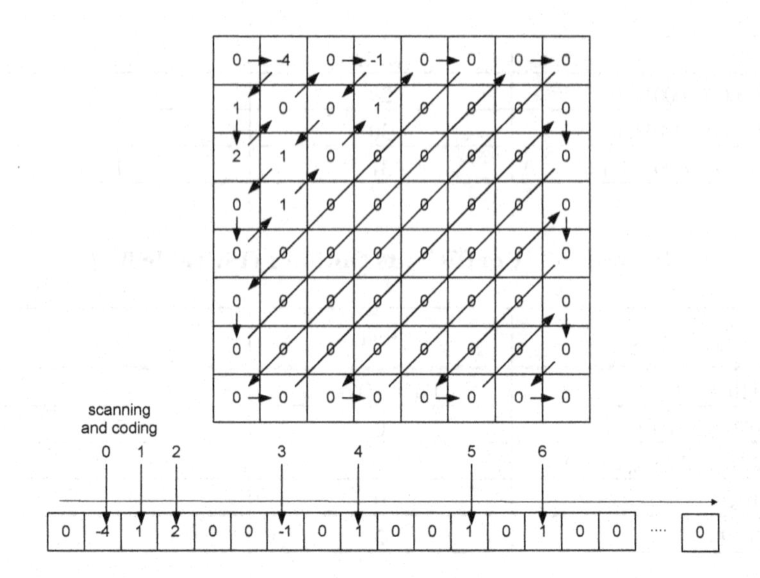


**Figure 9-7 Example of VLC Coding of a Block**

a. Zero-run and level combination: (1,-4), (0,1), (0,2), (2,-1), (1,1), (2,1), (1,1), (End of Block)

b. 00000011111,10,01000,01011,0110,01010,0110,10 (Table 9-7)

c. Final outcome: 00000011111100100001011011001010011010

## 9.3 Codes in VC-1

### Codes above MB-Level

This section is covered in Chapter 2.

### Codes below MB-Level

#### *Parsing and Decoding Process for Quantization Parameter*

In VC-1, the bitstream parsing and decoding process is heavily dependent on Quantization Parameter (Qp – particularly, PQUANT or PQINDEX). The value of Qp triggers a new set of Huffman codes to be used for MVD and residual decoding. In this sense, VC-1 coder is a kind of Context-Adaptive VLC (CA-VLC) based on Qp. In addition, VC-1 developed many useful representations of Qp for various scenarios, unlike other standards. Such scenarios and interpretations are covered in this section.

Picture level Qp syntax is described in Figure 9-8 and Figure 9-9.

```
......
        PQINDEX
    If((PQINDEX ≤8) {
            HALFQP
    }
    if(QUANTIZER==01b) {
            PQUANTIZER
    }
......
```

**Figure 9-8 Qp Syntax for All Pictures**

The interpretation of Picture level Qp information is based on DQUANT and QUANTIZER in the Entry Point header. DQUANT in the Entry Point header shall indicate whether or not the Quantization Step Size (QSS) may vary within a picture. If DQUANT==0, then only one QSS shall be used for a picture. If DQUANT==1 or 2, the QSS may vary within the picture. If DQUANT==1, there are four possibilities as follows:

- Those MBs located on the picture edge boundary shall be quantized with a second QSS (a.k.a., ALTPQUANT), while all other MBs shall be quantized with the picture QSS (a.k.a., PQUANT).
- Two adjacent edges are signaled. Those MBs located on the two edges of a picture shall be quantized with ALTPQUANT, while the rest of the MBs shall be quantized with PQUANT.
- One edge is signaled. Those MBs located on the picture edge are quantized with ALTPQUANT while the rest of the MBs are quantized with PQUANT.
- Every single MB may be quantized differently. In this case, it will be indicated whether each MB may be selected from PQUANT or ALTPQUANT, or whether each MB may be arbitrarily quantized using any QSS.

If DQUANT==2, the MBs located on the picture edge boundary shall be quantized with ALTPQUANT while the rest of the MBs shall be quantized with PQUANT.

The whole idea of VC-1 QSS design is to find efficient representations in object moving and/or camera panning scenarios. As explained in Chapter 6, when an object is moving or camera panning is happening, the edge has limited capability to represent MV and residual data. A MV can be represented effectively in VC-1 due to the Pullback algorithm. For the edge part of residual data, VC-1 introduces a policy to define QSSs that handle edges and the rest of the area independently. For example, if camera panning happens to the right direction, the left side of incoming video might have difficulty in finding best matches in reference pictures, thus producing relatively big residual errors. In this case, the left side edge might better be handled differently in terms of Qp (i.e., quality). Depending on direction of movement for objects or the camera, typically one or two edges show higher residual errors compared with the rest of the edges. This is why an independent ALTPQUANT is designed for specially specified edges, whose number is either 1 or 2, in VC-1.

QUANTIZER in the Entry Point header shall indicate the quantizer used for the sequence. If QUANTIZER==0, Quantizer is specified implicitly at picture level. If QUANTIZER==1, Quantizer is specified explicitly at picture level. If QUANTIZER==2, Non-uniform quantizer is used for all pictures. If QUANTIZER==3, Uniform quantizer is used for all pictures.

PQINDEX is a 5-bit syntax element that shall signal the quantizer scale index for the entire picture. If QUANTIZER==0, PQINDEX shall specify both the PQUANT and the quantizer (i.e., Uniform or Non-uniform) used for the picture. In this case, PQINDEX shall be translated to PQUANT as defined in Table 9-9. If QUANTIZER!=0, PQUANT shall be equal to PQINDEX for all values of PQINDEX except the value of 0.

**Table 9-9 PQINDEX to PQUANT/Quantizer Traslation – Implicit Case (IS: Table 36)**

| PQINDEX | PQUANT | Quantizer |
|---|---|---|
| 0 | NA | NA |
| 1 | 1 | Uniform |
| …… | …… | …… |
| 30 | 29 | Non-uniform |
| 31 | 31 | Non-uniform |

HALFQP is a 1-bit syntax element that shall be present in all picture types if PQINDEX is less than or equal to 8. The HALFQP syntax element allows the picture quantizer to be expressed in half step increments over the low PQUANT range. If HALFQP==1, the picture QSS shall be equal to PQUANT + ½. If HALFQP==0, the picture QSS shall be equal to PQUANT. Therefore, half step sizes are allowed up to PQUANT==9 with Uniform quantizer (i.e., 1, 1.5, 2, 2.5, …, 8.5,9, 10, 11, …), while half step sizes are possible up to PQUANT==7 with Non-uniform quantizer (i.e., 1, 1.5, 2, 2.5,…6.5,7, 8, 9, 10, …).

PQUANTIZER is a 1-bit syntax element that shall be present in all picture types if the syntax element QUANTIZER==1. In this case, the quantizer used for the picture shall be specified by PQUANTIZER. If PQUANTIZER==0, a Non-uniform quantizer shall be used for the picture. If PQUANTIZER==1, a Uniform quantizer shall be used for the picture.

```
VOPDQUANT( ) {
    If((DQUANT==2) {
           PQDIFF //quantizer can vary in frame
    }
    if(PQDIFF==7) {
           ABSPQ
    }
    else {
           DQUANTFRM
           If (DQUANTFRM==1){
                  DQPROFILE //quantizer can vary in frame
                  If(DQPROFILE== "Single Edge"){
                         DQSBEDGE
                  }
                  If(DQPROFILE== "Double Edge"){
                         DQDBEDGE
                  }
                  If(DQPROFILE== "All MBs"){
                         DQBILEVEL
                  }
                  If(!(DQPROFILE=="All MBs" &&
                        DQBILEVEL==0)){
                         PQDIFF
                         If(PQDIFF==7){
                                ABSPQ
                         }
                  }
           }
           else { //DQUANTFRM is 0
                  //same quantizer (PQUANT) is
                  //used for entire frame
           }
    }
}
```

**Figure 9-9 VOPDQUANT in P, B and AP I Pictures**

The VOPDQUANT as shown in Figure 9-9 shall be present in Progressive P and B pictures and in Advanced Profile I pictures when

DQUANT is non-zero. DQUANTFRM is a 1-bit syntax element that shall be present only when DQUANT==1 in the Entry Point header.

If DQUANTFRM==0, the current picture shall only be quantized with PQUANT. The value of DQUANTFRM shall be set to 0 when DQUANT is not 1.

The DQPROFILE is a 2-bit syntax element that shall be present only when DQUANT==1 and DQUANTFRM==1. This shall specify where it is allowable to change QSS within the current picture as shown in Table 9-10.

**Table 9-10 MB Quantization Profile (IS: Table 43)**

| DQPROFILE FLC | Location |
|---|---|
| 00 | All 4 edges |
| 01 | Double edges |
| 10 | Single edge |
| 11 | All MBs |

The DQSBEDGE is a 2-bit syntax element that shall be present when DQPROFILE== "Single edge." This shall specify which edge will be quantized with ALTPQUANT as shown in Table 9-11.

**Table 9-11 Single Boundary Edge Selection (IS: Table 44)**

| DQSBEDGE FLC | Location |
|---|---|
| 00 | Left |
| 01 | Top |
| 10 | Right |
| 11 | Bottom |

The DQDBEDGE is a 2-bit syntax element that shall be present when DQPROFILE== "Double edge." This shall specify which edge will be quantized with ALTPQUANT as shown in Table 9-12.

**Table 9-12 Double Boundary Edge Selection (IS: Table 45)**

| DQSBEDGE FLC | Location |
|---|---|
| 00 | Left and Top |
| 01 | Top and Right |
| 10 | Right and Bottom |
| 11 | Bottom and Left |

The DQBILEVEL is a 1-bit syntax element that shall be present only when DQPROFILE== "All MBs." If DQBILEVEL==1, each MB in the picture may only choose from PQUANT or ALTPQUANT. If DQBILEVEL==0, each MB in the picture may take on any QSS.

PQDIFF is a 3-bit syntax element that signals either the PQUANT differential or an ESC code. If PQDIFF!=7, PQDIFF signals the differential and ABSPQ syntax element shall not be present in the bitstream. In this case:

ALTPQUANT=PQUANT+PQDIFF+1 (9-10)

If PQDIFF==7, PQDIFF signals the ESC code and the ABSPQ syntax element shall be present in the bitstream. In this case:

ALTPQUANT=ABSPQ. (9-11)

ABSPQ is a 5-bit syntax element that shall be present in the bitstream only if PQDIFF equals 7. In this case, ABSPQ shall directly signal the value of ALTPQUANT as described above.

Once PQUANT and ALTPQUANT are derived, MB level processing continues. MQDIFF is a VLC syntax element that shall be present in P, B, and Advanced Profile I pictures only when DQPROFILE== "All MBs." The syntax depends on the DQBILEVEL, as described below. If DQBILEVEL==1, MQDIFF shall be a 1-bit syntax element. If MQDIFF

==0, MQUANT=PQUANT. If MQDIFF==1, MQUANT= ALTPQUANT.

If DQBILEVEL==0, MQDIFF shall be a 3-bit syntax element. In this case MQDIFF shall decode either to an MQUANT differential or to an ESC code as follows:

If MQDIFF!=7, MQDIFF shall decode to the differential and the ABSMQ syntax element shall not be present in the bitstream. In this case:

MQUANT=PQUANT+MQDIFF. (9-12)

MQUANT shall be in the range of 1 to 31 for the bitstream to be valid. If MQDIFF ==7, ABSMQ syntax element shall be present in the bitstream and MQUANT shall be decoded as:

MQUANT=ABSMQ. (9-13)

Here, ABSMQ is a 5-bit syntax element that shall be present in the bitstream only if MQDIFF==7. In this case, ABSMQ directly decodes to the value of MQUANT as defined in Figure 9-10.

```
If((DQPROFILE== "All MBs") {
   If(DQBILEVEL){
         //decode 1-bit flag MQDIFF
         if(MQDIFF==0)
               MQUANT=PQUANT
         else
               MQUANT=ALTPQUANT
   }
   else {
         //decode 3-bit syntax element MQDIFF
         If(MQDIFF!=7){
               MQUANT=PQUANT+MQDIFF
         }
         else {
               //decode 5-bit syntax element ABSMQ
               MQUANT=ABSMQ
         }
   }
}
```

**Figure 9-10 Derivation of MQUANT when DQPROFILE= "All MBs"**

### *Parsing Process for Motion Vector*

MV part syntax is described in Figure 9-11 and Figure 9-12, where two key data elements are MVDATA and BLKMVDATA for P pictures, and BMV1 and BMV2 for B pictures, respectively. The main decoding algorithm is described in Figure 9-13.

```
P_Simple/Main/Advanced_MB( ) {
    If((MVMODE== "Mixed-MV" || (MVMODE== "Intensity
    Compensation" && MVMODE2== "Mixed-MV") && MVTYPEMB
    Coding Mode== "Raw") {
        MVMODEBIT
    }
    if(SKIPMB Coding Mode== "Raw") {
        SKIPMBBIT
    }
    If(1-MV mode) {
        If (non-skipped MB){
            MVDATA
            If("hybridpred condition"){
                HYBRIDPRED
            }
    ...
        }
        else { //Skipped MB
            If("hybridpred condition"){
                HYBRIDPRED
            }
        }
    }
    else { //4-MV
        If (non-skipped MB){
            BLKMVDATA
            If("hybridpred condition"){
                HYBRIDPRED
            }
    ...
        }
        else { //Skipped MB
            If("hybridpred condition"){
                HYBRIDPRED
            }
        }
    }
}
```

**Figure 9-11 MV Syntax Parsing for P Pictures**

```
B_Main/Advanced_MB( ) {
    If((DIRECTMB Coding Mode == "Raw") {
           DIRECTBBIT
    }
    if(SKIPMB Coding Mode== "Raw") {
           SKIPMBBIT
    }
    If(!DIRECTBBIT) {
           If (!SKIPMBBIT){
                  BMV1
           }
           If (!("intra_flag of BMV1")||SKIPMBBIT){
                  BMVTYPE
           }
    }
    ...
    If(BMVTYPE== "Interpolative") {
           BMV2
    }
    ...
}
```

**Figure 9-12 MV Syntax Parsing for B Pictures**

To continue decoding MVDs, certain information from Picture level is also needed. Such information includes MVTAB and MVMODE. MVTAB is a 2-bit syntax element that shall be present only in P and B picture headers. The MVTAB VLC shall specify which one of four Tables is used to decode MVD as shown in Table 9-13~Table 9-18. MVD Tables are required for the MVD decoding process as shown in Figure 9-13. This adaptation is performed since the statistics of motion information is very different based on inter-prediction scenarios such as Frame Prediction or Field Prediction.

**Table 9-13 MVTAB and MVD Tables (IS: Table 51)**

| VLC | MVD Table |
|---|---|
| 00 | Table 9-14 (IS: Table 246) |
| 01 | Table 9-15 (IS: Table 247) |
| 10 | Table 9-16 (IS: Table 248) |
| 11 | Table 9-17 (IS: Table 249) |

**Table 9-14 MV Table 0 (IS: Table 246)**

| index | VLC |
|---|---|
| 0 | 000000 |
| 1 | 0000010 |
| …… | …… |
| 71 | 110111111 |
| 72 | 111 |

**Table 9-15 MVD Table 1 (IS: Table 247)**

| index | VLC |
|---|---|
| 0 | 00000 |
| 1 | 0000100 |
| …… | …… |
| 71 | 0010111101111 |
| 72 | 00101 |

**Table 9-16 MVD Table 2 (IS: Table 248)**

| index | VLC |
|---|---|
| 0 | 000 |
| 1 | 000100000000 |
| …… | …… |
| 71 | 00100111110 |
| 72 | 00100111111 |

**Table 9-17 MVD Table 3 (IS: Table 249)**

| index | VLC |
|---|---|
| 0 | 00000000000000 |

| | |
|---|---|
| 1 | 00000000001 |
| …… | …… |
| 71 | 00001110111 |
| 72 | 00100111111 |

Note that a different set of Tables are defined for Interlace tools as presented in Table 132~Table 143 in the standard.

MVMODE is a VLC syntax element that shall be present only in P and B picture headers. For P pictures, the MVMODE syntax element shall signal one of four MV coding modes, or the intensity compensation mode as shown in Table 9-18~Table 9-19. Note that VLC is context-adaptive in terms of Picture Quantization (PQUANT). VC-1 Inter Prediction tools are pre-combined and prioritized as explained in Chapter 6.

**Table 9-18 MVMODE for PQUANT>12 in P (IS: Table 46)**

| MVMODE VLC | mode |
|---|---|
| 1 | 1-MV Half-pel bilinear |
| 01 | 1-MV |
| 001 | 1-MV Half-pel |
| 0000 | Mixed-MV |
| 0001 | Intensity Compensation |

**Table 9-19 MVMODE for PQUANT≤12 in P (IS: Table 47)**

| MVMODE VLC | mode |
|---|---|
| 1 | 1-MV |
| 01 | Mixed-MV |
| 001 | 1-MV Half-pel |
| 0000 | 1-MV Half-pel bilinear |
| 0001 | Intensity Compensation |

**Table 9-20 MVMODE for PQUANT in B (IS: Table 48)**

| MVMODE VLC | mode |
|---|---|
| 1 | 1-MV |
| 0 | 1-MV Half-pel bilnear |

**Table 9-21 MVMODE2 for PQUANT>12 in P (IS: Table 46)**

| MVMODE VLC | mode |
|---|---|
| 1 | 1-MV Half-pel bilinear |
| 01 | 1-MV |
| 001 | 1-MV Half-pel |
| 0000 | Mixed-MV |

**Table 9-22 MVMODE2 for PQUANT ≤ 12 in P (IS: Table 47)**

| MVMODE VLC | mode |
|---|---|
| 1 | 1-MV |
| 01 | Mixed-MV |
| 001 | 1-MV Half-pel |
| 0000 | 1-MV Half-pel bilinear |

### *Decoding Process for Motion Vector*

The MVDATA or BLKMVDATA syntax elements are used to decode motion information for the block in the MB. 1-MV MBs have a single MVDATA syntax element and 4-MV MBs may have between 0 and 4 BLKMVDATA syntax elements. Each MVDATA or BLKMVDATA syntax element in the MB level jointly codes three parameters – the horizontal MVD component, the vertical MVD component and a binary flag indicating whether any Transform coefficients are present. Note that one of the VLC entries for MVD values indicates that the block is actually intra-coded. The MVDATA and BLKMVDATA syntax elements are a VLC, followed by a FLC, which is similar to that of MPEG-2. The value of the VLC determines the size of FLC. The MVTAB syntax element in the Picture header specifies the code Table used to decode the VLC. Figure 9-13 illustrates MVD decoding process, where k_x, k_y values are defined for fixed length for long MVs. The k_x and k_y values depend on the MVRANGE and shall be set according to following Table 9-23.

**Table 9-23 Interpretation of MVRANGE (IS: Table 75)**

| MVRANGE | k_x | k_y | range_x | range_y |
|---|---|---|---|---|
| 0 (default) | 9 | 8 | 256 | 128 |
| 10 | 10 | 9 | 512 | 256 |
| 110 | 12 | 10 | 2048 | 512 |
| 111 | 13 | 11 | 4096 | 1024 |

On top of this, the size_table and offset_table are arrays defined as follows:

size_table[6]={0,2,3,4,5,8} and

offset_table[6]={0,1,3,7,15,31}.

Decoding MVD is processed as shown in Figure 9-13, where index is obtained from Table 9-14, Table 9-15, Table 9-16 or Table 9-17.

```
index=vlc_decode( ) //use the table indicated by MVTAB in
the picture layer
index=index+1
If(index≥37) {
   more_present_flag=1
   index=index-37
} else
   more_present_flag=0
intra_flag=0
if(index==0){
         dmv_x=0
         dmv_y=0
} else if (index==35) {
   dmv_x=get_bits(k_x-halfpel_flag)
   dmv_y=get_bits(k_y-halfpel_flag)
```

```
} else if (index==36) {
    intra_flag=1
    dmv_x=0
    dmv_y=0
} else {
    index1=index%6
    if(halfpel_flag==1 && index1==5)
          hpel=1
    else  hpel=0

    val=get_bits(size_table[index1]-hpel)
    sign=0-(val&1)
    dmv_x=sign^((val>>1)+offset_table[index1])
    dmv_x=dmv_x-sign

    index1=index/6
    if(halfpel_flag==1 && index1==5)
          hpel=1
    else  hpel=0

    val=get_bits(size_table[index1]-hpel)
    sign=0-(val&1)
    dmv_y=sign^((val>>1)+offset_table[index1])
    dmv_y=dmv_y-sign
}
```

**Figure 9-13 MVD Decoding**

The subtraction operation requires that MVD be one bit more in representation compared with those of MV and MVP.

### *Parsing and Decoding Process for Residual Data*

Unlike MPEG-2, non-zero quantized AC coefficients shall be coded using a 3D run-length method. A set of Tables and constants are used to decode the combination of (*run*, *level*, *last_flag*) values. To continue decoding several terms are defined for Tables and constants as follows:

- CodeTable: The code Table used to decode the ACCOEF1 and ACCOEF2 VLC syntax elements.
- RunTable: The Table of run values indexed by the value decoded in the ACCOEF1 or ACCOEF2 syntax elements.

- LevelTable: The Table of level values indexed by the value decoded in the ACCOEF1 and ACCOEF2 syntax elements.
- NotLastDeltaRunTable: The Table of delta run values indexed by the level value as shown in Figure 9-14. It is used in ESC coding Mode 2.
- LastDeltaRunTable: The Table of delta run values indexed by the level value as shown in Figure 9-14. It is used in ESC coding Mode 2.
- NotLastDeltaLevelTable: The Table of delta level values indexed by the run value as shown in Figure 9-14. It is used in ESC coding Mode 1.
- LastDeltaLevelTable: The Table of delta level values indexed by the run value as shown in Figure 9-14. It is used in ESC coding Mode 1.
- Presence of Fixed Length Codes – Mode3 (a.k.a., first_mode3): This is used in ESC coding Mode 3 where Symbols are coded by FLC. It shall be set to one at the beginning of a frame, field or slice. It shall be set to zero, whenever Mode 3 is used for the first time.
- StartIndexOfLast: The VLC encodes index values from 0 to N. The index values are used to obtain the run and level values from RunTable and LevelTable, respectively. The first (StartIndexOfLast-1) of these index values correspond to run/level pairs that are not the last pair in the block. The next StartIndexOfLast to N-1 index values correspond to run/level pairs that are the last pair in the block. The last value, N, is the ESC index.
- ESC index (EscapeIndex): The last in the set of indices encoded by the VLC.

The following decoding process, as shown in Figure 9-14, shall be repeated until last_flag==1.

```
decode_symbol(&run, &level, &last_flag){
    last_flag=0
    index=vlc_decode( ) //use Code Table to decode VLC
(ACCOEF1)
    if(index!=ESCAPE){
            run=RunTable[index]
            level=LevelTable[index]
            sign=get_bits(1)
            if(sign==1)
                    level= -level
            if(index≥StartIndexOfLast)
                    last_flag=1
    }
    else {
            escape_mode=vlc_decode( )
            if(escape_mode==mode1){
                    index=vlc_decode( ) //decode VLC (ACCOEF2)
                    run=RunTable[index]
                    level=LevelTable[index]
                    if(index≥StartIndexOfLast)
                            last_flag=1
                    if(last_flag==0)

    level=level+NotLastDeltaLevelTable[run]
                    else

    level=level+LastDeltaLevelTable[run]
                    sign=get_bits(1)
                    if(sign==1)
                            level= -level
            }
            else if (escape_mode==mode2){
                    index=vlc_decode( ) //decode VLC (ACCOEF2)
                    run=RunTable[index]
                    level=LevelTable[index]
                    if(index≥StartIndexOfLast)
                            last_flag=1
                    if(last_flag==0)

    run=run+NotLastDeltaRunTable[level]+1
                    else
                            run=run+LastDeltaRunTable[level]+1
                    sign=get_bits(1)
                    if(sign==1)
                            level= -level
            }
            else if (escape_mode==mode3){
```

```
                    //decode FLC
                    last_flag=get_bits(1)
                    if(first_mode3==1)
                            first_mode3=0
                            level_code_size=vlc_decode( )
                            //use IS:Table 59 and Table 60
                            run_code_size=3+get_bits(2)
                    }
                    run=get_bits(run_code_size)
                    sign=get_bits(1)
                    level=get_bits(level_code_size)
                    if(sign==1)
                            level= -level
            }
        }
    }
```

**Figure 9-14 Coefficient Decoding**

To improve coding efficiency, there are eight AC coding sets. The coding sets are divided into two groups of four Intra/Inter coding sets. For Y blocks, one of the four Intra coding sets shall be used. For Cb and Cr blocks, one of the four Inter coding sets shall be used. For example, Table 9-24 and Table 9-25 are considered for I picture decoding and Intra blocks in P or B picture decoding, while Table 9-26 and Table 9-27 are used for Inter blocks in P or B picture decoding. Note that PQINDEX is also considered for Table selection as shown in Table 9-24~ Table 9-27.

For I picture decoding, the value decoded from the TRANSACFRM2 syntax element shall be used as the coding set index for Y blocks and the value decoded from the TRANSACFRM syntax element shall be used as the coding set index for Cb and Cr blocks. For P or B picture decoding, the value decoded from the TRANSACFRM syntax element shall be used to specify the coding set index used for decoding the Y, Cb and Cr. The value decoded from the TRANSACFRM syntax element shall be used to select the Intra coding set applied to decode the Y blocks, and it shall also be used to select the Inter coding set applied to decode the Cb and Cr blocks. The P or B picture header shall not contain the TRANSACFRM2 syntax element.

**Table 9-24 Coding Set Correspondence for PQINDEX ≤ 8 (IS: Table 71)**

| Y blocks | | Cb and Cr blocks | |
|---|---|---|---|
| index | Table | index | Table |
| 0 | High Rate Intra (IS: Table 219~Table 225) | 0 | High Rate Inter (IS: Table 226~Table 232) |
| 1 | High Motion Intra (IS: Table 177~Table 183) | 1 | High Motion Inter (IS: Table 184~Table 190) |
| 2 | Mid Rate Intra (IS: Table 205~Table 211) | 2 | Mid Rate Inter (IS: Table 212~Table 218) |

**Table 9-25 Coding Set Correspondence for PQINDEX>8 (IS: Table 72)**

| Y blocks | | Cb and Cr blocks | |
|---|---|---|---|
| index | Table | index | Table |
| 0 | Low Motion Intra (IS: Table 191~Table 197) | 0 | Low Motion Inter (IS: Table 198~Table 204) |
| 1 | High Motion Intra (IS: Table 177~Table 183) | 1 | High Motion Inter (IS: Table 184~Table 190) |
| 2 | Mid Rate Intra (IS: Table 205~Table 211) | 2 | Mid Rate Inter (IS: Table 212~Table 218) |

**Table 9-26 Coding Set Correspondence for PQINDEX ≤ 8 (IS: Table 78)**

| Y, Cb and Cr blocks | |
|---|---|
| index | Table |
| 0 | High Rate Inter (IS: Table 226~Table 232) |
| 1 | High Motion Inter (IS: Table 184~Table 190) |
| 2 | Mid Rate Inter (IS: Table 212~Table 218) |

**Table 9-27 Coding Set Correspondence for PQINDEX>8 (IS: Table 79)**

| Y, Cb and Cr blocks | |
|---|---|
| index | Table |
| 0 | Low Motion Inter (IS: Table 198~Table 204) |
| 1 | High Motion Inter (IS: Table 184~Table 190) |
| 2 | Mid Rate Inter (IS: Table 212~Table 218) |

In parsing residual data, ACCOEF1 is a VLC syntax element that may be present in both Intra and Inter blocks. The codeword shall decode to the run, level and last_flag for each non-zero AC coefficient. One of three code Tables shall be used to decode ACCOEF1. The Table is signaled by the TRANSACFRM or TRANSACFRM2 as was explained in this section. Examples of the Table groups include Table 9-32~Table 9-38. The group shown in Table 9-32~Table 9-38 is used to encode/decode for Low Motion Inter blocks.

ESCMODE is a VLC syntax element that may be present in both Intra and Inter blocks. It shall only be present if ACCOEF1 decodes to the ESC code. ESCMODE shall specify which of three ESC decoding methods are used as shown in Table 9-28. If Mode 1 or Mode 2 decoding mode is specified, the bitstream shall contain the ACCOEF2 element. If Mode 3 decoding mode is specified, the bitstream shall contain the ESCLR, ESCRUN, ESCLVL and VLVSIGN2 syntax elements and may contain the ESCLVLSZ and ESCRUNSZ elements.

**Table 9-28 AC Escape Mode (IS: Table 58)**

| ESC MODE VLC | AC Escape Decoding Mode |
|---|---|
| 1 | Mode 1 |
| 01 | Mode 2 |
| 00 | Mode 3 |

In parsing residual data, ACCOEF2 is a VLC syntax element that may be present in both Intra and Inter blocks. It shall only be present if ACCOEF1 decodes to the ESC code and if the ESCMODE syntax element specifies AC decoding Escape Mode 1 or 2. One of three code Tables shall be used to decode ACCOEF2. The Table is signaled by the TRANSACFRM or TRANSACFRM2 as was explained in this section.

ESCLR is a 1-bit syntax element that may be present in both Intra and Inter blocks. It shall only be present if ESCMODE specifies AC decoding ESC Mode 3. ESCLR shall specify whether the coefficient is the last non-zero coefficient in the block. If ESCLR==1, the coefficient shall be the last non-zero coefficient.

ESCLVLSZ is a VLC syntax element that may be present in both Intra and Inter blocks as shown in Table 9-29 and Table 9-30. It shall only be present if ESCMODE specifies AC decoding ESC Mode 3. If this is the first time to use Mode 3, then Mode 3 is signaled within the current picture. All subsequent instances of ESC Mode 3 coding within this picture do not have this syntax element. ESCLVLSZ shall specify the codeword size for the Mode 3 ESC coded "level" values for the entire picture.

**Table 9-29 Escape Mode 3 Level Codeword Size for Low Qp (IS: Table 59)**

| $1 \le PQUANT \le 7$ | |
|---|---|
| ESC MODE3 LEVEL SZ VLC | Level codeword size |
| 001 | 1 |
| 010 | 2 |
| 011 | 3 |
| …… | …… |
| 00010 | 10 |
| 00011 | 11 |

**Table 9-30 Escape Mode 3 Level Codeword Size for High Qp (IS: Table 60)**

| $8 \leq PQUANT \leq 31$ | |
|---|---|
| ESC MODE3 LEVEL SZ VLC | Level codeword size |
| 1 | 2 |
| 01 | 3 |
| 001 | 4 |
| …… | …… |
| 000001 | 7 |
| 000000 | 8 |

ESCRUNSZ is a 2-bit syntax element that may be present in both Intra and Inter blocks as shown in Table 9-31. It shall only be present if ESCMODE specifies AC decoding ESC Mode 3. If this is the first time, Mode 3 has been signaled within the current picture, ESCLRUNSZ shall specify the codeword size for the Mode 3 ESC coded "run" values for the entire picture.

**Table 9-31 Escape Mode 3 Run Codeword Size (IS: Table 61)**

| ESC RUN SZ VLC | Run codeword size |
|---|---|
| 00 | 3 |
| 01 | 4 |
| 10 | 5 |
| 11 | 6 |

ESCRUN may be present in both Intra and Inter blocks. It shall only be present if ESCMODE specifies AC decoding ESC Mode 3. The size of the ESCRUN codeword is fixed throughout the entire picture, where the size is specified in the ESCRUNSZ syntax element. In other words, the run is coded with FLC.

LVLSGN2 is 1-bit syntax element that may be present in both Intra and Inter blocks. It shall only be present if ESCMODE specifies AC decoding ESC Mode 3. If LVLSGN2==0, the level should be positive.

ESCLVL may be present in both Intra and Inter blocks. It shall only be present if ESCMODE specifies AC decoding ESC Mode 3. The size of the ESCLVL codeword is fixed throughout the entire picture, where the size is specified in the ESCLVLSZ syntax element. In other words, the level is coded with FLC.

In summary, the representation for ESC Mode 3 for the first time in a picture and the representation later on are, respectively, as follows:

ESCAPE ("000001101") + ESCAPE MODE ("00") + last_flag (1 bit-FLC) + ESCLVLSZ(VLC) + ESCRUNSZ(VLC) + Run (ESCRUNSZ bit-FLC) + Sign(1 bit-FLC) + Level (ESCLVLSZ bit-FLC)

ESCAPE ("000001101") + ESCAPE MODE ("00") + last_flag (1 bit-FLC) + Run (ESCRUNSZ bit-FLC) + Sign(1 bit-FLC) + Level (ESCLVLSZ bit-FLC). (9-14)

#### *Examples*

Let's assume the following 8x8 block to encode (or decode if reverse procedure applies) under the condition that QPINDEX is 9 and the block is Inter coded with TRANSACFRM being "0." Note that the zig-zag scanning pattern is different than that of MPEG-2.

a. Zero-run, level and last_flag combination: (1,1,0), (0,-4,0), (2,2,0), (1,1,0), (1,-1,0), (1,1,0), (1,1,1)

b. This example is to utilize Low Motion Inter group tables (IS: Table 198~Table 204) as shown in Table 9-32~Table 9-38.

c. Index mapping:

- (1,1,0)→ (index=14 in Table 9-33) and (sign=0)
- (0,-4,0)→ (index=3 in Table 9-33) and (sign=1)
- (2,2,0)→ (index=24 in Table 9-33) and (sign=0)
- (1,1,0)→ (index=14 in Table 9-33) and (sign=0)
- (1,-1,0)→ (index=14 in Table 9-33) and (sign=1)

- (1,1,0)→ (index=14 in Table 9-33) and (sign=0)
- (1,1,1)→ (index=86 in Table 9-34) and (sign=0)

d. Final outcome:

1011001111111101101101010110101111011001010

(index mapping in Table 9-32)

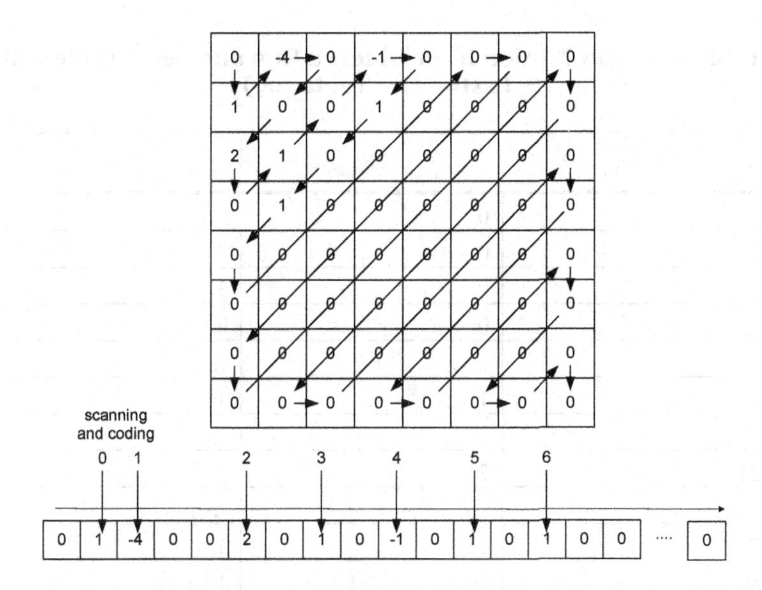


**Figure 9-15 Example of VLC Coding of a Block**

**Table 9-32 Low Motion Inter VLC Table (IS: Table 198)**

| index | VLC |
|---|---|
| 0 | 100 |
| 1 | 10100 |
| 2 | 0010111 |
| 3 | 01111111 |
| …… | …… |
| 14 | 1011 |
| …… | …… |

| | |
|---|---|
| 24 | 01101101 |
| ...... | ...... |
| 86 | 0101 |
| ...... | ...... |
| 147 | 1010110010100 |
| ESCAPE | 000001101 |

**Table 9-33 Low Motion Inter Indexed Run and Level Table with Last=0 (IS: Table 199)**

| index | Run | Level |
|---|---|---|
| 0 | 0 | 1 |
| 1 | 0 | 2 |
| 2 | 0 | 3 |
| 3 | 0 | 4 |
| ...... | ...... | ...... |
| 14 | 1 | 1 |
| ...... | ...... | ...... |
| 24 | 2 | 2 |
| ...... | ...... | ...... |
| 79 | 28 | 1 |
| 80 | 29 | 1 |

**Table 9-34 Low Motion Inter Indexed Run and Level Table with Last=1 (IS: Table 200)**

| index | Run | Level |
|---|---|---|
| 81 | 0 | 1 |
| 82 | 0 | 2 |
| ...... | ...... | ...... |
| 86 | 1 | 1 |
| ...... | ...... | ...... |
| 146 | 42 | 1 |
| 147 | 43 | 1 |

**Table 9-35 Low Motion Inter Delta Level Indexed by Run Table with Last=0 (IS: Table 201)**

| Run | Delta Level |
|---|---|
| 0 | 14 |
| 1 | 9 |
| 2 | 5 |
| 3 | 4 |
| …… | …… |
| 28 | 1 |
| 29 | 1 |

**Table 9-36 Low Motion Inter Delta Level Indexed by Run Table with Last=1 (IS: Table 202)**

| Run | Delta Level |
|---|---|
| 0 | 5 |
| 1 | 4 |
| 2 | 3 |
| 3 | 3 |
| …… | …… |
| 42 | 1 |
| 43 | 1 |

**Table 9-37 Low Motion Inter Delta Run Indexed by Level Table with Last=0 (IS: Table 203)**

| Level | Delta Run |
|---|---|
| 1 | 29 |
| 2 | 15 |
| 3 | 12 |
| 4 | 5 |
| …… | …… |
| 13 | 0 |
| 14 | 0 |

**Table 9-38 Low Motion Inter Delta Run Indexed by Level Table with Last=1 (IS: Table 204)**

| Level | Delta Run |
|---|---|
| 1 | 43 |
| 2 | 15 |
| 3 | 3 |
| 4 | 1 |
| 5 | 0 |

## Bitplane Coding

In each MB level, there are a few flags, each of which requires only one binary bit, to represent certain binary status as shown in Figure 9-16. The idea of Bitplane Coding is to form a big chunk of binary data in the Picture level to give another layer of compression efficiency. Note that if flags are described in the MB level, further compression is not possible since each flag occupies only one bit. The flags are independently coded in MB level when the Bitplane Coding mode is set to "Raw" mode. There are seven coding modes in Biplane Coding – Raw, Normal-2, $diff^{-1}$-2, Normal-6, $diff^{-1}$-6, Row-skip and Column-skip modes.

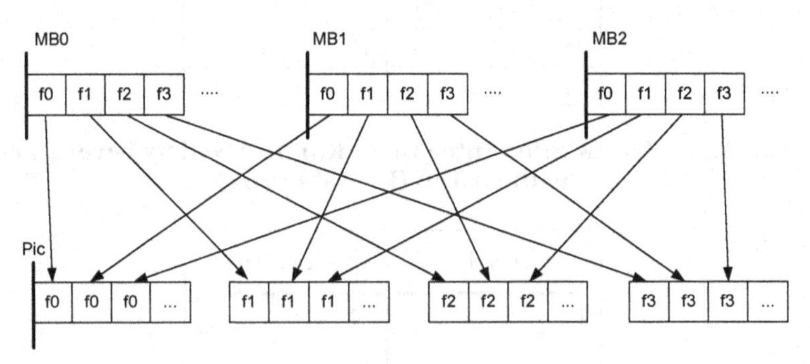


**Figure 9-16 Bitplane Coding**

When flags are grouped together, in the Picture header as shown in Figure 9-16, the grouped symbols correspond to a Huffman code. The Huffman code was designed in such a way that more "0"s are preferred. When two Symbols are jointly coded as shown in Table 9-39, it is called "Normal-2 mode." For example, symbol combination (0,0) is coded with "0," while symbol combination (1,0) is coded with "100." Note that the all zero case (0,0) is represented as the most compact code (i.e., 1-bit length), while (1,0) or (0,1) case is described with the longest codes (i.e., 3-bit length). This implies patterns of 0s are preferred in this Huffman code design to optimally represent. If somehow more "1"s are in the flag data, all flag data are first inverted from "1" to "0" and "0" to "1" to make the flag data pattern contain more 0s. When this binary inversion is performed on the flag data, INVERT flag is set equal to 1. Since more 0s remain in the flag data, compression efficiency gets higher on the inverted flag data than that on original flag data. The same idea is applied to a group of six Symbols. When six Symbols are jointly coded as shown in Figure 9-17, it is called "Normal-6 mode." In Normal-6 mode, these pixels shall be grouped into either 2x3 or 3x2 tiles. The bitplane shall be tiled maximally using a set of rules, and the remaining pixels shall be decoded using a variant of *Row-skip* and *Column-skip* modes that will be mentioned in the later part of this section. The 2x3 "vertical" tiles shall be used if and only if *rowMB* is a multiple of 3 and *colMB* is not. Else, the 3x2 "horizontal" tiles shall be used as shown in Figure 9-17. The algorithm to decode Normal-6 mode is shown in Figure 90 in the standard.

**Table 9-39 Norm-2/ Diff-2 Code Table (IS: Table 80)**

| Symbol (2N) | Symbol (2N+1) | VLC Codeword |
|---|---|---|
| 0 | 0 | 0 |
| 1 | 0 | 100 |
| 0 | 1 | 101 |
| 1 | 1 | 11 |

When pre-fixed differential process $diff^{-1}$ is performed on flag data before grouping 2 Symbols or 6 Symbols, such a coding mode is called "$diff^{-1}$-2 mode" or "$diff^{-1}$-6 mode," respectively. Again, the grouped symbols correspond to Huffman codes such as in Table 9-39. The INVERT mode can be applied to $diff^{-1}$-2 or $diff^{-1}$-6 coding modes. However, $diff^{-1}$-2 or $diff^{-1}$-6 coding modes let the value of INVERT syntax control the $diff^{-1}$ operation as below – the differential bitplane is generated using corresponding Normal modes. Then, differential bits shall be used to regenerate the original bitplane. The regeneration process, which is the inverse process of differential process, is defined in the standard. The regeneration process is a 2-D DPCM on a binary alphabet. In order to regenerate the bit at location (i,j), the predictor $b_p(i,j)$ shall be generated at positions (i,j) as follows:

$$b_p(i,j) = \begin{cases} A & i = j = 0, \quad or \quad b(i,j-1) \neq b(i-i,j) \\ b(0,j-1) & i = 0 \\ b(i-1,j) & otherwise. \end{cases} \tag{9-15}$$

In differential coding mode, the bit-wise inversion process based on INVERT shall not be performed. However, INVERT flag is still used to define A. The value of A is set equal to 0 if INVERT==0, while it is set equal to 1 if INVERT==1.

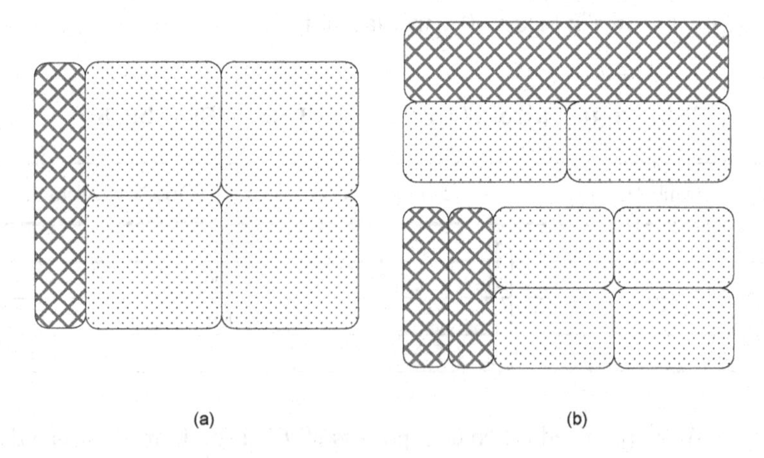


**Figure 9-17 (a) Vertical Tiles and (b) Horizontal Tiles**

The Row-skip mode uses 1-bit skip to signal rows for no set bits. In this mode, all-zero rows shall be skipped with 1-bit overhead. The process is as shown in Figure 9-18. For each row, ROWSKIP is a 1-bit syntax element that shall always be present. If ROWSKIP==0, the syntax element ROWBITS shall not be present, and the entire row of Symbols shall be set to zero. If ROWSKIP==1, the syntax element ROWBITS shall be present and shall be decoded as one bit per Symbol for the entire row. The Column-skip mode is the same as the Row-skip mode except that the direction is the transpose of Row-skip. Columns shall be scanned from the left to the right in the flag data of the picture.

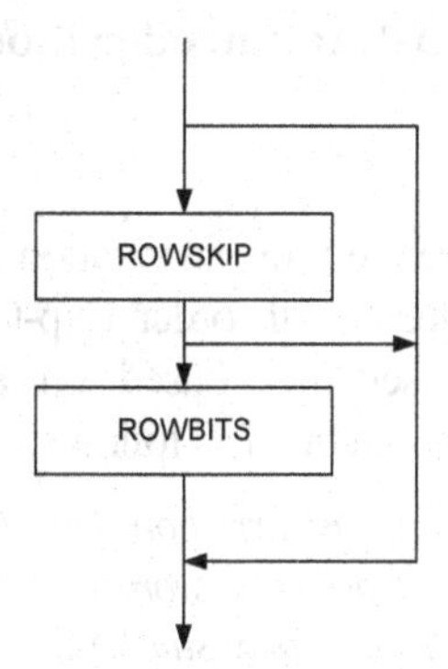


**Figure 9-18 Syntax of Row-skip Mode**

## 9.4 Codes in H.264

### Codes above MB-Level

This section is covered in Chapter 2.

### CA-BAC

The encoding process of CA-BAC is composed of three steps – binarization, context modeling, and binary arithmetic coding. In the 1st step, a given Symbol is uniquely mapped to a binary "bin string." The 2nd step utilizes assigned context models for bin and manages the context with input data on the fly, or so called "context modeling." The 3rd step is to apply BAC as explained in the previous section [marpe:context]. Figure 9-19 depicts the entire CA-BAC encoder diagram.

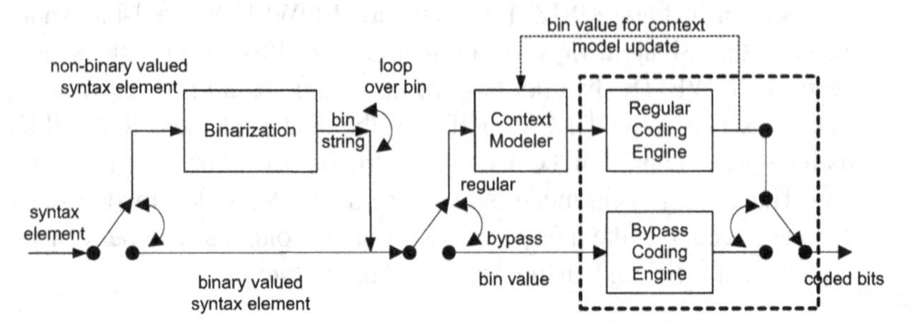


**Figure 9-19 CA-BAC Encoder Block Diagram**

### *Binarization*

In H.264 CA-BAC, there are four binarization schemes – the unary code, the truncated unary code, the *k*th order Exp-Golomb code, fixed-length code, and binarization schemes based on a concatenation of these elementary types. They are defined as follows:

*- Unary and Truncated Unary Binarization Scheme: for each unsigned integer valued symbol $x \geq 0$, the unary code word in CA-BAC consists of x "1" bits plus a terminating "0" bit. This may still be used to parse the code since the decoder already knows the entire length of the code.*

*- kth order Exponential Golomb Binarization Scheme: EG codes are constructed by a concatenation of a prefix and a suffix code word. Figure 9-20 shows a combination of Unary and kth order EG.*

*- Fixed Length (FL) Binarization Scheme.*

*- Concatenation of Basic Binarization Schemes: three more binarization schemes are defined in CA-BAC. The first one is a concatenation of a 4-bit FL prefix as a representation of the luminance related part of the coded block pattern and a TU suffix with S=2 representing the chrominance related part of coded_block_pattern. The second one is a concatenation of UT and the kth order EG, which is related with "small values" motion vector differences and absolute values of transform coefficient levels. The third one is a concatenation of TU and the kth order EG, which is related with "large values" motion vector differences and absolute values of transform coefficient levels. The cut off value for MVD is 9, while the cut off value for transform coefficients is 14. Note that the TU prefix part and the EG0 suffix are concatenated to compose codes when absolute values of MVD or transform coefficients are less than the corresponding cut off values. If those*

*absolute values are larger than the corresponding cut off values, the suffix is constructed as EGk values. Figure 9-20 illustrates about cut off explanation for transform coefficients.*

| abs_level | bin string | | |
|---|---|---|---|
| 1 | 0 | | |
| 2 | 1 0 | | |
| 3 | 1 1 0 | | |
| 4 | 1 1 1 0 | | |
| 5 | 1 1 1 1 0 | | |
| 6 | 1 1 1 1 1 0 | | |
| 7 | 1 1 1 1 1 1 0 | | |
| 8 | 1 1 1 1 1 1 1 0 | | Unary |
| 9 | 1 1 1 1 1 1 1 1 0 | | |
| 10 | 1 1 1 1 1 1 1 1 1 0 | | |
| 11 | 1 1 1 1 1 1 1 1 1 1 0 | | |
| 12 | 1 1 1 1 1 1 1 1 1 1 1 0 | | |
| 13 | 1 1 1 1 1 1 1 1 1 1 1 1 0 | | |
| 14 | 1 1 1 1 1 1 1 1 1 1 1 1 1 0 | | |
| 15 | 1 1 1 1 1 1 1 1 1 1 1 1 1 1 | 0 | |
| 16 | 1 1 1 1 1 1 1 1 1 1 1 1 1 1 | 1 0 0 | |
| 17 | 1 1 1 1 1 1 1 1 1 1 1 1 1 1 | 1 0 1 | TU prefix |
| 18 | 1 1 1 1 1 1 1 1 1 1 1 1 1 1 | 1 1 0 0 0 | EG suffix |
| 19 | 1 1 1 1 1 1 1 1 1 1 1 1 1 1 | 1 1 0 0 1 | |
| 20 | 1 1 1 1 1 1 1 1 1 1 1 1 1 1 | 1 1 0 1 0 | |
| ...... | ...... | | |
| bin | 1 2 3 4 5 6 7 8 9 10 11 12 13 14 | 15 16 17 18 19 | |

**Figure 9-20 Binarization Example**

Figure 9-20 shows a binarization scheme that is based on a concatenation of these elementary types. Note that the composition of codes is done in such a way that the probability of symbol "1" is made dominant.

***Context Modeling***

One of most unique characteristics of BAC is adaptation capability of the probability model. Probability model adaptation typically captures long-term statistics and hardly considers immediate local change of statistics. The problem of capturing locality in probability model adaptation can be resolved by context modeling. Based on neighboring context, the probability model is updated in different ways.

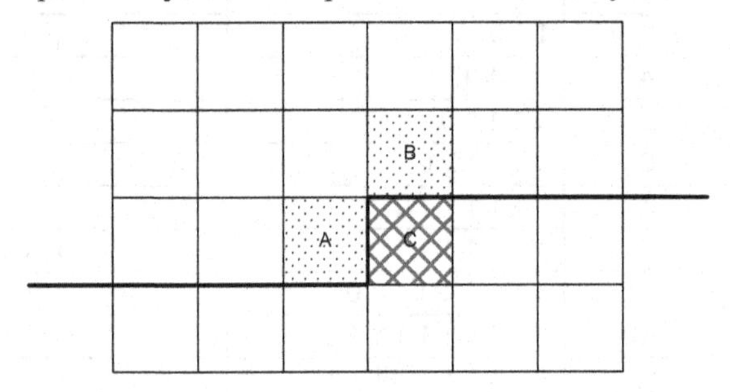


**Figure 9-21 Neighboring Context**

In CA-BAC, the model cost is reduced through two aspects: first, limited templates consisting of a few neighbors of the current symbol to encode are employed such that only a small number *C* of different context models are effectively used. Second, context modeling is restricted to selected bins of the binarized symbols.

Four basic context models are defined in CA-BAC. The first type of context model involves a context template with up to two neighboring syntax elements in the past of the current syntax element to encode, where the specific definition of the kind of neighborhood depends on the syntax element, as shown in Figure 9-21. For example, the probability model for MVD values is adaptively chosen based on evaluation of neighboring contexts as follows:

$e_k = |MVD_A| + |MVD_B|$ where A and B neighboring blocks are defined as shown in Figure 9-21.

Then, the rule for selection of context models is defined in the following Table 9-40.

**Table 9-40 3 Different Probability Models Assigned to bin1**

| $e_k$ | Context model (index) for bin1 |
|---|---|
| $0 \le e_k < 3$ | Model 0 |
| $3 \le e_k < 33$ | Model 1 |
| $33 \le e_k$ | Model 2 |

If $e_k$ is small, there is a high probability that the current MVD will have a relatively small value. If $e_k$ is large, there is a high probability that the current MVD will have a relatively big value. It is a good reason to adapt probability models based on $e_k$.

For overall assignment, the bins with MVD are coded based on seven context models, as defined in Table 9-41.

**Table 9-41 Different Probability Models Assigned to Higher bins**

| Bin | Context model (index) |
|---|---|
| 1 | Model 0, Model 1 or Model 2 depending on $e_k$ |
| 2 | Model 3 |
| 3 | Model 4 |
| 4 | Model 5 |
| 5 | Model 6 |
| 6 or higher | Model 6 |

The second type of context model is only defined for the syntax elements of mb_type and sub_mb_type. For this kind of context models, the values of prior coded bins ($b_0, b_1, b_2, \ldots, b_{i-1}$) are used for the choice of a model for a given bin with index i. Figure 9-22 shows an example for P or SP slices.

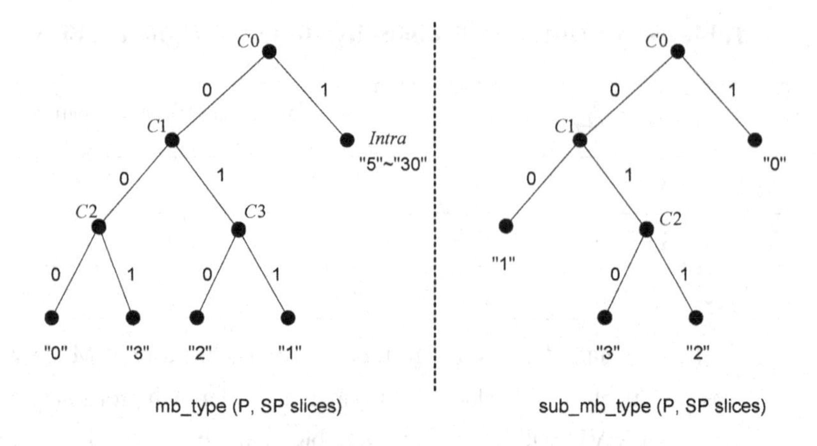


**Figure 9-22 Context Modeling Example for Mb_type and Sub_mb_type**

In Figure 9-22, the value "0" of mb_type means the MB type of "P_L0_16x16" in a P or SP slice. In the binary bin representation, the string is supposed to be "000." Here, the symbol probability $p("0")$ can be obtained based on $p^{(C0)}("0")$, $p^{(C1)}("0")$ and $p^{(C2)}("0")$, where *C0*, *C1* and *C2* denote the (binary) probability mode of the corresponding internal nodes as shown in Figure 9-22. In contrast, the value "2" of mb_type means the MB type of "P_L0_L0_8x16" in a P or SP slice. In the binary bin representation, the string is supposed to be "010," where the symbol probability $p("2")$ can be obtained based on $p^{(C0)}("0")$, $p^{(C1)}("1")$ and $p^{(C3)}("0")$. Note that the $p("0")$ and the $p("2")$ in the example proceed based on different context models (i.e., {C0→C1→C2} vs. {C0→C1→C3}). In CA-BAC these context models are only used to select different models for different internal nodes of the corresponding binary trees.

The third type and forth type of context models are applied to residual data only. The third type does not rely on past coded data, but on the position in the scanning path. For example, the coding of significance map requires 15 different probability models that are used for both the significant_coeff_flag and the last_significant_coeff_flag. The choice of the models for bins is dependent on position information. For a coefficient *coeff*[*i*], which is scanned at the *i*th position, the context model for the "significant map" is as follows:

Model *i*= *context model of coeff[i]* .

Depending on the maximum number of coefficients (MaxNumCoeff) for each context, this results in maxNumCoeff-1 different contexts. Thus, a total of 61 different models for both the significant_coeff_flag and the last_significant_coeff_flag are reserved.

**Table 9-42 Different Probability Models Assigned to Each bin**

| Bin | Context model (index) for mb_type | Context model (index) for sub_mb_type |
|---|---|---|
| 1 | Model 0 (C0) | Model 0 (C0) |
| 2 | Model 1 (C1) | Model 1 (C1) |
| 3 | Model 2 (C2) or Model 3 (C3) | Model 2 (C2) |
| .... | .... | .... |

**Table 9-43 Different Probability Models Assigned to Significant Map**

| The position transform coefficient scanned | Context model (index) |
|---|---|
| 1 | Model 0 |
| 2 | Model 1 |
| 3 | Model 2 |
| 4 | Model 3 |
| 5 | Model 4 |
| .... | .... |

The fourth type involves the evaluation of the accumulated number of encoded levels with a specific value prior to the current level bin to encode. Let NumT1(i) denote the accumulated number of already encoded/decoded trailing 1's, and let NumLgt1(i) denote the accumulated number of encoded/decoded levels with absolute value greater than 1, where both counters are related to the current scanning position i within the processed transform coefficient block. For a coefficient $coeff[i]$, which is scanned at the $i$th position, the context model for "level values" is as follows:

Model for $bin1$ in coeff[i]=

$$\left\{ Model \begin{matrix} 4, & if \quad NumLgt1(i) > 0 \\ clip3(0,3, NumT1(i)), & otherwise. \end{matrix} \right. \text{and}$$

(9-16)

Model for $bink$ in coeff[i]= $Model\ (5 + clip3(0,4, NumLgt1(i)))$, where $2 \leq bink \leq 14$. (9-17)

**Table 9-44 Different Probability Models Assigned to Each bin of Levels with coeff[i]**

| Bin at position of transform coefficient scanned at *i* | Context model (index) |
|---|---|
| 1 | Model 4 or Model (clip3(0,3, NumT1(i))) |
| 2 | Model (5+clip3(0,4, NumLgt1(i))) |
| 3 | Model (5+ clip3(0,4, NumLgt1(i))) |
| 4 | Model (5+ clip3(0,4, NumLgt1(i))) |
| …. | …. |
| 14 | Model (5+ clip3(0,4, NumLgt1(i))) |

***Combination of Hetero Context Modeling***

The probability models used in various parts of syntax in CA-BAC can be arranged in a linear fashion such that each model can be identified by a unique "context index $\gamma$." Figure 9-23 contains the entire syntax mapped with context index.

| Syntax element | Slice type | | |
|---|---|---|---|
| | SI/I | P, SP | B |
| Mb_type | 0/3-10 | 14-20 | 27-35 |
| Mb_skip_flag | | 11-13 | 24-26 |
| Sub_mb_type | | 21-23 | 36-39 |
| MVD (horizontal) | | 40-46 | 40-46 |
| MVD (vertical) | | 47-53 | 47-53 |
| Ref_idx | | 54-59 | 54-59 |
| Mb_qp_delta | 60-63 | 60-63 | 60-63 |
| Intra_chroma_pred_mode | 64-67 | 64-67 | 64-67 |
| Prev_intra4x4_pred_mode_flag | 68 | 68 | 68 |
| Rem_intra4x4_pred_mode | 69 | 69 | 69 |
| Mb_field_decoding_flag | 70-72 | 70-72 | 70-72 |
| Coded_block_pattern | 73-84 | 73-84 | 73-84 |
| Coded_block_flag | 85-104 | 85-104 | 85-104 |
| Significant_coeff_flag | 105-165, 277-337 | 105-165, 277-337 | 105-165, 277-337 |
| Last_significant_coeff_flag | 166-226,338-398 | 166-226,338-398 | 166-226,338-398 |
| Coeff_abs_level_minus1 | 227-275 | 227-275 | 227-275 |
| End_of_slice_flag | 276 | 276 | 276 |

**Figure 9-23 Syntax Elements and Context Indices ($\gamma$)**

Detailed description for context index $\gamma$ can be more efficiently represented by:

$$\gamma = \Gamma_S + \Delta_S(ctx\_cat) + \chi_S \tag{9-18}$$

where the context index offset $\Gamma_S$ and the context category $(ctx\_cat)$ dependent offset $\Delta_S(ctx\_cat)$ are employed. Note that the last term $\chi_S$ is a context index increment. $\Gamma_S$ is defined as the lowest value of the range given in Figure 9-23. The context category is as follows:

| BlockType | maxNumCoeff | Context category (ctx_cat) |
|---|---|---|
| Luma DC block for Intra16x16 | 16 | Luma-Intra16x16-DC (ctx_cat=0) |
| Luma AC block for Intra16x16 | 15 | Luma-Intra16x16-AC (ctx_cat=1) |
| Luma block for Intra4x4 | 16 | Luma-4x4 (ctx_cat=2) |
| Luma block for Inter | 16 | |
| U-Chroma DC block for Intra | 4 | Chroma-DC (ctx_cat=3) |
| V-Chroma DC block for Intra | 4 | |
| U-Chroma DC block for Inter | 4 | |
| V-Chroma DC block for Inter | 4 | |
| U-Chroma AC block for Intra | 15 | Chroma-AC (ctx_cat=4) |
| V-Chroma AC block for Intra | 15 | |
| U-Chroma AC block for Inter | 15 | |
| V-Chroma AC block for Inter | 15 | |

**Figure 9-24 Context Category (ctx_cat) Definition**

| Syntax element | Context category (cnt_cat) | | | | |
|---|---|---|---|---|---|
| | 0 | 1 | 2 | 3 | 4 |
| Codec_block_flag | 0 | 4 | 8 | 12 | 16 |
| Significant_coeff_flag | 0 | 15 | 29 | 44 | 47 |
| Last_significant_coeff_flag | 0 | 15 | 29 | 44 | 47 |
| coeff_abs_level_minus1 | 0 | 10 | 20 | 30 | 39 |

**Figure 9-25 Context Offset $\Delta_S(ctx\_cat)$ Definition**

The context index increment $\chi_S$ is taken as follows for each syntax element.

*Coding of Macroblock type, Prediction Mode and Control Info*

The context index increments used for mb_type, prediction mode and control information are as follows:

- $\chi_{MbSkip}(C) =$ ((mb_skip_flag($A$)!=0)?0:1)+(mb_skip_flag($B$)!=0)?0:1). (9-19)

- $\chi_{ChPred}(C) =$ ((ChPredInDcMode($A$)!=0)?0:1)+(ChPredInDcMode($B$)!=0)?0:1). (9-20)

- $\chi_{RefIdx}(C) =$ ((RefIdxZeroFlag($A$)!=0)?0:1)+ $2 \times$ (RefIdxZeroFlag($B$)!=0)?0:1). (9-21)

$$-\chi_{Mvd}(C, cmp) = \begin{cases} 0, & if \quad e(A,B,cmp) < 3 \\ 1, & if \quad 3 \le e(A,B,cmp) \le 32 \\ 2, & if \quad e(A,B,cmp) > 32 \end{cases} \qquad (9\text{-}22)$$

with $e(A,B,cmp) = |mvd(A,cmp)| + |mvd(B,cmp)|$.

- $\chi_{MbField}(C) =$ mb_field_decoding_flag($A$)+ mb_field_decoding_flag($B$). (9-23)

*Coding of Residual Data*

The context index increments used for residual data are as follows:

- $\chi_{CBP}(C, bin\_idx) =$ ((CBP_Bit($A$)!=0)?0:1)+ $2 \times$ (CBP_Bit($B$)!=0)?0:1). (9-24)

- $\chi_{CBFlag}(C) =$ coded_block_flag($A$)+ $2 \times$ coded_block_flag($B$). (9-25)

- $\chi_{SIG}(coeff[i]) = \chi_{SIG}(coeff[i]) = i.$ (9-26)

- $\chi_{AbsCoeff}(i, bin\_idx = 0) =$

$$\begin{cases} 4, & if \quad NumLgt1(i) > 0 \\ clip3(0,3, NumT1(i)), & otherwise. \end{cases} \quad (9\text{-}27)$$

- $\chi_{AbsCoeff}(i, bin\_idx) = 5 + clip3(0,4, NumLgt1(i))$, where $1 \leq bin\_idx \leq 13$. (9-28)

***Parsing Process for Residual Data***

The syntax parsing for residual data in H.264 is defined as shown in Figure 9-26.

```
If(coded_block_flag){
   numCoeff=maxNumCoeff
   i=0
   do{
         significant_coeff_flag[i]
         if( significant_coeff_flag[i]) {
               last_significant_coeff_flag[i]
               if( last_significant_coeff_flag[i]){
                     numCoeff=i+1
                     for(j=numCoeff; j<maxNumCoeff; j++)
                           coeffLevel[j]=0
               }
         }
         i++
   } while (i<numCoeff-1)
   coeff_abs_level_minus1[numCoeff-1]
   coeff_sign_flag[numCoeff-1]
   coeffLevel[numCoeff-
1]=(coeff_abs_level_minus1[numCoeff-1]+1)*
         (1-2*coeff_sign_flag[numCoeff-1])
   for (i=numCoeff-2;  i>=0; i--)
         if(significant_coeff_flag[i]){
               coeff_abs_level_minus1[i]
               coeff_sign_flag[i]
               coeffLeve[i]=
 (coeff_abs_level_minus1[numCoeff-1]+1)*
                     (1-2*coeff_sign_flag[numCoeff-1])
         } else
               coeffLevel[i]=0
```

**Figure 9-26 Representation in Residual Data in a Block**

As is listed, the order of information is to put "significant_coeff_flag" followed by "last_significant_flag." Note that the scanning order for these flags is from 0 to 15 -- forward order. When these are all listed, "coeff_sign_level_minus1" followed by "coeff_sign_flag" data are appended from 15 to 0 – reverse order. Note that coeff_abs_value_minus1 = abs_level-1 in Figure 9-26.

***Examples***

Let's assume the following 4x4 block to encode (or decode if reverse procedure applies).

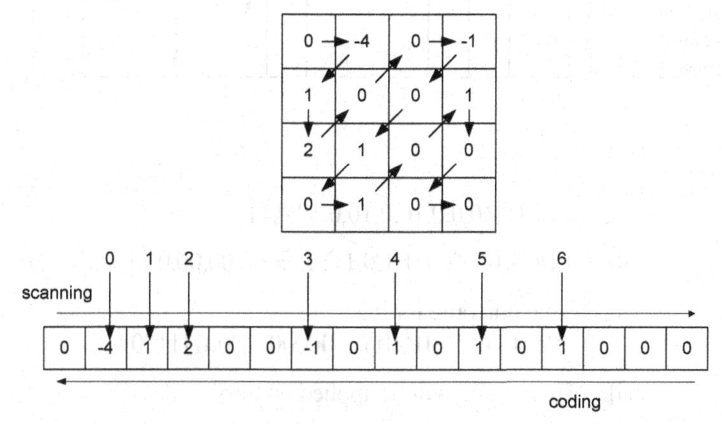


**Figure 9-27 Example of VLC Coding of a Block**

Table 9-45 and Table 9-46 illustrate lists of information to code for the example.

**Table 9-45 Example – Different Probability Models Assigned to significant_coeff_flag and last_significant_coeff_flag with coeff[i]**

| Scanning position i | 1 | 2 | 3 | 4 | 5 | 6 | 7 | 8 | 9 | 10 | 11 | 12 | 13 |
|---|---|---|---|---|---|---|---|---|---|---|---|---|---|
| Transform coeff. levels | 0 | 4 | 1 | 2 | 0 | 0 | -1 | 0 | 1 | 0 | 1 | 0 | 1 |
| significant_coeff_flag | 0 | 1 | 1 | 1 | 0 | 0 | 1 | 0 | 1 | 0 | 1 | 0 | 1 |
| last_significant_coeff_flag | | 0 | 0 | 0 | | | 0 | | 0 | | 0 | | 1 |

**Table 9-46 Example – Different Probability Models Assigned to Each bin of Levels with coeff[i]**

| Scanning position i | 1 | 2 | 3 | 4 | 5 | 6 | 7 | 8 | 9 | 10 | 11 | 12 | 13 |
|---|---|---|---|---|---|---|---|---|---|---|---|---|---|
| Transform coeff. levels | 0 | 4 | 1 | 2 | 0 | 0 | -1 | 0 | 1 | 0 | 1 | 0 | 1 |
| NumT1 | 3 | 3 | 3 | 3 | 3 | 3 | 3 | 3 | 2 | 2 | 1 | 1 | 0 |
| NumLgt1 | 2 | 1 | 1 | 0 | 0 | 0 | 0 | 0 | 0 | 0 | 0 | 0 | 0 |
| bin0 at position of transform coefficient scanned at i | | 4 | 4 | 3 | | | 3 | | 2 | | 1 | | 0 |
| bink at position of transform coefficient scanned at i | | 6 | 6 | 5 | | | 5 | | 5 | | 5 | | 5 |

a. 0,10,10,10,0,0,10,0,10,0,10,0,11

b. 0,1/0,1/0,1/0,-1/1,1/0,1/3,1 → 00,00,00,01,100,01,11100

c. Final outcome for bin:
0101010001001001001100000011000111100

d. Then, BAC will be applied on this.

***Probability Estimation***

In CA-BAC, the estimated probabilities of each context model can be represented by a sufficiently limited set of representative values. In this approach, an estimation ceiling for quantized probabilities exists – consecutive symbol of the same value more than 64 times cannot cause Least Probable Symbol (LPS) to be less than $p_{62}$. The set chosen for CA-BAC is composed of 64 different representative values in [0.01875, 0.5] for LPS with the following recursive equation:

$$p_\sigma = \alpha \times p_{\sigma-1} \text{ for all } \sigma = 1,....,63 \text{ with } \alpha = \left(\frac{0.01875}{0.5}\right)^{1/63} \text{ and}$$

$$p_0 = 0.5. \tag{9-29}$$

This approximation model for a consecutive LPS is taken with an exponential function to emulate behaviors of an inverse proportional function of such cases. Note that Prob(MPS)=1-Prob(LPS) and codes are composed to be biased to more "1"s. And, $p_0 = 0.5$ and $p_{62} = \frac{0.01875}{0.5 \times \alpha}$. Here, both the chosen scaling factor $\alpha \approx 0.95$ and the cardinality N=64 of the set of probabilities represent a good compromise between fast adaptation and an accurate estimation. The index $\sigma = 63$ is related to an autonomous non-adaptive state with a fixed value of MPS, which is only used for encoding binary decisions before termination of the arithmetic codeword.

The derivation of the transition rules for the LPS probability is based on the following relation between a given LPS probability $p_{hold}$ and its updated counter part $p_{new}$:

$$p_{new} = \begin{cases} \max(\alpha \times p_{hold}, p_{62}) & \textit{if a MPS occurs} \\ \alpha \times p_{hold} + (1-\alpha), & \textit{if a LPS occurs}. \end{cases} \tag{9-30}$$

The most efficient way to housekeep the models in CA-BAC is to use a 7-bit index for representing the 6-bit probability state index $\sigma_\gamma$ and the 1-bit value $\varpi_\gamma$ of the Most Probable Symbol (MPS). Thus, the pair ($\sigma_\gamma$, $\varpi_\gamma$) for each and every $\gamma$ ($0 \le \gamma \le 398$) and hence the models themselves can be efficiently represented by 7-bit unsigned integer values.

#### ***Table-Based Binary Arithmetic Coding***

In BAC, the current interval range R and the base (lower endpoint) L of the current code interval are continuously computed/updated with the probability model as a new binary input is applied. In H.264 CA-BAC, a novel approach is taken – for successive approximation, quantized range R, which is denoted with $Q_\rho$, and quantized probabilities $p_\sigma$ are proposed to be used for speed-up. Since values are quantized, limited and finite combinations exist for consideration in a pre-computed table. Figure 9-28 depicts inputs and outputs for Table-based BAC through TabRangeLPS and TransIdxLPS, where Prob(LPS) and MPS are represented with only 7-bit indices. Figure 9-29 illustrates the CA-BAC encoding process for a given binary value binVal using the regular coding mode. The precision needed to store these registers in the CA-BAC engine (both in regular and bypass mode) can be reduced by to 9 and 10 bits, respectively.

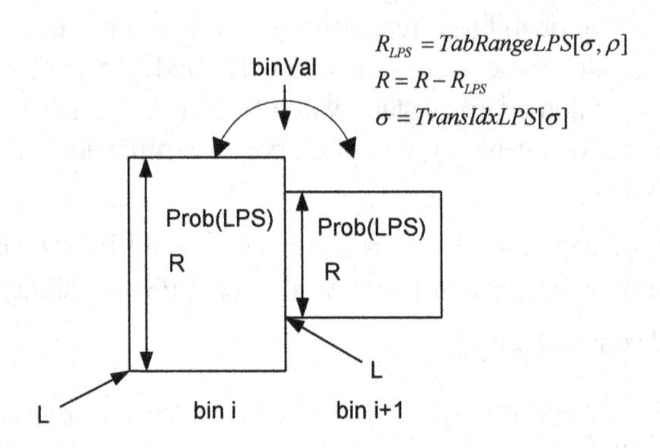


**Figure 9-28 Inputs and Outputs for Table-Based BAC**

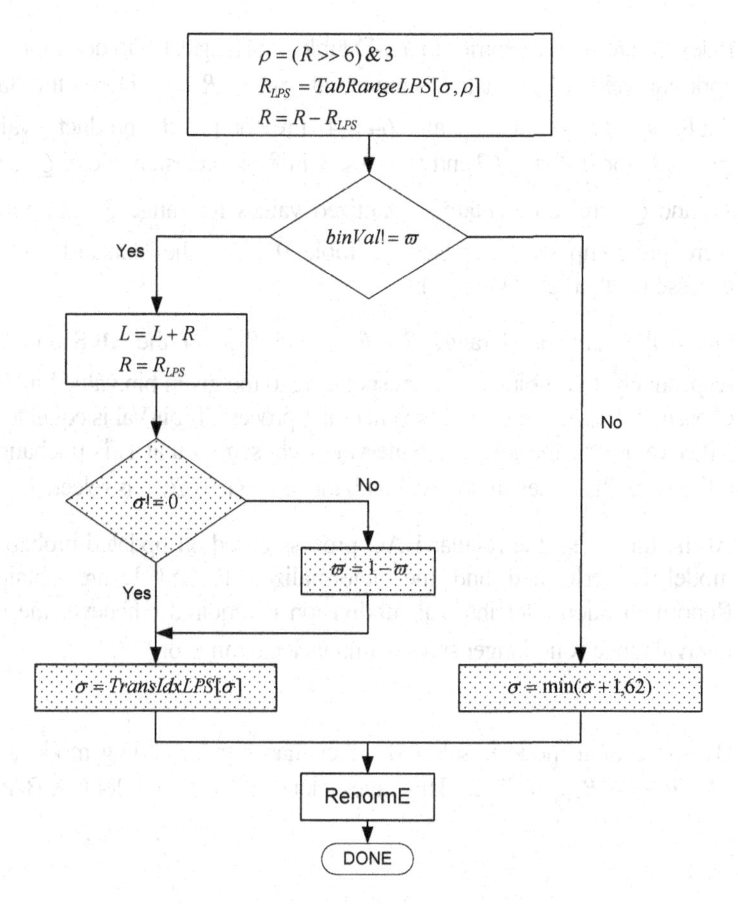


**Figure 9-29 BAC Encoding Process**

As the first step, the current interval is subdivided according to the given probability estimates. This interval subdivision process involves the three elementary operations in white boxes in Figure 9-29 – the range of R is quantized by using a equi-partition of the whole range $2^8 \leq R < 2^9$ into four cells. This can be simply performed by:

$$\rho = (R >> 6) \& 3,$$

which implies to take upper two bits of R. Note that the quantized value is only used for an input to seek for the next stage R. This does not mean that the accuracy of R is decreased. The index $\rho$ and the probability state index $\sigma$ are used as entries in a 2-D table TabRangeLPS to determine the approximated LPS related subinterval range $R_{LPS}$ . Here, the table TabRangeLPS contains all $64 \times 4$ pre-computed product values $p_\sigma \times Q_\rho$ for $0 \le \sigma < 63$ and $0 \le \rho < 3$ in 8-bit precision. Here, $Q_0$, $Q_1$, $Q_2$ and $Q_3$ are representative quantized values for range R. The values were pre-computed/ proposed in Table 9-35 in the standard and are accessed with a set of $(\sigma, \rho)$.

Given the subinterval range $R - R_{LPS}$ and $R_{LPS}$ for the MPS and LPS respectively, the subinterval corresponding to the given bin value binVal is chosen in the second step of the encoding process. If binVal is equal to the MPS value $\varpi$, the lower subinterval is chosen so that L is unchanged. Otherwise, the upper subinterval with range equal to $R_{LPS}$ is selected.

At the third step, the regular BAC process based on updated probability model is performed and the renormalized R and L are obtained. Renormalization after interval subdivision is required whenever the new interval range R no longer stays within its legal range of $[2^8, 2^9)$.

Bypass coding mode is selected rather than regular coding mode when $R - R_{LPS} \approx R_{LPS} \approx R/2$. This saves a lot of computation for CA-BAC.

## CA-VLC

The encoding and decoding process of CA-VLC in H.264 is unique in a sense that VLC and VLD are implemented through computation in obtaining Level values. Generally, VLC and VLD are implemented through Table look-up methods.

### *Parsing Process for TotalCoeffs and T1s (a.k.a., coeff_token)*

The TotalCoeffs and T1s are interpreted based on Table 9-5 in the standard and nC. The two values (TotalCoeffs, T1s) are defined with one value named "coeff_token." The nC value is used as a selected context to tell which VLC is to be used to decode coeff_token. Therefore, the nC computation should be performed before decoding coeff_token.

If the CA-VLC parsing process is invoked for ChromaDCLevel, nC is derived as follows:

- If chorma_format_idc is equal to 1, nC is set equal to –1.
- If chorma_format_idc is equal to 2, nC is set equal to –2.
- If chorma_format_idc is equal to 3, nC is set equal to 0.

For all other CA-VLC parsing processes, the following applies:

- When the CA-VLC parsing process is invoked for Intra16x16DCLevel, luma4x4BlkIdx is set equal to 0.
- The variables blkA and blkB are derived as follows:
    - o If the CA-VLC parsing process is invoked for Intra16x16DCLevel, Intra16x16ACLevel or LumaLevel, the process is invoked with luma4x4BlkIdx. The 4x4 luma block specified by mbAddrA/ luma4x4BlkIdxA is assigned to blkA, and the 4x4 luma block specified by mbAddrB/ luma4x4BlkIdxB is assigned to blkB.
    - o If the CA-VLC parsing process is invoked for ChromaACLevel, the process is invoked with chroma4x4BlkIdx. The 4x4 chroma block specified by mbAddrA/ chroma4x4BlkIdxA is assigned to blkA, and the 4x4 chroma block specified by mbAddrB/ chroma4x4BlkIdxB is assigned to blkB.
- Let nA and nB be the number of non-zero Transform coefficient levels in the block of Transform coefficient levels blkA located to the left of the current block and the block of Transform coefficient levels blkB located above the current block, respectively.
- With N replaced by A and B, in mbAddrN, blkN and nN the following applies:
    - o If any of the following conditions is true, nN is set equal to 0.
        - MbAddrN is not available.

- The current MB is coded using an Intra prediction mode. Constrained_intra_pred_flag is equal to 1. mbAddrN is coded using Inter prediction and slice data partitioning is in use (nal_unit_type is in the range of 2 to 4, inclusive).
- The MB mbAddrN has mb_type equal to P_Skip or B_Skip.
- All AC residual Transform coefficient levels of the neighboring block blkN are equal to 0 due to the corresponding bit of CodedBlockPatternLuma or CodedBlockPatternChroma being equal to 0.

  - If mbAddrN is an I_PCM MB, nN is set equal to 16.
  - Otherwise, nN is set equal to the value TotalCoeff(coeff_token) of the neighborhood block blkN.

- The value of nC is derived as follows:
  - If both mbAddrA and mbAddrB are available, the variable nC is set equal to the average value (nA+nB+1)>>1.
  - If mbAddrA is not available or mbAddrB is not available, the variable nC is set equal to nA+nB.

| TrailingOnes | TotalCoeff | $0 \le nC < 2$ | $2 \le nC < 4$ | $4 \le nC < 8$ | $8 \le nC$ | nC=-1 | nC=-2 |
|---|---|---|---|---|---|---|---|
| 0 | 0 | 1 | 11 | 1111 | 000011 | 01 | 1 |
| 0 | 1 | 000101 | 001011 | 001111 | 000000 | 000111 | 0001111 |
| 1 | 1 | 01 | 10 | 1110 | 000001 | 1 | 01 |
| …. | …. | …. | …. | …. | …. | …. | …. |

**Figure 9-30 coeff_token mapping to TotalCoeff and TrailingOnes (IS: Table 9-5)**

***Parsing Process for Level***

The Level values are interpreted based on "level_prefix," which is given in Table 9-47 below. The decoding algorithm for level_prefix is as follows:

*leadingZeroBits=-1;*

*for( b=0; !b; leading ZeroBits++)*

*b= read_bits(1);*

*level_prefix=leading ZeroBits.*

**Table 9-47 Codeword Table for level_prefix (IS: Table 9-6)**

| Level_prefix | Bit string | Level_prefix | Bit string |
|---|---|---|---|
| 0 | 1 | 9 | 0000000001 |
| 1 | 01 | 10 | 00000000001 |
| 2 | 001 | 11 | 000000000001 |
| 3 | 0001 | 12 | 0000000000001 |
| 4 | 00001 | 13 | 00000000000001 |
| 5 | 000001 | 14 | 000000000000001 |
| 6 | 0000001 | 15 | 0000000000000001 |
| 7 | 00000001 | .... | .... |
| 8 | 000000001 | | |

The Level decoding algorithm can be summarized as follows:

1. Initially an index i is set equal to 0. Then the following procedure is iteratively applied TrailingOnes times to decode the trailing one transform coefficient levels.

A 1-bit syntax element trailing_ones_sign_flag is decoded and evaluated as follows:

- If trailing_ones_sign_flag is equal to 0, the value +1 is assigned to Level[i].
- If trailing_ones_sign_flag is equal to 1, the value -1 is assigned to Level[i].

The index i is incremented by 1.

2. A variable suffixLength is initialized as follows:

- If TotalCoeff is greater than 10 and TrailingOnes is less than 3, suffixLength is set to 1.
- If TotalCoeff is less than or equal to 10 and TrailingOnes is equal to 3, suffixLength is set to 0.

3. The following procedure is then applied iteratively (TotalCoeff – TailingOnes) times to decode the remaining levels:

- The syntax element level_prefix is decoded.
- The variable levelSuffixSize is set equal to the variable suffxLength. The following two cases are exceptions:
    - If level_prefix==14 and suffixLength==0, levelSuffixSize is set equal to 4.
    - If level_prefix ≥ 15, levelSuffixSize is set equal to level_prefix-3.
- The syntax element level_suffix is decoded as follows:
    - If levelSuffixSize>0, the syntax element level_suffix is decoded as unsigned integer representation u(v) with levelSuffixSize bits.
    - If levelSuffixSize==0, the syntax element level_suffix shall be inferred to be equal to 0.
- A variable levelCode is set equal to (Min(15, level_prefix)<<suffixLength)+ level_suffix.
- When level_prefix ≥ 15 and suffixLength is equal to 0, levelCode is incremented by 15.
- When level_prefix ≥ 16, levelCode is incremented by (1<<(level_prefix-3))-4096.
- When the index i is equal to TrailingOnes and TrailingOnes is less than 3, levelCode is incremented by 2.

4. The variable level[i] is derived as follows:

- If levelCode is an even number, the value (levelCode+2)>>1 is assigned to level[i].
- If levelCode is an odd number, the value (-levelCode-1)>>1 is assigned to level[i].
- Examples –

(1|11) – level_prefix=0 and level_suffix=3 with suffixLength=2 leading to levelCode=(level_prefix<<suffixLength)+level_suffix=3, which is an odd number to perform (-3-1)>>1=-2.

(1|00) – level_prefix=0 and level_suffix=0 with suffixLength=2 leading to levelCode=(level_prefix<<suffixLength)+level_suffix=0, which is an even number to perform (0+2)>>1=1.

(1|01) – level_prefix=0 and level_suffix=1 with suffixLength=2 leading to levelCode=(level_prefix<<suffixLength)+level_suffix=1, which is an odd number to perform (-1-1)>>1=-1.

(001|0) – level_prefix=2 and level_suffix=0 with suffixLength=1 leading to levelCode=(level_prefix<<suffixLength)+level_suffix=4, which is an even number to perform (4+2)>>1=3.

(001|1) – level_prefix=2 and level_suffix=1 with suffixLength=1 leading to levelCode=(level_prefix<<suffixLength)+level_suffix=5, which is an odd number to perform (-5-1)>>1=-3.

(0001) – level_prefix=3 and level_suffix=0 with suffixLength=0 leading to levelCode=(level_prefix<<suffixLength)+level_suffix=3, which is an odd number to perform (-3-1)>>1=-2.

5. SuffixLength modification:

- When suffixLength==0, suffixLength is set equal to 1.
- When |level[i]|>(3<<(suffixLength-1)) and suffixLength<6, suffixLength is incremented by 1 as shown in Table 9-48.

**Table 9-48 Rule for suffixLength Increment**

<table>
<tr><th>Current suffixLength</th><th>Level triggering suffixLength+1</th></tr>
<tr><td>0</td><td>If(|level|>0), suffixLength++;</td></tr>
<tr><td>0,1</td><td>If(|level|>3), suffixLength++;</td></tr>
<tr><td>0,1,2</td><td>If(|level|>6), suffixLength++;</td></tr>
<tr><td>0,1,2,3</td><td>If(|level|>12), suffixLength++;</td></tr>
<tr><td>0,1,2,3,4</td><td>If(|level|>24), suffixLength++;</td></tr>
<tr><td>0,1,2,3,4,5</td><td>If(|level|>48), suffixLength++;</td></tr>
<tr><td>0,1,2,3,4,5,6</td><td>If(|level|>96), suffixLength++;</td></tr>
</table>

The index i is incremented by 1.

### *Parsing Process for Run*

The Run values are interpreted based on Table 9-7, Table 9-8, Table 9-9 and Table 9-10 in the standard.

1. Initially, an index i is set equal to 0. The variable zeroLeft is derived as follows:

- If the number of non-zero transform coefficient levels TotalCoeff is equal to the maximum number of non-zero transform coefficient levels maxNumCoeff, a variable zerosLeft is set equal to 0.
- If the number of non-zero transform coefficient levels TotalCoeff is less than the maximum number of non-zero transform coefficient levels maxNumCoeff, total_zeros is decoded and a variable zerosLeft is set equal to its value.

2. The VLC used to decode total_zeros is derived as follows:

- If maxNumCoeff is equal to 4, one of the VLCs specified in Table 9-9 (a) in the standard is used.
- If maxNumCoeff is equal to 8, one of the VLCs specified in Table 9-9 (b) in the standard is used.

- If maxNumCoeff is not equal to 4 or 8, one of the VLCs specified in Table 9-7 and Table 9-8 in the standard are used.

3. The variable run[i] is derived as follows:
   - If zeroLeft is greater than zero, a value run_before is decoded based on Table 9-10 in the standard and zerosLeft. The run[i] is set equal to run_before.
   - If zeroLeft==0, The run[i] is set equal to 0.
   - The value of run[i] is subtracted from zerosLeft and the result assigned to zerosLeft. The result of the subtraction shall be greater than or equal to 0.

The index i is incremented by 1. Finally, the value of zerosLeft is assigned to run[i].

| total_zeros | TotalCoeff | | | | | | |
|---|---|---|---|---|---|---|---|
| | 1 | 2 | 3 | 4 | 5 | 6 | 7 |
| 0 | 1 | 111 | 0101 | 00011 | 0101 | 000001 | 000001 |
| 1 | 011 | 110 | 111 | 111 | 0100 | 00001 | 00001 |
| .... | .... | .... | .... | .... | .... | .... | .... |

**Figure 9-31 total_zeros for 4x4 Blocks with TotalCoeff 1 to 7 (IS: Table 9-7)**

| total_zeros | TotalCoeff | | | | | | | |
|---|---|---|---|---|---|---|---|---|
| | 8 | 9 | 10 | 11 | 12 | 13 | 14 | 15 |
| 0 | 000001 | 000001 | 00001 | 0000 | 0000 | 000 | 00 | 0 |
| 1 | 0001 | 000000 | 00000 | 0001 | 0001 | 001 | 01 | 1 |
| .... | .... | .... | .... | .... | .... | .... | .... | .... |

**Figure 9-32 total_zeros for 4x4 Blocks with TotalCoeff 8 to 15 (IS: Table 9-8)**

| total_zeros | TotalCoeff | | |
|---|---|---|---|
| | 1 | 2 | 3 |
| 0 | 1 | 1 | 1 |
| 1 | 01 | 01 | 0 |
| 2 | 001 | 00 | |
| 3 | 000 | | |

**Figure 9-33 Chroma DC2x2 Blocks (4:2:0 chroma sampling) (IS: Table 9-9 (a))**

| total_zeros | TotalCoeff | | | | | | |
|---|---|---|---|---|---|---|---|
| | 1 | 2 | 3 | 4 | 5 | 6 | 7 |
| 0 | 1 | 000 | 000 | 110 | 00 | 00 | 0 |
| 1 | 010 | 01 | 001 | 00 | 01 | 01 | 1 |
| …. | …. | …. | …. | …. | …. | …. | …. |

**Figure 9-34 Chroma DC2x4 Blocks (4:2:2 chroma sampling) (IS: Table 9-9 (b))**

| run_before | zerosLeft | | | | | | |
|---|---|---|---|---|---|---|---|
| | 1 | 2 | 3 | 4 | 5 | 6 | 6< |
| 0 | 1 | 1 | 11 | 11 | 11 | 11 | 111 |
| 1 | 0 | 01 | 10 | 10 | 10 | 000 | 110 |
| …. | …. | …. | …. | …. | …. | …. | …. |

**Figure 9-35 Tables for run_before (IS: Table 9-10)**

***Examples***

Let's assume the following 4x4 block to encode (or decode if reverse procedure applies). Let's assume that nC=2.

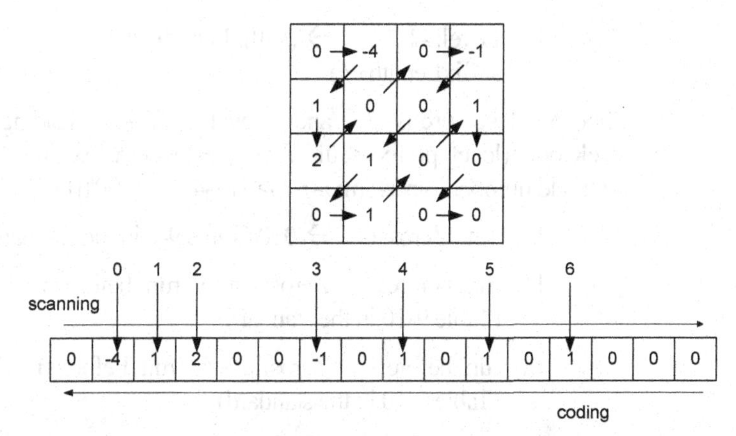


**Figure 9-36 Example of VLC Coding of a Block**

a. TotalCoeff=7 and TotalZeros=6.

b. T1s=3.

c. coeff_token (TotalCoeff=7, T1s=3) → 000000100 (Table 9-5 in the standard)

d. T1 sign: "+" → 0

e. T1 sign: "+" → 0

f. T1 sign: "+" → 0

g. Level: –1 → (suffixLength=0; suffixLength++)

choose level_prefix=1 and level_suffix=0 leading to levelCode=(level_prefix<<suffixLength)+level_suffix=1, which is an odd number to perform (-1-1)>>1=-1. --> (01)

h. Level: 2 → (suffixLength=1)

choose level_prefix=1 and level_suffix=0 leading to levelCode=(level_prefix<<suffixLength)+level_suffix=2, which is an even number to perform (2+2)>>1=2. --> (01|0)

i. Level: 1→ (suffixLength=1)

choose level_prefix=0 and level_suffix=0 leading to levelCode=(level_prefix<<suffixLength)+level_suffix=0, which is an even number to perform (0+2)>>1=1. --> (1|0)

j. Level: -4 → (suffixLength=1; suffixLength++)

Choose level_prefix=3 and level_suffix=1 leading to levelCode=(level_prefix<<suffixLength)+level_suffix=4, which is an odd number to perform (-7-1)>>1=-4. --> (0001|1)

k. TotalZeros: 6 → 010 (Table 9-7 in the standard)

l. run_before: ZerosLeft=6/ run_before=1 → 000 (Table 9-10 in the standard)

m. run_before: ZerosLeft=5/ run_before=1 → 10 (Table 9-10 in the standard)

n. run_before: ZerosLeft=4/ run_before=1 → 10 (Table 9-10 in the standard)

o. run_before: ZerosLeft=3/ run_before=2→ 01 (Table 9-10 in the standard)

p. run_before: ZerosLeft=1/ run_before=0→ 1 (Table 9-10 in the standard)

q. run_before: ZerosLeft=1/ run_before=0→ 1 (Table 9-10 in the standard)

r. run_before: ZerosLeft=1/ run_before=0→ 1 (Table 9-10 in the standard)

s. Final outcome: 0000001000000010101000011000101001111

# References


[ahmed:dct] N. Ahmed, T. Natarajan and K. R. Rao, "Discrete cosine transform," *IEEE Trans. on Compt.*, pp. 90-93, January 1974

[ahmed:orth] N. Ahmed and K. R. Rao, *Orthogonal transforms for digital signal processing*, Springer Verlag, New York, 1975

[bhaskaran:image] V. Bhaskaran and K. Konstantinides, *Image and video compression standards: Algorithms and architectures*, Springer, MA, 1997

[chen:dct] C.-F. Chen and K. –K. Pang, T., "The optimal transform of motion-compensated frame difference images in a hybrid coder," *IEEE Trans. on Circuits and Systems-II*, pp. 393-397, June 1993

[chen:fastdct] W. H. Chen, C. H. Smith and S. C. Fralick, "A fast computational algorithm for the discrete cosine transform," *IEEE Trans. on Commun.*, Vol. COM-25, pp. 1004-1009, September 1977

[chiang:newrate] T. Chiang and Y. Q. Zhang, "A new rate control scheme using quadratic rate control model," *IEEE Trans. on CSVT*, Vol. 7, No. 1, pp. 246-250, February 1997

[chung:percept] T.-Y. Chung and K.-H. Jung and Y.-N. Oh and D.-H. Shin, "Quantization control for improvement of image quality compatible with MPEG-2," *IEEE Trans. on Consumer Electronics*, Vol. 40, No. 4, pp. 821-825, November 1993

[cover:IT] T. Cover and J. Thomas, *Elements of information theory*, John Wiley and Sons, New York, 1991

[ding:ratecontrol] W. Ding and B. Liu, "Rate control of MPEG video coding and recoding by rate-quantization modeling," *IEEE Trans. on CSVT*, Vol. 6, No. 1, pp. 12-20, February 1996

[ebrahimi:VVM] Touradj Ebrahimi, "MPEG-4 Video Verification Model: A video encoding/decoding algorithm based on content representation," *Signal Processing: Image communication*, Vol. 3, No. 1, pp. 26-40, June 1997

[eleftheriadis:auto1] A. Eleftheriadis and A. Jacquin, "Automatic face location detection for model-assisted rate control in H.261 compatible coding of video," *Signal Processing: Image Communication* (Special Issue on Coding Techniques for Very Low Bit rate Video), Vol. 7, No. 4-6, pp. 435-455, November 1995

[flierl:motion] M. Flierl and B. Girod, "Generalized B pictures and the draft H.264/ AVC video compression standard," *IEEE Trans. on CSVT*, Vol. 13, No. 7, pp. 587-596, July 2003

[gersho:VQ] A. Gersho and R. M. Gray, *Vector quantization and signal compression*, Kluwer Academic Publishers, Boston, 1992

[gordon:simplified] JVT, "Simplified Use of 8x8 Transforms – Proposal," JVT-J029, Hawaii, USA, December 2003

[haskell:MPEG2] B. G. Haskell, A. Puri and A. N. Netravali, *Digital video: Introduction to MPEG-2*, Chapman & Hall, New York, NY 10003, 1997

[ISO:MPEG1] ISO/IEC JTC1, "Coding of moving pictures and associated audio for digital storage media up to about 1.5 Mbits/s: video," ISO/IEC 11172, November 1993

[ISO:MPEG2systems.amd] ISO/IEC JTC1, "Generic coding of moving pictures and associated audio: systems," ISO/IEC 13818-1:2000/Amd.3:2004, November 2004 (with several subsequent amendments and corrigenda)

[ISO:MPEG2-TM5] ISO/IEC JTC1/SC29/WG11, "Motion Picture Expert Group Test Model 5, Draft," April 1993

[ISO:MPEG4.VM6] ISO/IEC JTC1/SC29/WG11, "Motion Picture Expert Group MPEG-4 Video Verification Model version 6.0, MPEG96/N1582, Sevilla," February 1996

[ISO:MPEG4.VM8] ISO/IEC JTC1/SC29/WG11, "MPEG-4 Video Verification Model version 8.0," February 1997

[ISO:MPEG4] ISO/IEC JTC1 "Coding of audio-visual objects – Part 2: Visual," ISO/IEC 14496-2 (MPEG-4 Part 2), Jan. 1999 (with several subsequent amendments and corrigenda)

[ITU:H.261] ITU-T, "Video codec for audio visual services at $p \times 64$ kbit/s," ITU-T Rec. H.261 version 1: November 1990, version 2: March 1993

[ITU:H.263] ITU-T and ISO/IEC JTC1, "Video coding for low bitrate communication," ITU-T Rec. H.263 version 1: July 1995, version 2: 1998, version 3: 2000

[ITU:MPEG2] ITU-T and ISO/IEC JTC1, "Generic coding of moving pictures and associated audio: video," ISO/IEC 13818-2, November 1994 (with several subsequent amendments and corrigenda)

[ITU:MPEG2systems] ITU-T and ISO/IEC JTC1, "Generic coding of moving pictures and associated audio: systems," ITU-T Rec. H.222.0| ISO/IEC 13818-1, November 1994 (with several subsequent amendments and corrigenda)

[jain:DIP] A. K. Jain, *Fundamentals of digital image processing*, Prentice Hall, Englewood Cliffs, NJ 07632, 1989

[JVT:H.264] JVT of ITU-T and ISO/IEC JTC1 "Draft ITU-T Recommendation and Final Draft International Standard of Joint Video Specification (ITU-T Rec. H.264| ISO/IEC 14496-10 AVC)," May 2003 (with several subsequent amendments and corrigenda)

[lee:fastdct] B. G. Lee, "A new algorithm for the discrete cosine transform," *IEEE Trans. on Acoust., Speech, and Signal Prcoess.*, Vol. ASSP-32, pp. 1243-1245, December 1984

[lee:improved] Y.-L Lee, K.-H. Han and G. Sullivan, "Improved lossless intra coding for H.264/ MPEG-4 AVC," *IEEE Trans. on CSVT*, Vol. 15, No. 9, pp. 2610-2615, September 2007

[lee:ratecontrol] Jae-Beom Lee, *Model assisted activity and assisted coding for content-based video*, Columbia University, New York, NY 10025, 2000

[legall:MPEG] D. Le Gall, "MPEG: A video compression standard for multimedia applications," *Commun. ACM*, Vol. 34, No. 4, pp. 47-58, April 1991

[lim:rate] JVT, "Test Description of Joint Model Reference Encoding Methods and Decoding Concealment Methods," JVT-O079, Busan, Korea, April 2005

[list:adaptive] P. List, A. Joch, J. Lainema, G. Gjontegaard and M. Karczewicz, "Adaptive deblocking filter," *IEEE Trans. on CSVT*, Vol. 13, No. 7, pp. 614-619, July 2003 [malvar:low] H. S. Malvar, A. Hallapuro, M. Karczewicz and L. Kerofsky, "Low-complexity transform and quantization in H.264/ AVC," *IEEE Trans. on CSVT*, Vol. 13, No. 7, pp. 598-603, July 2003

[marpe:context] D. Marpe, H. Schwarz and T. Wiegand, "Context-based adaptive binary arithmetic coding in the H.264/ AVC video compression standard," *IEEE Trans. on CSVT*, Vol. 13, No. 7, pp. 620-635, July 2003

[marpe:H.264] T. Wiegand and G. Sullivan, "The H.264/ AVC video coding standard," *IEEE Signal Processing Magazine*, Vol. 24, No. 2, pp. 148-153, March 2007

[martucci:zerotree] S. A. Martucci, I. Sodagar and T. Chiang and Y.-Q. Zhang,"A zerotree wavelet video coder," *IEEE Trans. on CSVT*, Vol. 7, No. 1, pp. 109-118, February 1997

[mitchell:MPEG] J. Mitchell, W. Pennebaker, C. Fogg and D. Le Gall, *MPEG video compression standard*, Chapman and Hall, New York, 1997

[netravali:DIP] A. Netravali and B. Haskell, *Digital pictures: Representation, compression, and standards*, Plenum Press, New York, 1994

[ortega:optimal] A. Ortega, K. Ramchandran and Martin Vetterli, "Optimal Trellis-based buffered compression and fast approximations," *IEEE Trans. on Image Processing*, Vol. 3, No. 1, pp. 26-40, January 1994

[puri:mobile] A. Puri and A. Eleftheriadis, "MPEG-4: An object-based multimedia coding standard supporting mobile applications," *Mobile Networks and Applications*, Vol. 3, pp. 5-32, March 1998

[ramchandran:optimal] K. Ramchandran, A. Ortega and M. Vetterli, "Bit allocation for dependent quantization with applications to multiresolution and MPEG video," *IEEE Trans. on Image Processing*, Vol. 3, No. 5, pp. 533-545, September 1994

[rao:DCT] K. R. Rao and P. Yip, *Discrete cosine transform*, Academic Press, San Diego, 1990

[rao:techniques] K. R. Rao and J. J. Hwang, *Techniques, standards for image/ video and audio coding*, Prentice Hall, Upper Saddle River, NJ 07458, 1996

[ribas-corbera:generalized] J. Ribas-Corbera, P. A. Chou and S. L. Regunathan, "A generalized hypothetical reference decoder for H.264/ AVC," *IEEE Trans. on CSVT*, Vol. 13, No. 7, pp. 674-687, July 2003

[richardson:H.264] I. E. G. Richardson, *H.264 and MPEG-4*, Wiley, West Sussex, England, 2003

[richardson:white] I. E. G. Richardson, *White papers: H.264/ MPEG-4 Part 10*, www.vcodex.com, 2002

[rijkse:H.263] K. Rijkse, "ITU standardisation of very low bitrate video coding algorithms," *Signal Processing: Image Communication*, pp. 553-565, July 1995

[ronda:multiple] J. Ronda, M. Eckert, S. Rieke, F. Jaureguizar and A. Pacheco, "Advanced rate control for MPEG-4 coders," *SPIE VCIP-97*, Vol. 3309, pp. 383-390, February 1997

[schafer:digital] R. Schafer and T. Sikora, "Digital video coding standards and their role in video communication," *Proceedings of the IEEE*, Vol. 83, No. 6, June 1995

[SMPTE:VC1] SMPTE, "Proposed SMPTE standard for television: VC-1 compressed video bitstream format and decoding process," SMPTE 421M, August 2005

[SMPTE:VC1systems] SMPTE, "VC-1 bitstream transport encodings," SMPTE RP227, March 2005

[srinivasan:WMV9] S. Srinivasan, P. Hsu, T. Holcomb, K. Mukerjee, S. L. Regunathan, B. Lin, J. Liang, M.-C. Lee and J. Ribas-Corbera, "Windows Media Video 9: overview and applications," *Signal Processing: Image Communication*, Vol. 19, pp. 851-875, October 2004

[sullivan:new] G. Sullivan, H. Yu, S. Sekiguchi, H. Sun, T. Wedi, S. Wittmann, Y.-L Lee, A. Segall and T. Suzuki, "New standardized extensions of MPEG-4 AVC/ H.264 for professional-quality video applications," *ICIP-2007* Vol. 1, pp. I.13-I.16, September 2007

[sullivan:video] G. Sullivan and T. Wiegand, "Video compression: From concepts to the H.264/ AVC standard," *Proceedings of the IEEE*, Vol. 93, pp. 18-31, January 2005

[sun:joint] H. Sun, W. Kwok, M. Chien and C. H. J. Ju, "MPEG coding performance improvement by jointly optimizing coding mode decisions and rate control," *IEEE Trans. on CSVT*, Vol. 7, No. 3, pp. 449-458, June 1997

[vetro:multiple] A. Vetro, H. Sun and Y. Wang, "MPEG-4 rate control for multiple video objects," *IEEE Trans. on CSVT*, Vol. 9, No. 1, pp. 186-199, February 1999

[vetterli:wavelet] M. Vetterli and J. Kovacevic, *Wavelets and subband coding*, Prentice Hall, Englewood Cliffs, NJ 07632, 1995

[wang:multiple] L. Wang and A. Vincent and P. Corriveau, "Muliti-program video coding with joint rate control," *IEEE ICIP-96*, Vol. 3, pp. 1516-1520, September 1996

[wedi:motion] T. Wedi and H. G. Musmann, "Motion- and aliasing-compensated prediction for hybrid video coding," *IEEE Trans. on CSVT*, Vol. 13, No. 7, pp. 577-586, July 2003

[wiegand:H.264] D. Marpe and T. Wiegand, "The H.264/ AVC advanced video coding standard and its applications," *IEEE Communication Magazine*, Vol. 44, No. 8, pp. 134-143, August 2006

[wiegand:overview] T. Wiegand and G. Sullivan, "Overview of the H.264/ AVC video coding standard," *IEEE Trans. on CSVT*, Vol. 13, No. 7, pp. 560-576, July 2003

[wiegand:rate] T. Wiegand, H. Schwarz, A. Joch, F. Kossentini and G. Sullivan, "Rate-constrained coder control and comparison of video coding standards," *IEEE Trans. on CSVT*, Vol. 13, No. 7, pp. 688-703, July 2003

[wiegand:ratedistortion] T. Wiegand, M. Lightstone and D. Mukherjee, "Rate-distortion optimized mode selection for very low bit rate video coding and the emerging H.263 standard," *IEEE Trans. on CSVT*, Vol. 6, No. 2, pp. 182-190, April 1996

[yu:MPEG2systems] S. R. Yu, K. H. Chang, B. U. Lee, J. I. Kim and H. M. Jung, *MPEG systems*, Daeyoung Press, Seoul, Korea, 1997

[yu:MPEGvideo] S. R. Yu, *MPEG video*, Yonam Press, Seoul, Korea, 1995

# Index